Developmental Biology of the Axolotl

AXOLOTL GENETICS 1/25 Diane Malacinski

Developmental Biology of the Axolotl

Edited by

JOHN B. ARMSTRONG
GEORGE M. MALACINSKI

New York Oxford
OXFORD UNIVERSITY PRESS
1989

Oxford University Press

Oxford New York Toronto
Delhi Bombay Calcutta Madras Karachi
Petaling Jaya Singapore Hong Kong Tokyo
Nairobi Dar es Salaam Cape Town
Melbourne Auckland

and associated companies in
Berlin Ibadan

Published by Oxford University Press, Inc.,
200 Madison Avenue, New York, New York 10016

Oxford is a registered trademark of Oxford University Press

Library of Congress Cataloging-in-Publication Data

Developmental biology of the axolotl / edited by John B. Armstrong and George M. Malacinski.
p. cm.
Bibliography: p.
Includes index.
ISBN 0-19-505073-8 (alk. paper)
1. Axolotls—Development. 2. Amphibians—Development.
I. Armstrong, John B. II. Malacinski, George M.
QL668.C23D48 1989
597.6′5—dc19

9 8 7 6 5 4 3 2 1

Printed in the United States of America

Preface

How did an obscure salamander from Mexico become one of the leading experimental animals in modern developmental biology? Many readers undoubtedly will be familiar with the story (related in Chapter 1) of how General Forey's expeditionary force in the Emperor Maximilian's Mexico sent a few dozen of these curious animals to Paris. Somehow, most survived the trip to become a major popular and zoological curiosity. Naturalists of the day were concerned with understanding how a salamander could reproduce in an apparent larval state. Some even thought that if it did metamorphose, it became another species!

Developmental biologists soon found it more than a curiosity. The axolotl is easily bred in the laboratory. Its embryos are large and easily manipulated. The animal soon began to supplant locally caught species as the choice for embryological experiments, and an extensive literature began to accumulate (reviewed in Chapter 1). By the one-hundredth anniversary of its captivity, it is fair to say that the axolotl had become one of the focal species in developmental biology. It has been and continues to be used for a wide variety of studies, including embryonic induction, pattern formation, organogenesis, and regeneration. In addition, it has also been a major character in a novel (Peltonen, 1978), and at least one book has been written about keeping axolotls as pets (Scott, 1981). Surprisingly, however, there has been no treatise documenting its history and use in developmental biology, aside from the comprehensive review of the axolotl literature by H. M. Smith and R. B. Smith (1971). Considering the vast literature that exists, writing such a volume would be a daunting task for one individual. Consequently, when we first discussed preparing a book about the axolotl, we quickly agreed that it would be best to have several contributors. What we have tried to accomplish is to present a cross section of the developmental studies currently underway. We hope that our selections will stimulate the interest of readers unfamiliar with the axolotl. We must, however, ask the indulgence of our colleagues who were not included in this volume due to space limitations. We have also tried, in Part III, to present all the necessary information required for a novice to start working with the axolotl.

We dedicate this volume to the memories of Robert Briggs and Rufus Humphrey. Surely they would have been the two persons best qualified to write this kind of book. Humphrey could probably be called the North American champion of the axolotl. Most of his over 50-year scientific career was devoted to various studies on the axolotl (Malacinski, 1978a). Briggs, a pioneer in the study of amphibian tumors and nuclear transplantation (DiBernardino and Brothers, 1984), brought

Humphrey to Indiana University in 1956, and together they created a new field of amphibian developmental genetics (see Chapter 10). Both men were generous in sharing their knowledge and hospitality. They brought an enthusiasm to their work that rubbed off on all who knew them.

In addition to our cocontributors, Kathleen Armstrong and Steven Smith deserve special thanks for their assistance with the editing, and Diane Malacinski for her preparation of the fine art print used as the frontispiece. The print is based on Fig. 3 from Humphrey's 1967 paper, "Albino Axolotls from an Albino Tiger Salamander through Hybridization."

Contents

Contributors

David Able
Department of Biology
Indiana University
Bloomington, Indiana 47405

John B. Armstrong
Department of Biology
University of Ottawa
Ottawa, Ontario K1N 6N5, Canada

Makoto Asashima
Department of Biology
Yokohama City University
Yokohama 236, Japan

Jean-Claude Beetschen
Centre de Biologie du Développement
Université Paul Sabatier
31062, Toulouse Cedex, France

N. P. Bordzilovskaya
Institute of Developmental Biology
Academy of Sciences of the U.S.S.R.
Moscow, U.S.S.R.

Ronald A. Brandon
Department of Zoology
Southern Illinois University
Carbondale, Illinois 62901

G. Wayne Brodland
Department of Civil Engineering
University of Waterloo
Waterloo, Ontario N2L 3G1, Canada

Susan V. Bryant
Developmental Biology Center
University of California, Irvine
Irvine, California 92717

Robert Cuny
Department of Anatomy
University of Saskatchewan
Saskatoon, Saskatchewan, S7N 0W0, Canada

T. A. Dettlaff
Institute of Developmental Biology
Academy of Sciences of the U.S.S.R.
Moscow, U.S.S.R.

Susan T. Duhon
Department of Biology
Indiana University
Bloomington, Indiana 47405

Hans H. Epperlein
Department of Anatomy
Freiburg University
D-7800 Freiburg, West Germany

Sally K. Frost
Department of Physiology and Cell Biology
University of Kansas
Lawrence, Kansas 66045

David M. Gardiner
Developmental Biology Center
University of California, Irvine
Irvine, California 92717

Jean Gautier
Centre de Biologie du Développement
Université Paul Sabatier
31062, Toulouse Cedex, France

Richard Gordon
Department of Botany
University of Manitoba
Winnipeg, Manitoba R3T 2N2, Canada

William A. Harris
Department of Biology
University of California, San Diego
La Jolla, California 92093

G. F. M. Jacobs
Laboratory of Comparative Endocrinology
Zoological Institute
Catholic University of Leuven
B-3000 Leuven, Belgium

E. R. Kühn
Laboratory of Comparative Endocrinology
Zoological Institute
Catholic University of Leuven
B-3000 Leuven, Belgium

Michael J. Lannoo
Department of Anatomy
University of Ottawa
Ottawa, Ontario K1H 8M5, Canada

Jan Löfberg
Department of Zoology
Uppsala University
S-751 22 Uppsala, Sweden

George M. Malacinski
Department of Biology
Indiana University
Bloomington, Indiana 47405

Pat G. Model
Department of Neuroscience
Albert Einstein College of Medicine
Kennedy Center
Bronx, New York 10461

Ken Muneoka
Department of Biology
Tulane University
New Orleans, Louisiana 70118

Anton W. Neff
Anatomy Section
Medical Sciences Program
Indiana University
Bloomington, Indiana 47405

Roberto Perris
Department of Zoology
Uppsala University
S-751 22 Uppsala, Sweden

Jacques Signoret
Laboratoire d'Embryologie
Faculté des Sciences
Université de Caen
14 Caen, France

Jonathan M. W. Slack
ICRF Developmental Biology Unit
Department of Zoology
University of Oxford
Oxford OX1 3P5, England

Hobart M. Smith
Department of Environmental, Population, and Organismic Biology
University of Colorado
Boulder, Colorado 80209-0334

Steven C. Smith
Department of Biology
University of Ottawa
Ottawa, Ontario K2E 7A9, Canada

Michael Stigson
Department of Zoology
Uppsala University
S-751 22 Uppsala, Sweden

I

SCIENTIFIC AND NATURAL HISTORY

1

Discovery of the Axolotl and Its Early History in Biological Research

HOBART M. SMITH

For an animal on which several thousand scientific and numerous popular articles and books have been written (H. M. Smith and R. B. Smith, 1971), and which has become one of the most widely useful experimental subjects in the world, the axolotl, *Ambystoma mexicanum* (Shaw, 1978), (Fig. 1–1), remains uniquely enigmatic, at least in nature if not in the laboratory. For example, some of the misconceptions and problems pertaining to the axolotl were reviewed briefly nearly 20 years ago (H. M. Smith, 1969a), and although most were properly interpreted, at least one misconception was perpetuated there: that spontaneous transformation can occur naturally, under environmental stress such as that induced by shipment. The quite unanticipated basis for that misconception is that the axolotl exists macrosympatrically (presumably not microsympatrically) with a deceptively similar local race of tiger salamander, *Ambystoma tigrinum* (Green, 1825) (Chapter 2), which of course does readily undergo transformation, whereas the axolotl is a true obligatory neotene (in the sense used in Gould, 1977) that can be induced to transform only under experimental conditions with hormone treatment.

Thus, to avoid confusion, restricting the common name axolotl to the species *Ambystoma mexicanum,* rather than using it to refer to any of the group of related species (including the tiger salamander and 15–18 other species in Mexico alone, plus a nearly equal number of species and two more genera in the United States), or to refer to the larval stages of any or all of these species, as has been the practice in the past, is highly desirable. The imprecise use of the name would eventually require deciding on some alternative name for *A. mexicanum;* therefore, every effort should be made to curb the use of the name axoltl for any species other than *A. mexicanum.*

Native to certain permanent lakes near Mexico City, where extinction is now said to be imminent (Gonzalez et al., 1986), the axolotl was well known to local Indians for thousands of years as an important source of food (which it still is to a lesser degree), substituting for fish because few large species are native to the waters of the Mexican plateau. Indeed, the animal came to bear the náhuatl name *axolotl* in reference, at least in part, to the Aztec god Xolotl, to which it may have contributed some characteristics. The name, rendered *ajolote* in Spanish, means water

Fig. 1–1. Adult axolotls. The top animal is a white male (note enlargement around the cloaca); the bottom animal is a wild-type female. ×0.5 See Chapter 12 for a discussion of pigmentation and color variants. (Photograph from Armstrong, 1985.)

slave, water servant, water sprite, water monstrosity, water twin, or, preferably, water dog.

Xolotl (Fig. 1–2) was a god of exceptionally broad surveillance, over the dead and resurrected (in which role he took the form of a dog); games (hence water sprite, water player); monstrous or grotesque aberrations and beings (e.g., congenital and other deformities, any repulsive-appearing objects); and twins.

The close relation of the deity Xolotl with the axolotl is apparent in (1) the derisive expression meaning "ugly" that is used now even as in early times *(parece un ajolote),* and (2) the myth that the god Xolotl, in attempting to escape banishment from the earth (death), assumed many forms, all of which were either monstrous or "double" (e.g., corn or agave plants with double stalks) and finally assumed the form of the axolotl (whereupon Xolotl was captured, killed, and used to nourish the sun and moon). The axolotl's supposedly ugly features and its "double" existence in an aquatic and terrestrial form may be regarded as fitting the pattern of both the ugly and the paired disguises assumed by the god as he attempted to escape. The two forms seem to have been recognized even then as stages in the life cycle of a single species.

It is remarkable that, despite the close association in the Indian mind of the axolotl and Xolotl, the salamander has never been revered as an embodiment of the god. The association in no way interfered with capturing and consuming the salamander freely.

Foreign knowledge of the axolotl came in 1519 as Cortés's troops reached Mexico City, initially as Montezuma's guests, and later as his conquerors. The role of the axolotl in Aztec life and thought in Cortés's time was documented in several codices (e.g., Florentine, Madrid, Vatican) written by Spanish friars (e.g., Bernardino de Sahagún) in the period 1558–1577 (Dibble and Anderson, 1963). The first printed reference to the axolotl appeared in Ximénez' natural history book of 1615, and it was first given a scientific name conforming to generally agreed-on nomenclatural rules nearly 200 years later (Shaw, 1798). However, it was not until 1865 that it became the focal point of a scientific revolution unparalleled in biology. In that year a flood of publications on the axolotl began in France, and has continued since then in ever-increasing volume throughout the world (Fig. 1–3).

The impetus for the explosion of interest in the axolotl was the arrival in Paris in 1863, at the Natural History Museum, of 33 black and 1 white live axolotls, shipped by General Forey of the French Expeditionary Forces in Mexico, presumably in containers of water or wet moss. Although the animals are hardy, it is miraculous that they survived the several weeks they would have had to spend en route

Fig. 1–2. The Aztec god Xolotl, as depicted in the *Codex Borgia*. (From *Manuel d'archéologie américaine* by H. Beauchat. Copyright © 1912 Editions A. & J. Picard, Paris. Reprinted with permission.)

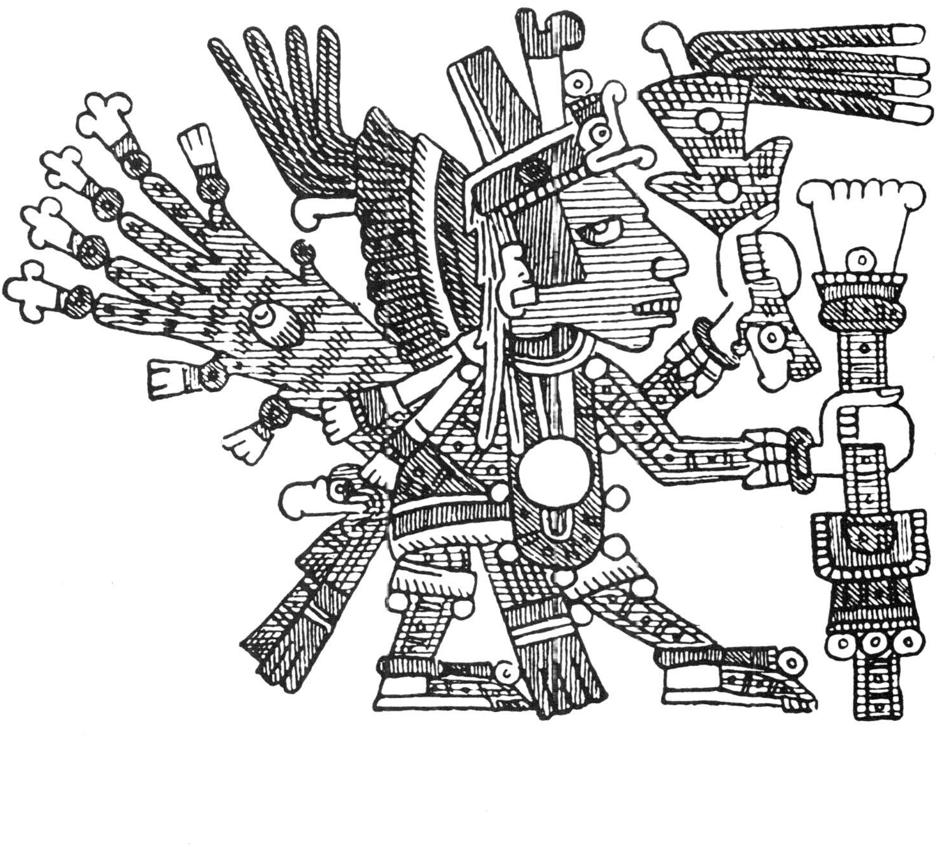

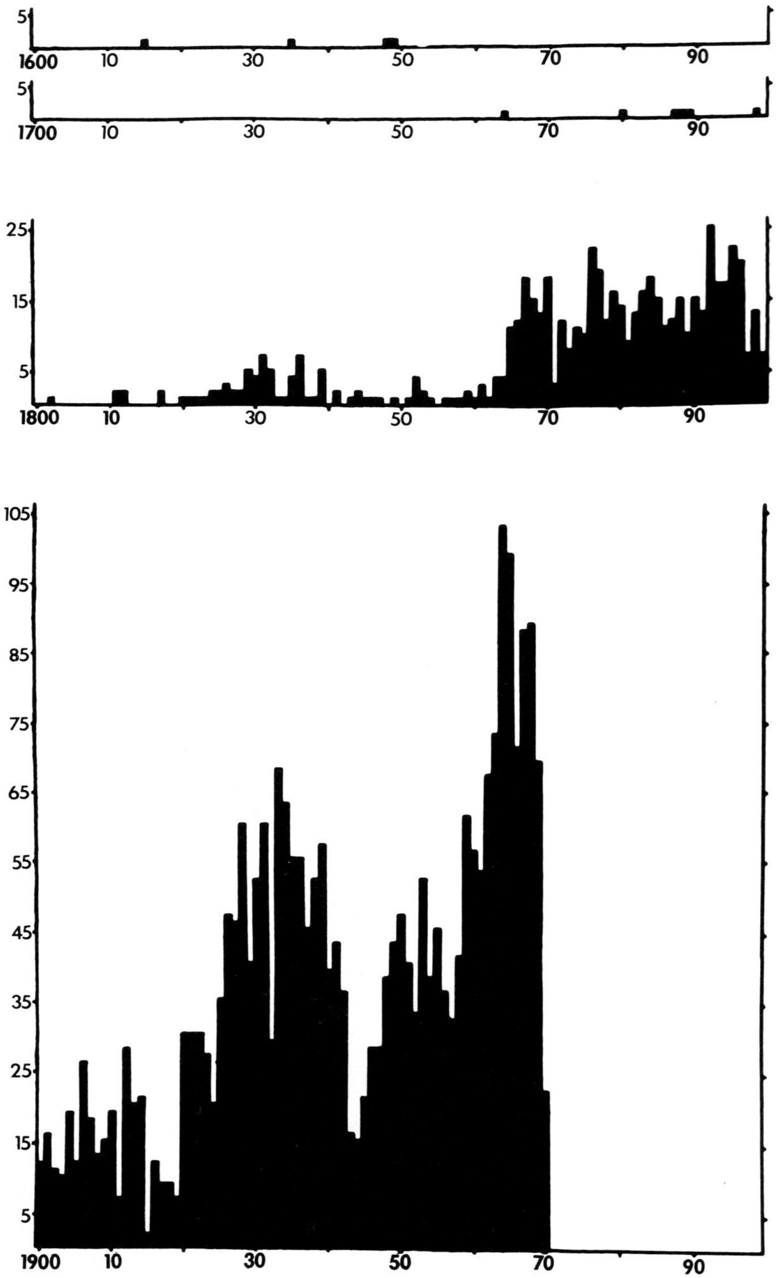

Fig. 1–3. Worldwide publication on the axolotl, 1600–1971. (Adapted from H. M. Smith and R. B. Smith.)

from their lakes of origin in the Valley of Mexico to Paris. Only 6 of these 34 specimens (five males and one female)—supplemented with a live white specimen ("albino") later (November 26, 1866) donated by Méhédin—were turned over to Auguste Duméril for scientific study (Duméril, 1870a), and from these 7 came the many thousands of axolotls sent all over Europe (Duméril, 1872). A large proportion of the experimental stocks now in existence throughout the world can be traced in a direct line to the shipment received in 1863 in Paris—surely one of the longest pedigrees of experimental animals of any kind!

What excited such immediate interest in this case was what then seemed an amazing phenomenon: This animal in sexual maturity bred successfully in a seemingly terminal gilled, aquatic form, thought to represent one genus; then it transformed into an equally successfully breeding terrestrial form thought to represent another genus. Duméril's discovery of this propensity met with skepticism in the scientific community, leading to intensive reexamination over the next several decades by other respected workers (e.g., Chauvin, 1885; Weismann, 1882). The literature on the axolotl during the period 1865–1885 is dominated by experiments and speculations concerning the cause and significance—in the axolotl itself and in zoology in general—of the still inadequately understood phenomenon of the transformation. Gould (1977) cogently discusses the conceptual quandary of that era.

Despite important advances in knowledge, the phenomenon of transformation in amphibians is still a general problem. Specifically in the axolotl it continues to be even more perplexing, mostly because of uncertainty in identifying the animals observed and recorded. Apparently the original stock received in Paris represented both the axolotl and its cryptic mimic, the black race of the tiger salamander; that would explain the exceedingly common observance of transformation in the offspring of that stock. Thus it is uncertain which of these two species was the subject of much of the early work—all purportedly on the axolotl. The tiger salamander genome was apparently gradually weeded out of experimental stocks; reproduction of the neotenic form has always been far easier to induce in the laboratory than reproduction of transformed individuals. Since all transformed individuals carried the *A. tigrinum* genome, obviously pure *A. mexicanum* eventually remained. For that reason many workers during this century have maintained, on the basis of experience with tens of thousands of axolotls (e.g., Brunst, 1955b; Humphrey, 1946, 1962), that they never, or at least rarely, transform in nature. That conclusion is borne out by recent observations (R. A. Brandon, personal communication), despite the claims in past years that any stress is likely to induce transformation (e.g., H. M. Smith, 1969a)—a conclusion unwittingly based upon examples of the dark race of *A. tigrinum.*

It is tempting to speculate that the source of contamination with *A. tigrinum* in the Paris culture of *A. mexicanum* was the white example received in 1866. In any event, obviously the two species readily hybridized there and subsequently, in captivity, whereas they presumably do not do so in nature in spite of their sympatric occurrence. Indeed, Humphrey (1967a) has recorded laboratory hybridization of axolotls with an albino tiger salamander.

By around 1885 the excitement over transformation had abated—partly because it occurred progressively less frequently in laboratory animals, and partly because the problem remained unsolved despite consideration of every conceivable causative factor (see especially Weismann, 1882). Attention then turned gradually to other topics of investigation, simply because the axolotl had proved to be such an excellent experimental animal: very hardy, easily induced to reproduce in captivity, reasonably quick to mature, moderately large, and relatively fecund. Gradually the axolotl became the most widely popular amphibian experimental animal in the world, utilized in every inhabited continent. While other amphibian species have attained equal or surpassing status locally for experimental study, none have attained the long-enduring popularity of the axolotl. The ultimate reason is the remarkably unselfish largesse and zeal with which Duméril, back in the 1860s and 1870s, unstintingly shared with his colleagues all over the world samples of his exciting laboratory animals.

SCIENTIFIC NAME

The scientific name of the axolotl, *Ambystoma mexicanum* (Shaw, 1798), now appears to be firmly established in perpetuity, through two actions of the International Commission on Zoological Nomenclature (ICZN, 1963, 1974). The first ICZN action conserved the name (and spelling) *Ambystoma* for the genus, and *mexicanum* for the species. The second also crossed the bridge of any future determination of conspecificity of the axolotl and the tiger salamander *Ambystoma tigrinum* (Green, 1825; conserved by the ICZN in a separate 1963 action)—an eventuality that now seems extremely remote—by ruling that the latter would have nomenclatural priority despite its later date, and hence would have to be accepted as the valid name of the species. If the axolotl were to be regarded as subspecifically related to the tiger salamander, its scientific name would be *Ambystoma tigrinum mexicanum.* Otherwise, without that ruling, all of the some 8–10 subspecies now assigned to *Ambystoma tigrinum* would become subspecies of *A. mexicanum.* Fortunately that relationship—of conspecificity of the axolotl and the tiger salamander—seems conclusively out of the question.

The nomenclatural stability the axolotl now enjoys contrasts sharply with its chequered history, however. At least 48 different—or differently spelled—scientific names have been applied to it, involving at least 27 different generic names or erroneous spellings thereof (Tihen, 1969; Smith and Smith, 1976: C-C-2-3). The most frequently used alternative generic name was *Amblystoma* Agassiz (1847), once accepted but now long rejected by taxonomists; persistence in the experimental literature led finally to official rejection by the ICZN 1963 ruling. The name *Siredon* Wagler (1830, actually antedating *Ambystoma* Tschudi, 1839, but rejected by the ICZN in 1963) was also popular for many years, applied in the context of obligatory neoteny; some workers (e.g., Brunst) went so far as to distinguish neotenes by the name *Siredon mexicanum* from the metamorphosed *Ambystoma mexicanum,* while accepting that the two forms are different developmental stages of the same species.

Curiously, six other generic names, all based on *A. mexicanum,* also antedate *Ambystoma* Tschudi, 1839: *Axolotus* Jarocki, 1822; *Philhydrus* Brookes, 1828; *Phyllhydrus* Gray, 1831; *Axolot* Bonaparte, 1831; *Sirenodon* and *Stegoporus* Wiegmann, 1832. All were either long overlooked or were superseded in popularity by

Ambystoma; thus, resurrection of any of these names for the genus through application of the Law of Priority would result in unnecessary confusion. Hence all were suppressed by the ICZN in 1963.

Since 1838, at least 20 other generic names, emendations, erroneous subsequent spellings, or erroneous generic allocations have been used for the axolotl: *Acholotes* Cope, 1868; *Amblystoma* Agassiz, 1847; *Axelotes* Landois, 1874; *Axoloteles* Wood, 1863; *Axolotes* Owen, 1844; *Axolotl* Mivart, 1870; *Axolotus* Gray, 1850 (a junior homonym of Jarocki's name, but proposed as new); *Gyrinus* Shaw, 1798 (a junior homonym of *Gyrinus* Geoffroy, 1762, Insecta); *Hypochthon* Merren, 1820 (a junior synonym of *Proteus,* in a different family; first used for the axolotl by Gravenhorst, 1829); *Lusus* Duméril et al., 1854 (a nomen nudum via citation in synonymy of a pre-linnaean name); *Menobranchus* (a junior synonym of *Necturus,* in a different family; first used for the axolotl by Harlan, 1835); *Phylhydrus* Swainson, 1839; *Phyllidrus* Agassiz, 1845; *Piscis* Duméril et al., 1854 (a nomen nudum via citation in synonymy of a pre-Linnaean name); *Protei* Humboldt and Bonpland, 1811 (probably intended as the plural of *Proteus*); *Proteus* (a valid genus of another family; first used for the axolotl by Green, 1818); *Salamandra* (a valid genus in another family; first used for the axolotl by Peters, 1863); *Siren* (a valid genus in another family; first used for the axolotl by Shaw, 1802); and *Sirena* (a junior synonym of *Siren,* in another family; first used for the axolotl by Velasco, 1879).

Nine different specific names, all proposed after Shaw's *mexicanus* in 1798, have been used in combination with the various generic names previously mentioned: *alba* Duméril, 1870a; *aquarum* Duméril et al., 1854 (a nomen nudum); *axolotl* Wagler, 1830; *edulis* Duméril et al., 1854; *guttata* Owen, 1844; *humboldtii* Duméril et al., 1854; *lubricus* Duméril et al., 1854 (a nomen nudum); *pisciformis* Shaw, 1802; and *weismanni* Wiedersheim, 1879. The scientific names formed in combination of these specific names with different generic names in various works differ not only in the particular combination chosen but in the different endings of the specific names, made to agree in gender with the generic name used with them, or used optionally with the double or single *-i* genitive ending. It is not surprising that as many as 48 different name combinations or spellings have been used for the scientific name of the axolotl in the past. Through the 1963 and 1974 actions of the ICZN, however, the nomenclatural chaos that has afflicted the history of the axolotl need no longer haunt the future progress in its study.

THE AXOLOTL LITERATURE

The most complete bibliography of the axolotl literature, covering all publications up to 1970 (although representation of the later years was necessarily incomplete) is the volume by H. M. Smith and R. B. Smith (1971), listing 3311 titles. Later bibliographies, derived from BIOSIS, are contained in issues 8 (1979), 12 (1983), and 15 (1987) of the *Axolotl Newsletter.* Copies of a pre-1978 search are also available from the *Newsletter* editor (G. M. Malacinski).

The 1971 bibliography contains a thorough subject index (pp. 199–211) and thereby permits illuminating time analyses of research emphases on the axolotl up to that time. For example, the general subject of greatest interest was embryology, numbering 830 titles. Its chronology (Fig. 1–4) exhibits an initial surge of activity from 1892 to 1895 (28 titles), a greater one from 1929 to 1942 (238 titles), and the peak from 1948 to 1968 (383 titles). Similar chronological analyses have been presented separately, though based largely on the 1971 compilation, for axolotl endocrinology (R. B. Smith et al., 1971), stemming from the early concern with transformation; for various aspects of descriptive anatomy (osteology, myology, visceral/lymphatic system, and the blood vascular system, in R. B. Smith and H. M. Smith, 1970a,b,c,d, respectively).

All of these subject chronologies more or less parallel each other, but those sorted by nationality reflect also the specific history of each nation. The effects of both world wars are evident in each country, but the more limited upheavals such as the Franco-Prussian war of 1871, and the U.S. Civil War of 1860–1864, show their effects in their respective nations. It is sobering indeed to observe that virtually all peaks and valleys in the chronology of axolotl research can be correlated most meaningfully with sociopolitical events, not with scientific discoveries.

Other historic generalities can be derived from the 1971 literature survey by Smith and Smith. In the pretaxonomic era, prior to 1798 when the species was first formally and acceptably named, all accounts were sketchy descriptions of appearance, habits, utility, and folklore. As occasional additional specimens arrived in museums and universities, the first truly investigative works appeared, and they pertained to anatomy. Indeed anatomy was the only field of investigation available to workers who were limited to preserved material; not until propagation was suc-

Fig. 1–4. Chronology of publications pertaining to the embryology of the axolotl, contrasting frequencies of descriptive (solid black) and experimental (open) works. (Adapted from H. M. Smith and R. B. Smith, 1971.)

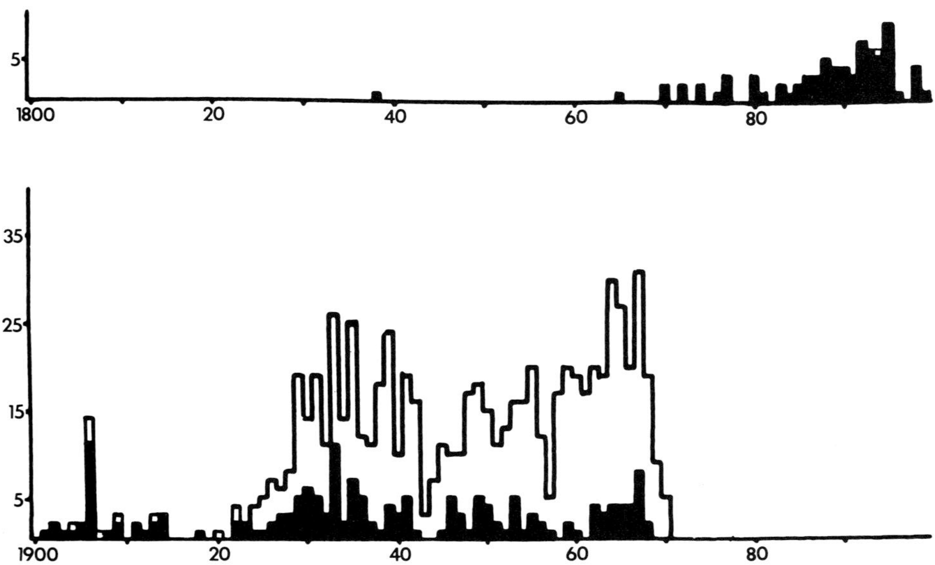

cessfully carried out, about 1870, did experimentation become possible. Scientists of the Western Hemisphere did not become involved importantly in axolotl research until several years after World War I, and at least initially were dependent on stocks derived from Europe, not directly from Mexico.

Evidence of the steady interest in the axolotl may be found in the fact that not since 1855 has a year passed without at least one publication appearing on it. Through 1970, publications on the axolotl had appeared in 29 countries (some now not independent), with the greatest numbers coming from Germany, the United States, France, Russia, and England. The earliest work appeared, appropriately, in Mexico (Ximénez, 1615). Marked proliferation of works on the axolotl occurred in several European countries in the 1860s and 1870s, undoubtedly due to Duméril's largesse. Not until the 1920s, however, did the spectacular increase in publication on the axolotl begin in Russia and the United States.

As of 1971, the most popular general subject areas were embryology, regeneration, metamorphosis, endocrinology, cultivation, radiation, transplantation, and anatomy. Amelanism received early attention through receipt of a white specimen in Paris in 1866, establishing this genetic variant in stocks widely distributed throughout the world. Curiously, little explicit attention was given to it after the initial announcements in 1870, until 1908.

Cultivation naturally was an exceptionally popular topic, partly because of the amazing popularity in central Europe of home aquariums, and partly because of the ease of axolotl propagation and experimentation in the laboratory. The popular journals *Wochenschrift* and *Blätter für Aquarien- und Terranenkunde* reflect the abundance of axolotls in home and zoo aquariums and terrariums. Together, these articles provide a little-tapped mine of information on care, diseases, treatment, and habits of the species.

The most prolific worker on the axolotl—according to the number of publications with him cited as an author—was Victor V. Brunst, with 79. Rufus Richard Humphrey, with 70, was the next most prolific.

As of 1971, a total of 525 journals had carried articles on the axolotl. Two that rank high—the *Wochenschrift für Aquarien- und Terrarienkunde* (109) and *Blätter für Aquarien- und Terrarienkunde* (98)—reflect popular interest in aquarium animals. The other journals most frequently used as publication outlets more accurately reflect the pattern of interest in the axolotl as a research subject.

The BIOSIS literature searches kindly made available by Dr. George M. Malacinski, partly in the *Axolotl Newsletter,* list 713 articles for the period 1970–1986. The coverage is obviously incomplete, primarily because all the popular journals were not included, but also because certain years were not fully canvased. However, of the 12 journals with the most articles after 1969, only 3 (indicated by an asterisk in the following list) are among the most popular journals prior to that time. The journals with 15 or more post-1969 articles on the axolotl are ranked as follows: *Developmental Biology,* 59; *Journal of Embryology and Experimental Morphology,* 53; *American Zoologist,* 44; *Journal of Experimental Zoology,** 34; *Ontogenez,* 29; *Anatomical Record,** 25; *Journal of Cell Biology,* 23; *General and Comparative Endocrinology,* 18; *Journal of Physiology,* 17; *Cell Differentiation,* 16; *Wilhelm Roux's Archiv für Entwicklungsmechanik der Organismen,** 16; and *Society of Neuroscience Abstracts,* 15.

A total of 150 journals have carried articles on the axolotl since 1969, according to the BIOSIS searches; of these, 85 were not among the 525 journals involved prior to 1970. A major shift of research from descriptive to experimental endeavors is evident.

ACKNOWLEDGMENTS

I am much indebted to Rozella B. Smith for the extensive data retrieval that permitted the analyses incorporated in our 1971 compilation, to Ronald A. Brandon for generous sharing of his knowledge of Mexican ambystomatids with me, and to George M. Malacinski for data on the axolotl literature since 1970.

2

Natural History of the Axolotl and Its Relationship to Other Ambystomatid Salamanders

RONALD A. BRANDON

ZOOGEOGRAPHY AND HABITAT

The axolotl, *Ambystoma mexicanum,* appears to be indigenous to the spring-fed Lakes Xochimilco and Chalco along the southern edge of the Basin of Mexico. Today, suitable habitat is greatly reduced, through drainage, from what it was only 100 years ago. About a quarter of the 7800-sq km (3000 sq mi) basin (Coe, 1964) was dominated by a series of shallow lakes that, in times of heavy rain, formed a continuous body of water some 60 km (37 mi) south to north (Fig. 2–1). These were remnants of the large Late Pleistocene lake formed when tectonic activity blocked drainage of the valley to the south (Sanders et al., 1979), called by the Aztecs "Lake of the Moon." The closed basin received water from springs (particularly near Xochimilco, Chalco, and Zumpango), meltwater from nearby snow-capped volcanos (Ixtaccíhuatl and Popocatépetl), and runoff. The slightly higher, spring-fed lakes Xochimilco and Chalco were fresh water, Zumpango was slightly saline, and Lake Texcoco at the bottom of the basin was highly saline (Sanders et al., 1979). At the time of the Spanish conquest in 1521, these relatively shallow lakes remained well formed (Martínez, 1961), but Lakes Chalco and Xochimilco had become separated from the more saline Lake Texcoco by the great Aztec dike just to the east of Tenochtitlán, and contained large areas of intensively cultivated, man-made strips of land called *chinampas* (Armillas, 1971; Coe, 1964; Sanders et al., 1979), which were separated by canals. This Xochimilco–Chalco basin—a continuous expanse (200 sq km or 77 sq mi) of marshes, swamps, and lagoons until 85 years ago when drainage works were completed (Armillas, 1971)—seems to have been the natural habitat of *Ambystoma mexicanum,* and what little remains still is.

Four of the lakes mentioned in the *Ambystoma* literature are Zumpango, Texcoco, Xochimilco, and Chalco. *Ambystoma mexicanum* has been reported from the last two, *A. velasci/lacustris* from the first two. Today, nearly all of this area is covered by rapidly growing Mexico City, now the largest city in the world with a pop-

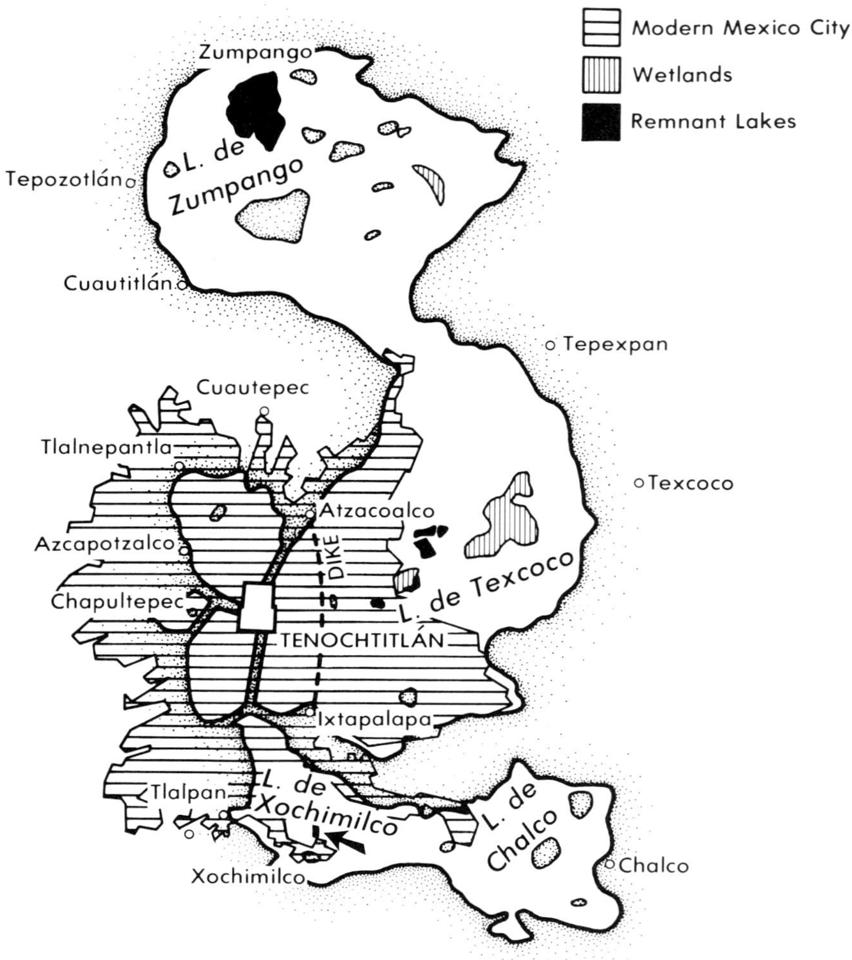

Fig. 2–1. Configuration of the Basin of Mexico lake system before the Spanish conquest, showing the extent of Lakes Xochimilco and Chalco, where spring-fed marshes apparently provided the axolotl's natural habitat. Present limits of Mexico City taken from "Ciudad de México, Mexico NE-14-2," in the Americian 1:250,000 series by the Instituto Panamericano de Geografía e Historia. Arrow indicates the Pista Olímpica, the largest body of water in what formerly was Lake Xochimilco. [Map modified from *La Ciudad de México (1325–1960)* (Martínez, 1961).]

ulation of about 8 million. Over an area of some 2000 sq km (772 sq mi), including Mexico City and the metropolitan Federal District, live some 16 million inhabitants (González et al., 1986; Marlin et al., 1986; McDowell, 1984). The progress of urbanization is vividly illustrated by the expansion of Nezahualcoyotl, itself a city of 2.3 million inhabitants, across the Texcoco lowlands (R. W. Fox, 1984, p. 176). Lake Chalco now essentially is dry, Lake Texcoco greatly reduced, Lake Zumpango small and isolated, and Lake Xochimilco reduced to remnant swamps and canals. Although the remaining canals, except for the Pista Olimpica, are too small to show on Fig. 2–1, they appear in detail on modern city street maps of Mexico City, such as the Guia Roji 1980 city map. The degree to which the lakes that surrounded Mexico City 300 years ago have been lost is illustrated dramatically by the following description in Marlin et al (1986): "With no river flowing through the city, and without a large body of water nearby, the city is unique among most populous world cities, which ordinarily are located on oceans, bays, lakes, or rivers." This is a telling statement about a city that began on a small island surrounded by lakes and regularly flooded during wet years (Robles, 1941). Photographs in McDowell (1980, pp. 716–717, 724–725) show the past and present extent of Lake Texcoco, and remnant habitat of Lake Xochimilco.

Thus, most of the original habitat of these animals is gone and remaining populations surely are in danger of extirpation (González et al., 1986). Like the axolotl, most of the other lentic species of *Ambystoma* in Mexico are highly vulnerable to extinction, having been subjected to several hundred years of habitat destruction. The availability of spring water and popularity of the canals and gardens of Xochimilco, however, may assure survival of at least some axolotl habitat. Photographs of this habitat may be found in McDowell (1980), Pan American Union (1964, p. 32), and the Enciclopedia de México (1985). Apparently, there are introduced populations of the axolotl in the lakes of Chapultepec and perhaps other spots around Mexico City as well (Enciclopedia de México, 1985).

The unfortunate fact is that current distributions of the axolotl and other species of *Ambystoma* within the Basin of Mexico remain poorly known. Current survey work such as that being carried out by José L. Camarillo R. of the Escuela Nacional de Estudios Profesionales Iztacala undoubtedly will prove helpful. Neither is there enough information on morphological and genetic variation among known populations to allow satisfying taxonomic treatment of known populations.

LIFE HISTORY IN NATURE

Virtually all information on life history of the axolotl has been obtained from laboratory colonies. Apparently there has been no ecological study of a wild population—a situation that is incredible considering the popularity of the axolotl as a research animal, but is easily understood considering the limited and relatively inaccessible natural populations, the ready availability of live animals in local markets for those who want them, and the ease with which breeding colonies have been maintained for over 100 years in laboratories around the world. However, this lack of basic information from natural populations contributes to current uncertainties about metamorphic response and about relationships with tiger salamander populations in the Basin of Mexico.

CONFUSION OVER THE NAME "AXOLOTL"

A common source of confusion in identifying individuals of the several large, neotenic species in central and eastern Mexico (and confusion in much of the axolotl literature as well; see Chapter 1) is that neither in Mexico nor in laboratories has the term axolotl (or *ajolote* in Spanish) been tied neatly to the species *Ambystoma mexicanum.* Dunn (1931) pointed out long ago that the term axolotl was being applied generally to branchiate adults of several different species. This practice now is much less common in the English-language literature but, as H. M. Smith and R. B. Smith (1971) have pointed out, in much of central and eastern Mexico the term *ajolote* remains a generalized common name for any large branchiate salamander, and that usage continues even in recent literature. In the interest of preventing further confusion, however, the common name axolotl (and *ajolote*) should be limited absolutely to the nontransforming species *Ambystoma mexicanum,* which occurs only in remnants of Lake Xochimilco and Lake Chalco at the southeastern edge of Mexico City. It may seem pedantic—even futile—to attempt to control the use of a common name, but in this instance the benefit would be great.

GENERAL RELATIVES

The Mexican axolotl, *Ambystoma mexicanum* (Shaw, 1798), is a member of the largest family of salamanders found exclusively in the New World, the family Ambystomatidae. This family of around 30 species ranges throughout North America from southern Canada in the east and Alaska in the west to the transverse volcanic belt along the southern edge of the Mexican Central Plateau (D. R. Frost, 1985; Gehlbach, 1967; Tihen, 1958).

Other than the marbled *(Ambystoma opacum),* spotted *(A. maculatum),* and small-mouthed salamanders *(A. texanum),* and the various diploid, triploid, tetraploid, and hybrid forms of the *A. jeffersonianum/A. laterale* complex, which all have relatively large ranges, the ranges of most ambystomatid salamanders are small (Conant, 1975; Stebbins, 1985). This is particularly true of the Mexican species (H. M. Smith and E. H. Taylor, 1948; Shaffer, 1984), most of which are known virtually only from their type localities. Part of the reason is that their ranges are poorly known and much further fieldwork remains to be carried out; in addition, however, the level of endemism really seems to be greater across central Mexico, and the ambystomatid radiation there is relatively young, probably dating from only 10–12 million years ago (Shaffer, 1984).

Only the tiger salamander *(A. tigrinum)* occurs over nearly the entire range of the family (Gehlbach, 1967), making it one of the most widely distributed species of salamander known. It is likely, however, that what currently is considered one species of tiger salamander eventually will be shown to be a complex of several closely related, allopatric or parapatric species (Brandon et al., 1981; Collins et al., 1980; Shaffer, 1983). As should become clear below, satisfactory understanding of the status of the axolotl as a species rests upon resolution of the diversity of tiger salamaders in Mexico, because the axolotl seems very closely related to some of these "tiger salamander-like" forms. Comparable increases in knowledge of species diversity have resulted in the recognition of several North American leopard frogs, formerly all called *Rana pipiens;* and over a dozen species of slimy salamanders, formerly all called *Plethodon glutinosus.*

CLOSEST RELATIVES

Ambystoma mexicanum is one of the approximately 17 recognized species (Table 2–1) of what Shaffer (1984) has termed the "*Ambystoma tigrinum-mexicanum-dumerilii-Rhyacosiredon* complex"—the *A. tigrinum* species group of Tihen (1958) as subsequently broadened (Brandon, 1972, 1977; Brandon et al., 1974). Distinctive features of these species are (1) a series of osteological traits pointed out by Tihen (1958), (2) absence of balancers on newly hatched larvae, (3) neoteny (maturation without metamorphosis) as a species-specific or dimorphic trait in some or all wild populations (see Table 2–1), (4) large size of larvae at metamorphosis, (5) large size of the heavy-bodied adults, particularly neotenic ones, (6) rapid growth to maturity in about 1 year regardless of metamorphosis, and (7) high degree of reproductive intercompatibility (similar courtship behaviors, and high viability of hybrid offspring produced in the laboratory).

Morphologically and geographically, some Mexican congeners (*A. andersoni, A. dumerilii, A. ordinarium, A. taylori,* and the four nominal species usually placed in the genus *Rhyacosiredon*) are quite distinct from *A. mexicanum. Ambystoma ordinarium* and the species of *Rhyacosiredon* are lotic species of cold, high-elevation streams where they are found in and along the banks and in spring seeps; they are variable in metamorphic response, with branchiate and transformed adults often found side by side. *Ambystoma andersoni* is a distinctive neotenic species known only from streams draining a lake at Zacapu, Michoacán; *A. dumerilii* is the quite distinctive and neotenic *achoque* of Lake Pátzcuaro, Michoacán, noted for its hyperfilamentous gills; and *A. taylori* is a neotenic species restricted to the highly saline waters of Lake Alchichica, Puebla.

Other species living primarily but not exclusively in lentic habitats, particularly *A. velasci* and several named and unnamed populations similar to *A. tigrinum,* may be difficult to distinguish from axolotls, and the taxonomic treatment of these forms remains tentative.

From datings of important geological events and a phylogenetic analysis of genetic distances (32 electromorphic characters obtained through starch gel electrophoresis of tissue extracts) among 21 populations and species of Mexican *Ambystoma,* Shaffer (1984) estimated that (1) the ambystomatid radiation in central Mexico is a relatively recent event, beginning 10–12 million years ago; (2) uplift of the Neovolcanic Axis isolated the montane, lotic species *(A. ordinarium* and the species of *Rhyacosiredon*) from plateau populations 7–11 million years ago; and (3) subsequent differentiation of isolated plateau populations within the three major basins of internal drainage probably dates from Late Pliocene–Pleistocene volcanic activity. Shaffer (1984) presents convincing support for the view that reproduction by branchiate adults (neoteny) has evolved independently within each of the three major basins, and has itself promoted greater genetic divergence between populations than would be expected otherwise. Barbour's analysis (1973a,b) of the distribution of atherinid fishes of the genus *Chirostoma* provides another useful zoogeographic perspective of geological events along the southern end of the Mexican Central Plateau.

Thus, the Xochimilco axolotl *(A. mexicanum)* is closely related to the tiger

Table 2–1 Currently Recognized Species of the Salamander Genus *Ambystoma* in Central Mexico

Species	Type Locality	Metamorphic Response[a]
A. altamirani[b]	Manantial de los Axolotes, Serrania de las Cruces, Vallee de México [Distrito Federal] (probably includes *A. zempoalaensis,* Lakes of Zempoala, Morelos) (Dugès, 1895)	V
A. amblycephalum	15 km west of Morelia, Michoacán (Taylor, 1940a)	V
A. andersoni	Laguna de Zacapu, 2000 m [6562 ft] elevation, Michoacán (Krebs and Brandon, 1984)	N
A. bombypellum	Near Rancho Guadalupe, 14 km [8.7 mi] east of San Martín, (Asunción) Mexico (Taylor, 1940a)	T
A dumerilii	La laguna de Pátzcuaro [Michoacán] (Dugès, 1870)	N
A. flavipiperatum	One mile [1.6 km] north of Santa Cruz, Jalisco (Dixon, 1963)	T
A. granulosum	Km 74, about 12 mi [19 km] northwest of Toluca, México (Taylor, 1944)	T
A. leorae[b]	Near Rio Frío, México (Taylor, 1943b)	T
A lermaensis	Lake Lerma, east of Toluca, México (Taylor, 1940a)	V
A. mexicanum	Mexico, Mexico (Shaw, 1798); Recorded from Lakes Xochimilco and Chalco (Smith and Taylor, 1948); restricted to Xochimilco, Distrito Federal, by Smith and Taylor (1950)	N
A. ordinarium	Four miles [6.5 km] west of El Mirador near Puerto Hondo, Michoacán (Taylor, 1940a)	V
A. queretarensis[c]	San Juan del Rio (Qro.) (Maldonado-Koerdell, 1948)	N
A. rivularis[b]	About 13 km [8 mi] west of Villa Victoria, México (Taylor, 1940b)	T
A. rosaceum	Mojárachic, Chihuahua (Taylor, 1941)	V
A. schmidti[d]	10 miles [16 km] east of San Martín (Asunción) at Rancho Guadalupe [Mexico] (Taylor, 1939)	T
A. subsalsum[e]	At Lake Alchichica, Puebla (Taylor, 1943a)	V
A. taylori	Laguna Alchichica, Puebla (Brandon et al., 1981)	N
A. velasci	Laguna Santa Isabel, Guadalupe Hidalgo, Distrito Federal (Dugès, 1888); includes *A. lacustris,* Lake Texcoco, Mexico[f]	V
A. zempoalaensis	Lakes of Zempoala, Morelos, Mexico (Taylor and Smith, 1945)	T

[a]The species is characterized by transformation (T), neoteny (N); or it is variable in metamorphic response (V).

[b]Following Dunn (1928), these three species usually have been placed in a separate genus, *Rhyacosiredon,* but this has been questioned (Brandon, 1977) and is not supported by Shaffer's data (1984) on genetic divergence.

[c]The status of this nominal species, originally described as a subspecies of *A. dumerilii,* is unknown, as no additional specimens have been collected and the types have not been available for study.

[d]Because this purported species so closely resembles the small-mouthed salamander, *Ambystoma texanum,* of the central United States; is in a different subgenus from all other Mexican species; and no specimen other than the holotype has ever been found, its validity is doubtful.

[e]This name does not apply to the neotenic species in Lake Alchichica, as Taylor intended, but to the *A. tigrinum*-like transforming species in the vicinity of the lake (Brandon et al., 1981).

[f]The published type locality of *A. lacustris* is "Lake Zumpango" (Taylor and Smith, 1945), but the holotype USNM 117410 is part of a series collected at Lake Texcoco (Cochran, 1961; personal check of museum records).

salamander-like populations distributed across the southern end of the Central Plateau (see photographs in Brandon, 1977, Fig. 5B; and Brandon et al., 1981, Fig. 1), but particularly to those in the Basin of Mexico. Shaffer (1984) obtained Nei (1978) genetic distances (quantitative estimates of the proportion of unshared electromorphic alleles) of only 0.054 –0.118 between the axolotl and tiger salamander populations of Hidalgo and Puebla. Genetic distances of this magnitude are small compared with distances of 0.221, 0.427, and 0.404 between *A. mexicanum* and *A. rosaceum,* California tiger salamanders, and Durango tiger salamanders, respectively (Shaffer, 1984). Unfortunately, no genetic comparison was made with tiger salamanders from the Basin of Mexico. Shaffer may have assumed that all animals from Xochimilco would have been *A. mexicanum,* since he asserted (1984, p. 1195) "no examples of sympatry are known for Mesa Central ambystomatids. . . . "

SPECIES INTEGRITY OF THE LABORATORY AXOLOTL

Evidence has accumulated to show that *A. mexicanum* and a species of tiger salamander both occur in the vicinity of Xochimilco. Branchiate *Ambystoma* of two phenotypes are sold at the market in Xochimilco (Brandon, 1976); one a long-toed form *(A. mexicanum)* and the other a broad-toed form (*A. tigrinum* complex) easily confused with it. The broad-toed form is similar to larvae and neotenic adults in other populations of the *A. tigrinum* complex in central and eastern Mexico, and in the laboratory transforms.

Many of the reports of metamorphosis in the axolotl may have resulted from failure to distinguish between these two forms. Both have been reared in our laboratory, where they have bred true and have been hybridized to yield intermediates (Brandon, 1977, Fig. 3A). The larval phenotypes can be distinguished by toe morphology, body pigmentation, and metamorphic response. Those few axolotls that have transformed in the laboratory (Fig. 2–2A) are black with flecks of white and are not nearly as hardy as postmetamorphic tiger salamanders. Postmetamorphic Xochimilco tiger salamanders (Fig. 2–2B) have a brown ground coloring and a variable (individually and ontogenetically) pattern of cream or yellow markings. Branchiate adults usually transform after being moved to the laboratory, and all of their offspring reared in the laboratory transform well before maturity. Whether these two forms occur in sympatry in nature or simply are sold together in the market is unknown.

Branchiate animals obtained at the market in Xochimilco in May 1970 and March 1971 were of both phenotypes. Other collectors have had similar experiences. For example, over half of 22 branchiate animals obtained at the market in Xochimilco in 1967 and shipped to the University of Illinois promptly transformed (H. M. Smith and R. B. Smith, 1971), and one of them—misidentified as an axolotl—subsequently was illustrated on the cover of a journal (H. M. Smith, 1969b). This specimen, representing the *tigrinum*-like transforming species, and another similar animal from the same collection were kept in our laboratory where they produced F_1 offspring, all of which transformed before maturation and developed tiger salamander-like body pigmentation. In addition, all F_1-hybrid offspring, obtained through artificial insemination, between these animals and branchiate *A.*

Fig. 2–2. Spontaneously transformed individuals of the two species of *Ambystoma* purchased as branchiate adults at Xochimilco, DF, May 21, 1970. (A) *A. mexicanum* (RAB A5-8), adult female, measuring 125 mm snout to vent, transformed within 6 months and lived 19 months longer (Brandon, 1976, Table 1). (B) Tiger salamander (RAB A5-7), adult male, measuring 120 mm snout to vent, transformed within 5 months and stayed alive 4 years longer. Note the strikingly longer and more slender toes in the axolotl, and the darker, more uniform ground coloring.

mexicanum from Xochimilco likewise transformed before maturation while *mexicanum* × *mexicanum* controls matured, remained branchiate, and gave rise to subsequent neotenic stocks (Brandon, 1977, cross H11). Thus, the 1967 collection contained animals of both species.

It is not clear yet whether the second species is identifiable as *Ambystoma velasci* or is an undescribed form. This question will be dealt with elsewhere (R. A. Brandon, in preparation). It is almost certain, however, that some of them have been imported into laboratory axolotl colonies and have hybridized there with axolotls; see, for example, Taylor's illustration of Emil Witschi's transformed *"mexicanum,"* which looks more like a tiger salamander (1940a, plate XLV, Fig. 3). A broad capacity for interspecific hybridization within the genus *Ambystoma* is well documented, particularly among members of the *tigrinum* species group (Brandon, 1972, 1977; Humphrey, 1967a; Nelson and Humphrey, 1972).

It is possible the original group of animals transported to the Natural History Museum in Paris in 1863 (H. M. Smith and R. B. Smith, 1971), as the foundation of a colony from which many others arose, was a mixture of the two forms. For some experimental studies of mechanisms of gene action or morphogenesis, the taxonomic source of material may seem unimportant; however, for systematists seeking comparative data from "natural" species, the likelihood of mixed genomes in various laboratory axolotl colonies could pose a serious problem. Given the potential for interspecific hybridization among species of *Ambystoma,* their apparent dependence on ecological and geographic separation for reproductive isolation in Mexico, and the greatly changed habitats in the Basin of Mexico over the past few hundred years, one must consider the possibility that populations in the Basin of Mexico may be as mixed as some laboratory colonies.

Why the transforming species commonly (usually?) is neotenic in nature but almost always transforms in the laboratory is unclear, but neoteny is common in tiger salamander populations of western United States (except California) and Mexico, sometimes with neotenic and transformed adults occurring in the same population.

ACKNOWLEDGMENTS

Without the kind and generous support of Hobart M. Smith, neither my studies of Mexican ambystomatids nor this manuscript would have been possible.

II

DEVELOPMENTAL BIOLOGY

3
Oogenesis

JEAN-CLAUDE BEETSCHEN AND JEAN GAUTIER

Unlike those of *Xenopus,* axolotl oocytes have been studied only sporadically. Reports of these studies are found mostly in the early literature. The various stages of oocyte growth and maturation have not been well documented in the axolotl, although its eggs were studied and compared to those of the indigenous *Triturus* species soon after the axolotl was imported into Europe in 1864. Van Bambeke (1870), working in Belgium, published a polar view of the freshly laid secondary oocyte. His drawing clearly shows the spot on the animal pole with the first polar body in the center. An interpretation of this spot was published some years later (Schultze, 1887). Van Bambeke (1870) was also the first to describe the multiple sperm entry points (or pits) in the urodele egg, although he did not hazard an interpretation in view of the absence of such phenomena in anuran eggs.

The oocytes and eggs of the axolotl have probably been studied less than those of other species as a result of the histological complications caused by their relatively large size. These problems were noted very early (Fick, 1893; Jenkinson, 1904) and still do not seem to have been fully resolved. For example, in a relatively recent ultrastructural and cytochemical study of oogenesis in various anurans and urodeles, Wartenberg (1962) used oocytes from both *Triturus alpestris* and the axolotl, but published no figures of the latter. In comparison to the oocytes of *Xenopus* and another common laboratory urodele, *Pleurodeles waltl,* however, the oocyte of the axolotl has several features that make it particularly suitable for in vitro experimentation. It would be useful, therefore, to have at one's disposal a data base equivalent to those available for other amphibians.

OOCYTE GROWTH

The origin of the primordial germ cells and the formation of the genital ridges in the axolotl were first investigated by Spehl and Polus (1912), and in greater detail by Schapitz (1912). There was no description of the differentiation of the gonadal primordium into ovarian structures. Burns (1932), however, briefly described the ovarian and follicular structures as "essentially like . . . the ovary of *Triton*" before investigating the effects of hypophysectomy on these structures. Similarly, Humphrey, in a number of publications, compared normal and abnormal ovarian con-

Table 3–1 Stages in Oogenesis

	Stage	Size (μm)	External Appearance
Previtellogenesis	I	≤500	Cytoplasm transparent; germinal vesicle visible
Vitellogenesis	II	500–700	Opalescent
	III	700–1000	Opaque and white
	IV	1000–1300	Uniformly gray
	V	1300–1600	Brown, except for a light-gray area at the vegetal pole that gradually expands
End of vitellogenesis	VI	1600–2100	Animal hemisphere brown; vegetal hemisphere depigmented

ditions, but included no comprehensive description of the former (see Humphrey and Fankhauser, 1946).

In a normal adult female the different kinds of oocytes, in various stages of growth, can be distinguished by their outward appearance. We shall adapt the classification proposed for oocytes of the salamander *Pleurodeles waltl* (Bonnanfant-Jais and Mentré, 1983), since the general characteristics of axolotl oocytes are essentially the same. The features are summarized in Table 3–1. Our observations generally agree with those of Callan (1966, p. 96), who described nucleolar changes in the nuclei of various-stage axolotl oocytes. The final size seems to be slightly more than 2 mm in diameter. Unlike *Pleurodeles,* no detailed ultrastructural study of axolotl oogenesis has been reported.

The germinal vesicle (nucleus) of the growing and full-grown oocyte has been extensively studied by Callan (1966), who described both the lampbrush chromosomes and somatic karyotype of the axolotl. He states that the nuclear sap in the axolotl germinal vesicle is more viscous than in other salamanders such as *Triturus cristatus* and *T. viridescens,* and must therefore be treated with a special saline solution in order to spread out the chromosomes. This viscous nature has tended to make injections of nuclear sap technically difficult, though still feasible.

OOCYTE MATURATION

Morphological Features of Maturation in Vitro

It has long been recognized that migration of the germinal vesicle to the animal pole is one of the primary events in amphibian oocyte maturation (see Fick, 1893, for observations in the axolotl). Unlike the situation in *Xenopus,* this migration is externally visible in the oocytes of the axolotl and *Pleurodeles.* It is preceded by the appearance of an unpigmented area at the animal pole (Fig. 3–1A), which should not be confused with the true maturation spot observed after germinal vesicle breakdown (GVBD). In the axolotl, this unpigmented area is transient, and is followed by the appearance of a large translucent nucleus (Fig. 3–1B). After GVBD the nuclear area gradually disappears as the yolk-filled cytoplasm mixes with nucleoplasm, and the genuine maturation spot forms at the animal pole (Fig. 3–

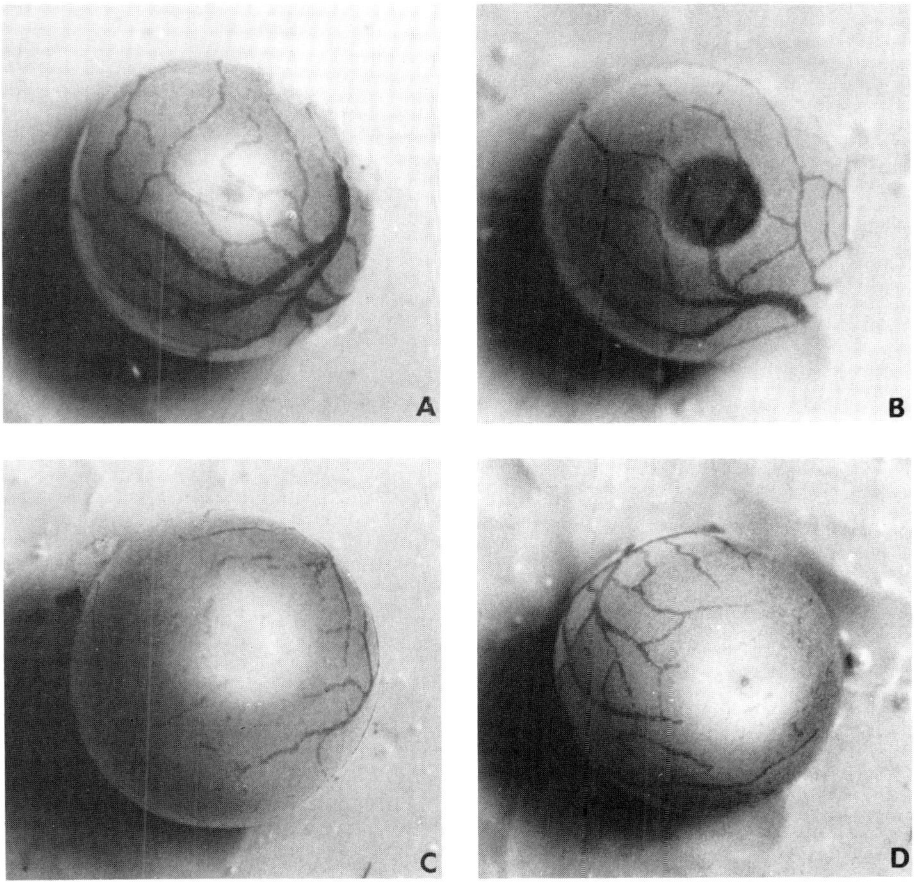

Fig. 3–1. Four stages of axolotl oocyte maturation at 18°C, seen through the transparent follicle. (A) At 5 hours, a depigmented spot is visible on the animal pole; germinal vesicle migration is indicated by a darker central area. (B) At 7.5 hours, the germinal vesicle is visible. (C) At 10 hours, maturation spot appearance following GVBD. (D) At 15 hours, first polar body has formed.

1C). The maturation spindle appears as a small pigmented circle, and the first polar body forms in the clear area and can be seen as a dark point (Fig. 3–1D). Compared to *Xenopus*, this assists greatly in determining the progress of maturation. Pseudomaturation (with or without GVBD) can be easily distinguished from the normal situation by examination with a dissecting microscope. Some possible mechanisms of germinal vesicle migration were first discussed in Fick's early work (1893). Precise data only recently became available. It has been shown that the cytoskeleton is directly involved in this phenomenon, which occurs even in inverted oocytes. However, migration can be prevented by inverting the oocytes and treating them simultaneously with colcemid or nocodazole, which disrupt microtubule (MT) organization (J. Gautier, unpublished data).

The successive steps of maturation can thus be easily followed in vitro. Maturation of full-grown oocytes can be induced by exposure to progesterone. A concentration of 1 μM seems adequate, although 10-fold higher concentrations have been used by some workers. Axolotl oocyte maturation in vitro was first demonstrated by Brachet (1974), followed by Vilain (1978) and Beetschen (1979). Vilain (1978) established that the proportion of oocytes reaching maturity was maximal in Merriam's modification (1971) of Barth's medium at pH 8.5. The oocytes could then be cultured at pH 7.5 or 7.8. These conditions have been used to study the artificial formation of a gray crescent (GC) in maturing oocytes (Grinfeld and Beetschen, 1982; Gautier and Beetschen, 1985). OR2 medium (Wallace et al., 1973) has also been used (Vilain et al., 1980b; Baud and Barish, 1984). We currently use Ford and Gurdon's modification (1977) of Barth's saline, which is buffered with HEPES (MBS-H).

In *Xenopus,* the follicle cells must be eliminated by an enzymatic pretreatment before the oocytes may be treated with progesterone. In the axolotl, various collagenases have proved toxic (Vilain et al., 1980b; J. C. Beetschen and J. Gautier, unpublished data), but dispase—a neutral protease—has been used by Baud and Barish (1984), after which the oocytes are isolated mechanically with fine forceps. Unlike *Xenopus,* an enzymatic pretreatment is not required for the induction of maturation in the axolotl. Progesterone is quite effective through the follicle wall, which also provides protection against injuries. Moreover, its transparent nature allows for the observation of events occurring in the oocyte. Among follicles isolated manually with fine scissors and treated with progesterone, up to 95% mature normally. Using a simplified technique (Gautier and Beetschen, 1985) adapted from Ford and Gurdon (1977), full-grown oocytes can also be enucleated through the follicle wall. The follicles are placed in healing solution (Ford and Gurdon, 1977) for 1–1.5 hour and then transferred into OR2 medium. Experiments involving enucleation are discussed below.

Cytological Data on Meiotic Divisions

Nuclear maturation events in the axolotl and other amphibian oocytes were outlined a century ago by Schultze (1887). He noticed that the first spindle was initially formed with its long axis parallel to the animal pole. Fick (1893) published a painstakingly detailed study of the axolotl that included both maturation and fertilization. The plates drawn by this author correctly show the main steps of the two meiotic divisions, even though he was only able to obtain a small number of coelomic and intraoviductal oocytes. At that time he was unable to interpret either chromosome pairing or disjunction. Fine cytological observations of meiotic divisions were described by Jenkinson (1904) some years later. Since then it seems that no detailed study on the completion of oocyte meiosis has been published.

The need for a precise timetable of meiotic events led us to carry out a kinetic analysis of these processes in intrafollicular oocytes maturing in vitro at 18°C. GVBD normally occurs 8–9 hours after progesterone has been added to the culture medium. It is followed by the formation of a mushroomlike structure with an axial stalk and an enlarged discoidal head in the animal hemisphere (Fig. 3–2A). This feature does not seem to be present in another urodele, *Cynops pyrrhogaster,* nor is it found in *Xenopus* (Imoh and Miyazaki, 1984). The condensed chromosomes

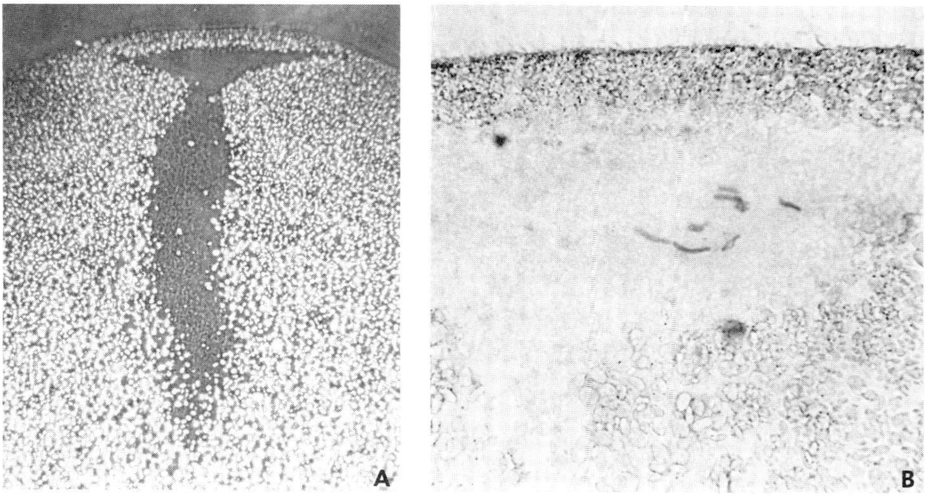

Fig. 3–2. Meridional sections through maturing oocytes (phase contrast). (A) At 9 hours, remnants of the germinal vesicle following GVBD. (B) At 10 hours, oocyte chromosomes in the upper part of the animal hemisphere prior to spindle formation.

Fig. 3–3. Timing of meiotic divisions as observed in oocytes maturing in vitro. Abscissa, number of hours at 18°C. Ordinate, successive stages, defined as follows: prometaphase I, the chromosomes are still dispersed around the spindle that is forming; metaphase I, the spindle is formed, at first parallel with and later perpendicular to the animal pole; anaphase I; telophase I, first polar body forms and is then extruded (arrow, 1PB); prophase II, partly decondensed chromosomes; prometaphase II, the chromosomes condense again but the equatorial plate has not yet formed; metaphase II, second spindle with chromosomes, at first oblique and then perpendicular to the animal pole surface.

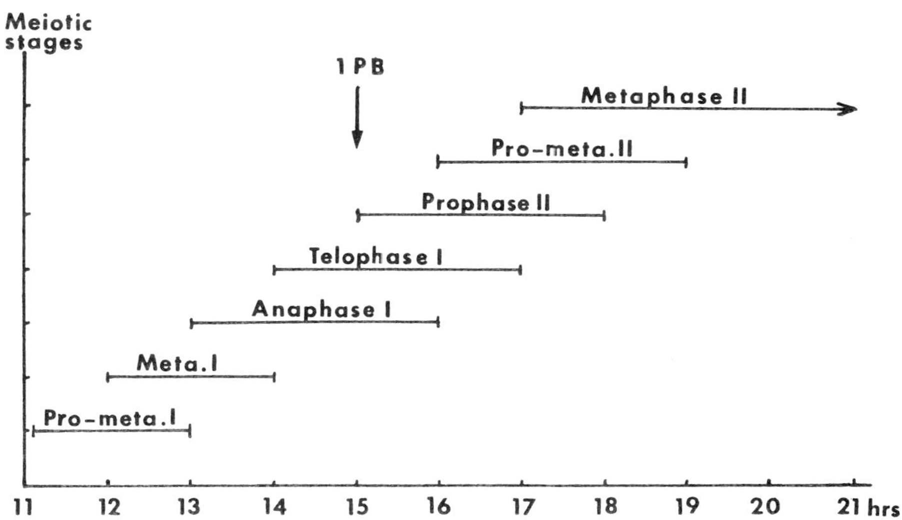

migrate into the head, where the first meiotic spindle appears (Fig. 3–2B). The stalk, corresponding to the central area of the germinal vesicle, mixes with yolk-laden cytoplasm and finally vanishes. After 12 hours most oocytes are in metaphase I and the spindle is first parallel with, and then perpendicular to, the animal pole. Histological sections of 130 oocytes from three different females were examined between 12 and 22 hours. Batches of 4–5 oocytes from each female were fixed every hour. Between 12 and 20 hours meiosis was not strictly synchronized in all the oocytes from each batch. There always appeared to be some overlap between the successive division stages at any given time due to the presence of a delayed (or advanced) oocyte. The first polar body is completely extruded at 15–16 hours, although it is visible much earlier. In some cases metaphase II is reached before 18 hours, but after 20 hours all oocytes stop at this stage. Figure 3–3 summarizes the cytological observations.

Not much data are available on the ultrastructural changes taking place during axolotl oocyte maturation. Taghy-Sadak and Vilain (1985) have reported the disappearance of annulate lamellae after GVBD, following an initial increase of these organelles during the early phases of progesterone action. On the other hand, glycogen granules tend to be extruded during maturation. These processes are found in both the axolotl and *Pleurodeles waltl.*

SOME ASPECTS OF MATURATION-PROMOTING FACTOR (MPF) ACTIVITY IN THE AXOLOTL OOCYTE

Involvement of the Germinal Vesicle in MPF Activity

In amphibian oocytes, meiosis can be reinitiated by various steroid hormones, especially progesterone. This hormonal stimulation causes the appearance of a cytoplasmic factor, the sо́-called maturation-promoting factor, or MPF (Masui and Markert, 1971; L. D. Smith and R. E. Ecker, 1971). When a small amount of cytoplasm from a maturing oocyte is microinjected into a full-grown oocyte that was not treated with progesterone, GVBD occurs earlier and maturation proceeds up to metaphase II. The cytoplasmic maturing properties are related to the presence of active MPF (which may be a result of the activation of an inactive precursor). MPF activity has since been found in a wide variety of maturing oocytes, and in dividing somatic cells, from yeast to humans. For this reason, MPF is more appropriately termed "M phase-promoting factor" (Gerhart et al., 1985).

We have studied the time course of MPF activation in both nucleated and enucleated axolotl oocytes (Gautier, 1987). Cytoplasm was taken from either nucleated or enucleated oocytes at different times following progesterone treatment. It was then transferred into untreated nucleated oocytes to test for the ability to induce maturation. MPF activity was detected in nucleated oocytes between 5 and 6 hours after the addition of progesterone to the culture medium. In contrast, enucleated oocytes had an additional lag of 2–3 hours. This difference in appearance of MPF activity might be a result of an enhancing property of the germinal vesicle. If nucleoplasm is injected into a previously enucleated oocyte before hormone treatment, MPF activity appears at the same time as in a nucleated oocyte (i.e., 6 hours after the addition of progesterone; Fig. 3–4).

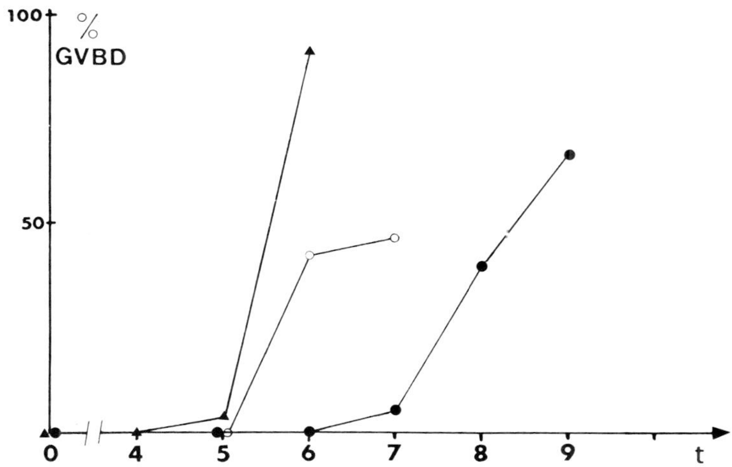

Fig. 3–4. Timing of appearance of MPF activity after progesterone treatment (time 0), revealed by GVBD in recipient oocytes after transfer of cytoplasm from (▲) nucleated oocytes, (●) enucleated oocytes, and (○) enucleated reinjected oocytes (time in hours).

It must be emphasized that MPF activity was not detected in the nucleus itself before GVBD. However, whereas the transfer of cytoplasm from a nucleated oocyte 5 hours after the addition of progesterone was ineffective, transfer of both cytoplasm and nucleoplasm at 5 hours was able to promote maturation in 50% of the recipient oocytes. Thus, MPF activity could be detected at an early stage when components from both cellular compartments were combined.

These detailed investigations on the induction kinetics of MPF activity in nucleated and enucleated (or in enucleated and reinjected) oocytes have not yet been carried out in other amphibians. In *Xenopus,* the species more commonly used for studies on maturation, difficulties may arise due to the shorter time course of MPF activation and the rather asynchronous nature of oocyte maturation.

Effects of Heterospecific MPF Preparations

Although MPF has not yet been completely purified, or even chemically defined, active MPF preparations have been reported to induce GVBD in full-grown *Xenopus* oocytes. Morever, this has also been found with MPF extracts from various other cell systems, including other amphibian species (Reynhout and Smith, 1974), invertebrate oocytes (Meijer and Guerrier, 1984), cultured mammalian cells (Sunkara et al., 1979; Nelkin et al., 1980), and even cell division cycle *(cdc)* temperature-sensitive yeast mutants, blocked in G_2 to M phases (Weintraub et al., 1982).

In an attempt to characterize any maturation processes that might be independent of MPF production, we tried to induce axolotl oocyte maturation with heterospecific MPF preparations (J. Gautier, M. Buscaglia, E. E. Baulieu, and J.-C. Beetschen, submitted). Those extracts would be expected to contain few axolotl-related cytoplasmic components. Some are inevitably introduced along with MPF when crude cytoplasm is microinjected into an axolotl oocyte.

Crude cytoplasm from maturing *Xenopus* oocytes actually induced maturation in 82.5% of axolotl oocytes, and supernatants from high-speed centrifugation were even more effective (97% of axolotl oocytes matured). However, the reactivity of axolotl and *Xenopus* oocytes was quite different when extracts from *cdc* yeast mutants were used (Pringle and Hartwell, 1981). Two extracts from the *cdc 20* mutant, the second partially purified by chromatography on arginine–agarose columns, induced 97% and 77% maturation, respectively, in *Xenopus,* but only 0 and 2% in the axolotl. A third extract induced 63.5% maturation in *Xenopus* and 3% in the axolotl. Positive results were obtained in both species using a fourth partially purified extract from another mutant *(cdc 16),* but while *Xenopus* still reacted very strongly (63.5% of the oocytes matured), GVBD was observed in only 26% of the axolotl oocytes. The maturation spot was much smaller than normal in the axolotl oocytes, although it appeared quite normal in *Xenopus.* Histological examination of five axolotl oocytes with such a maturation spot revealed that the maturation spindle lay deep in the animal cytoplasm in three of them, while the first polar body and the second metaphase spindle were present in the other two.

These abnormalities clearly indicate unusual maturation conditions in a significant number of axolotl oocytes. In contrast, a normal second meiotic division was seen in *Xenopus* oocytes injected with yeast MPF and in axolotl oocytes matured with *Xenopus* MPF. Cytological anomalies reminiscent of "pseudomaturation," as described by Brachet (1974) in *Pleurodeles* oocytes, were also found in several axolotl oocytes injected with yeast MPF. There were no maturation spots and no GVBD; a huge nucleolar body was enclosed by the shrunken nuclear membrane. These anomalies are not observed when a small amount of MPF-active cytoplasm from a maturing axolotl oocyte is injected into a full-grown axolotl oocyte.

It seems, therefore, that the apparent lack of species specificity in MPF activity with respect to *Xenopus* oocytes might not apply to other amphibians. In the axolotl, normal GVBD might require an association between MPF activity and other oocyte cytoplasmic factor(s) that yeast cells lack. We suggest that these factor(s) are present in *Xenopus* but are not essential for GVBD in this species. The axolotl may therefore be used as a new model system for the investigation of GVBD during maturation.

CHANGES IN PROTEIN PATTERNS DURING OOCYTE MATURATION

Protein accumulation patterns during axolotl oocyte maturation were first described by Jäckle and Eagleson (1980). Overall protein distribution was determined using two-dimensional (2D) gels stained with silver nitrate. The study concerned mainly the protein pattern along the animal–vegetal axis of the oocyte, but the authors did demonstrate the appearance of new proteins after oocyte maturation. They proposed that these new proteins were due to posttranslational modification of preexisting polypeptides, rather than de novo synthesis. Protein phosphorylation and synthesis during maturation were investigated, after incorporation of [^{32}P]orthophosphate and [^{35}S]methionine into polypeptides, by 2D gel electrophoresis (Gautier and Tencer, 1986). Two main phosphorylation events were detected. The first is a wave of phosphorylation during GVBD similar to that observed in *Xenopus* (Maller et al., 1977; Maller and Smith, 1985). More significant is an early dephosphorylation of a polypeptide within the first 3 hours of maturation. This is the only report of this process so far, and it should be noted that this protein is rephosphorylated during the wave of phosphorylation.

The axolotl oocyte would seem to be a good model for investigation of de novo protein synthesis during progesterone- or MPF-induced maturation. Studies of protein synthesis during oocyte maturation have produced somewhat conflicting results in *Xenopus* (Ballantine et al., 1979; Wasserman et al., 1982; Younglai et al., 1982), but in the axolotl have yielded highly reproducible differences in protein patterns during progesterone-induced maturation. Between 4 and 8 hours after exposure to the hormone there was evidence of de novo synthesis of several proteins. We also observed a marked decrease in the synthesis of four polypeptides. These became almost undetectable as maturation proceeded. In contrast, a recent study of alterations in protein synthesis patterns during hormone-induced maturation did not report any changes in protein patterns during the first 5 hours (Meuler and Malacinski, 1985). We looked over a longer period after exposure to hormone, however, and this may explain the discrepancy. As in the case of the studies on MPF production, the longer duration of oocyte maturation in the axolotl allows for more detailed investigations of these phenomena.

EXPERIMENTAL EARLY GC FORMATION DURING OOCYTE MATURATION

The first indication of dorsalization and of bilateral symmetry in many amphibian eggs is the formation of a cytoplasmic area, the GC, prior to first cleavage. The GC appears to be a topographic marker of the future dorsal side of the embryo. Axolotl eggs display a GC 4–5 hours after fertilization, whereas *Pleurodeles* and *Triturus* eggs display only a clear depigmented area. Moreover, early GC formation can be elicited immediately by heating fertilized or unfertilized axolotl eggs within the first hour after laying (Benford and Namenwirth, 1974). Since the physiological mechanisms of the cytoplasmic rearrangements leading to GC formation are still poorly understood (Brachet, 1977), it was of interest to find out when the capacity to form a GC appeared (i.e., the stage at which it could be artificially expressed) before egg formation.

GC formation can be induced by heat in maturing intraoviductal or intrafollicular oocytes (Beetschen, 1979). The conditions necessary for GC formation are likely established during maturation. We examined the possible role of protein synthesis as a prerequisite for this phenomenon. Cycloheximide and diphtheria toxin were used as inhibitors of protein synthesis. An unexpected effect was observed with both agents. They induced early appearance of GC in unheated maturing oocytes. This was more marked when a 45° to 90° rotation along the animal–vegetal axis was used to combine the effects of gravity with those of the inhibitor (Grinfeld and Beetschen, 1982; Gautier and Beetschen, 1983). Inhibition of protein synthesis thus could be related to GC formation in oocytes at an advanced stage of maturation (later than 12 hours at 18°C). During earlier stages of maturation, oocytes did not react to heat shock or to inhibitors, no GC was formed, and—as expected—the inhibitors of protein synthesis prevented GVBD. It was further shown that an early GC formation did not take place in late oocytes that had been enucleated before the initiation of maturation. However, the ability to form a GC could be restored in these oocytes by injecting nucleoplasm from a normal prophase-arrested oocyte either before or after injecting the drug. It has also been shown that this nuclear-cytoplasmic interaction can only occur if cytoplasm has "matured" at least 10 hours after exposure to progesterone.

A three-step model for artificial GC formation has been put forward (Gautier and Beetschen, 1985):

1. Cytoplasmic maturation proceeds until a reactive state is attained, enabling interactions with nuclear factor(s).
2. Nuclear factor(s) interact(s) with matured cytoplasm.
3. Inhibition of protein synthesis triggers GC formation.

This last step can be replaced either by a heat shock or by injection of an MT-depolymerizing agent (nocodazole, colcemid), after completion of the first two steps. It should be noted that maturation could still proceed up to the block at metaphase II if low concentrations of cycloheximide were used. This stage persisted several hours after GC formation, indicating that the latter phenomenon could occur in a unactivated oocyte (Grinfeld and Beetschen, 1982). In this way, the situation appears to be different from normal GC formation in a fertilized (previously activated) egg.

Several observations in anuran eggs indicate that MT play a key role in the final steps leading to GC formation (Ubbels et al., 1983; Elinson, 1985). All the changes relating to MT during hormone-induced maturation or artificial GC formation therefore need to be defined. Maturation would appear to be a period when the specific cytoplasmic conditions for GC formation are established. Many biochemical changes (e.g., in cAMP and Ca^{2+} levels, protein synthesis, and phosphorylation) occur at this time. Concomitant structural rearrangements such as germinal vesicle migration and breakdown probably also involve MT (Lessman, et al., 1986; J. Gautier, unpublished data). On the other hand, nuclear-cytoplasmic mixing following GVBD seems to regulate MT cycling in the oocyte, since it is followed in *Xenopus* by the formation of a special MT apparatus (Huchon et al., 1985; Jessus et al., 1986) and by changes in MT properties (Heidemann et al., 1985, Jessus et al., 1986). Thus, oocyte maturation involves a reorganization of the MT system, which then can be followed by cytoplasmic rearrangements.

We have proposed a mechanism to account for the triggering effect of MT inhibitors on early GC formation (Gautier and Tencer, 1987). At the end of maturation, an artificial GC would be the result of an MT cycle, that is, from depolymerization followed by repolymerization. Addition of MT inhibitors might artificially speed up this phenomenon, which normally takes place either during late maturation or during the subsequent fertilization and activation periods in a normal egg. In *Xenopus,* the symmetry reaction in the fertilized egg is actually accompanied by the appearance of a high level of polymerized tubulin (Elinson, 1985).

The effects of protein synthesis inhibitors (PSI) on artificial GC formation in the axolotl require clarification. This phenomenon is also related to changes in polymeric tubulin content (J. Gautier and J.-C. Beetschen, unpublished results). PSI might therefore act on the synthesis of one or several protein(s) involved in regulation of the MT cycle.

Heat shock-induced early GC formation is a very rapid phenomenon, occurring within a few minutes. Direct action on cytoplasmic structures (through the MT system or not) is more likely than inhibition of protein synthesis, although there is some inhibition during this time (Beetschen and Gautier, 1987).

CONCLUSION

The axolotl oocyte provides a useful alternative model for the study of amphibian oocyte maturation, although the details of MPF formation and its effects still have to be worked out. The changes leading up to the initial events of early morphogenesis have been shown to depend on nuclear-cytoplasmic interactions, while the cytoplasm itself is conditioned by hormone-induced processes. The characteristics of the oocyte cytoskeleton deserve greater attention. As well, the overall molecular bases of these gross phenomena still remain to be described in detail.

ACKNOWLEDGMENTS

We thank Mrs. C. Daguzan and Mrs. C. Ardourel for their invaluable technical assistance in these studies.

4

Spermatogenesis

JOHN B. ARMSTRONG

In urodeles, each testis is made up of individual lobes linked together by slender cords of tissue (Lofts, 1984). Their number is variable and seems to increase with age. Each lobe is composed of saclike structures, termed "lobules," which drain into the efferent ducts. In turn, each lobule is composed of a number of "cysts," each of which starts out as a stem cell or primary spermatogonium surrounded by follicle (Sertoli) cells (Fig. 4–1a). The lobules are surrounded by lobule boundary cells, which are steroidogenic and have an activity correlated with androgen secretion and the seasonal development of secondary sex characteristics.

In most temperate species, the reproductive cycle can be closely correlated with the seasons of the year. Norris et al. (1985) followed spermatogenesis in neotenic *Ambystoma tigrinum,* and their data can be taken as representive of most urodeles. Spring is the time of renewal of germinal tissue in the male gonad. It is also the active breeding season, and sperm stored in the vas deferens from the previous winter are laid in spermatophores. In May and June, a new generation of secondary spermatogonia arises from primary spermatogonia that have proliferated in regions where old cysts ruptured and degenerated. Meiosis begins in late June or early July.

Spermiogenesis begins in July, and spermatozoa are present by August. The testis is swollen and has nearly doubled in weight. In the autumn and early winter months, sperm are released into the vas deferens, and ruptured cysts degenerate. The weight of the gonad declines as most of the degenerating tissue is resorbed by the salamander in preparation for a new reproductive cycle.

Spermatogenesis is not perfectly coordinated in all cysts. It is initiated in new cysts over a 2- to 3-month period and, consequently, the testis generally contains a variety of cysts in different stages. In most urodeles there is synchrony within a zone of the testis such that, in cross section, all cysts are approximately at the same stage. This is termed zonation. Spermatogenesis therefore proceeds through the testis in a wave traveling along a cephalocaudal axis. The most mature stages will be posterior to the less mature ones. However, for the axolotl raised in captivity, there is no apparent zonation (Miltner and Armstrong, 1983).

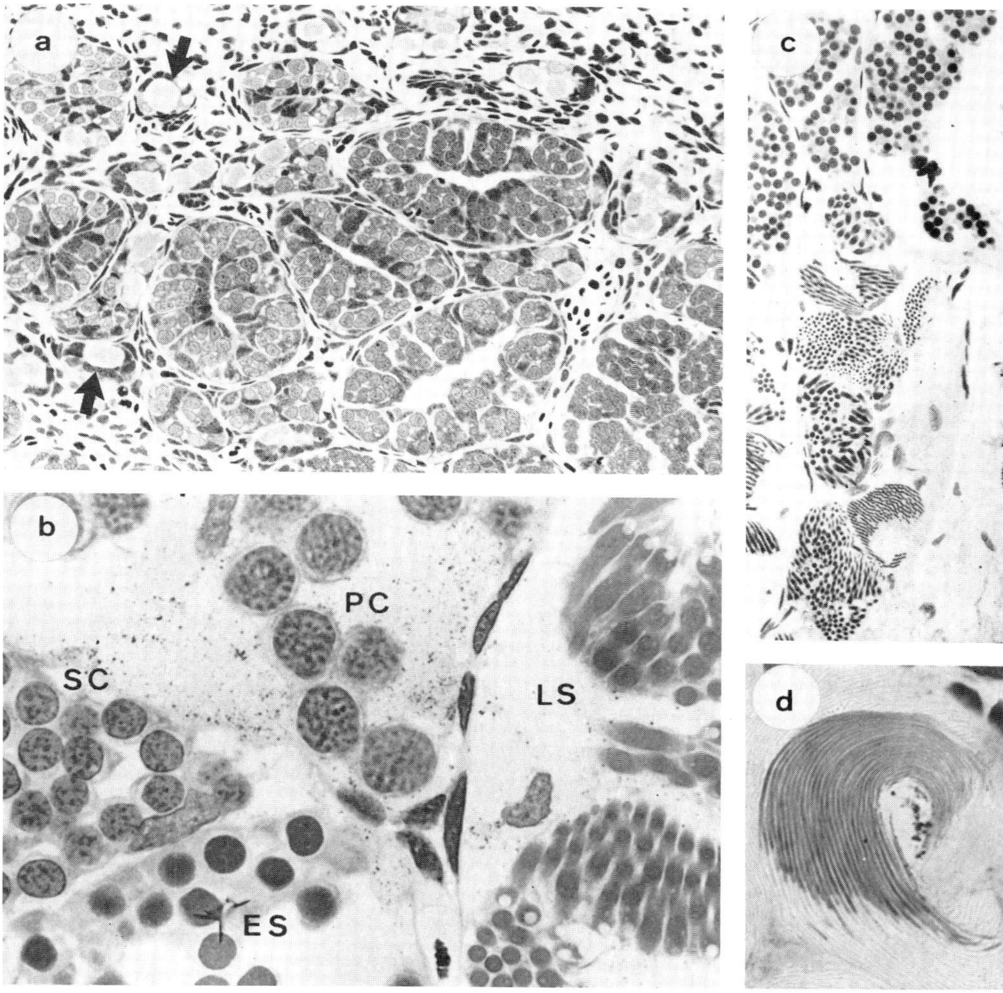

Fig. 4–1. (a) Spermatogonial cells showing the progression from primary spermatogonia (two of which are indicated by arrows) through to doughnut-shaped secondary spermatogonial cysts. ×110. (b) Section showing the size relationships between primary spermatocytes (PC), secondary spermatocytes (SC), early spermatids (ES), and spermatids that have matured further (LS). ×470. (c) Section showing secondary spermatocytes (top) and spermatids at various stages of elongation, some of which are seen in cross section and some in longitudinal section. ×110. (d) Mature sperm. ×470. (Adapted from Miltner and Armstrong, 1983.)

HORMONAL CONTROL

In amphibians, as in other vertebrates, testicular function seems to be controlled by pituitary gonadotropins similar to mammalian follicle-stimulating hormone (FSH) and luteinizing hormone (LH) (reviewed by Lofts, 1984). FSH stimulates spermatogenesis, while LH appears to regulate steroid synthesis by lobule boundary cells. Hypophysectomy leads to cessation of spermatogenesis and, in the long term, to regressed testes. In the eyeless mutant of the axolotl, the hypophysis is atrophied and the gonads remain immature (Chapter 13). Although axolotl gonadotropins have not been characterized, two distinct hormones have been demonstrated in *A. tigrinum* (Licht et al., 1983).

The secretion of gonadotropins is regulated by hypothalamic releasing factors, which in turn may respond to the extrinsic factors that provide timing cues (Lofts, 1984). There is also a negative feedback by the sex steroids that inhibits gonadotropin production.

The major steroids produced by urodeles are testosterone (T) and 5α-dihydrotestosterone (DHT). In seasonally breeding *A. tigrinum,* Norris et al. (1985) found that the plasma levels of both steroids were highest in September and November, when spermatogenesis had essentially finished and testes had begun to regress, and lowest in June following spawning and initiation of a new round of spermatogenesis. The peak coincided with hypertrophy of the lobule boundary cells. However, at present, it is not clear whether gonadal development correlates with levels of gonadotropin secretion or, instead, with fluctuations in gonadal sensitivity.

In the axolotl, Lazard (1979) has determined the activity of two steroidogenic enzymes. In the lobule boundary (Leydig) cells, he found the greatest activity at the very end of spermiogenesis, when sperm were being released into the vas deferens and plasma levels of T and DHT were presumably increasing. In contrast, the Sertoli cells showed maximum activity just before or during spermatogonial division. Since the levels of T and DHT should be low at that time, either the Sertoli cells are producing other steroid hormones or their products do not reach the circulatory system. Conceivably, this synthesis by the Sertoli cells could be under the control of FSH, as in higher vertebrates, and be responsible for the initiation of a new round of spermatogenesis.

NORMAL MORPHOLOGY

In the axolotl testis, each cyst starts out as a single stem cell (primary spermatogonium) that averages 24 μm in section. When stained with toluidine blue, they are characterized by large, lightly staining, irregularly shaped nuclei, and are surrounded by many smaller, darkly staining follicle cells (Fig. 4–1a). At some point, incomplete cytokinesis occurs after routine mitosis of a primary spermatogonium. The two daughter cells remain attached by a cytoplasmic bridge and secondary spermatogonia are formed.

All cells with the cyst divide synchronously and remain interconnected through several additional mitoses. Secondary spermatogonia are slightly smaller but similar in appearance to the primary spermatogonia. After additional mitoses they become more darkly staining, and numerous heterochromatic patches appear. A "mature" secondary spermatogonial cyst will often have the cells arranged around the perimeter with a clear area at its center (Fig. 4–1a).

The primary spermatocyte is the stage of first meiotic division. The various stages of prophase I are more easily distinguished in squash preparations than in sections (see Miltner and Armstrong, 1983, Figs. 4–6). Leptotene cells are densely packed with chromatin, but its fibrillar pattern is not easily distinguished. In zygotene, the chromatin appears less tightly packed and its fibrillar nature can be made out in some cells. In pachytene, the chromosomes can be distinguished as thickened fibers and, where these are not too tightly packed, synapsis of the chromatids can be distinguished. Finally, in diplotene, the paired chromosomes can be clearly distinguished. Diakinesis and metaphase I are easily recognized in both squash preparations and sections, as the chromosomes are fully condensed.

The secondary spermatocyte is the stage of second meiotic division. The cells are clearly much smaller than in the previous stages (Fig. 4–1b).

However, unlike *Xenopus* (Kalt, 1976), there appears to be no marked size reduction at the end of the secondary spermatocyte stage. Thus, very early spermatids are difficult to distinguish from secondary spermatocytes. Early spermatids stain more uniformly than the cells classified as secondary spermatocytes and appear more oval as they begin to elongate. As elongation progresses, spermatid diameter decreases (Fig. 4–1c). Mature sperm are recognized by the characteristic swirl of extremely elongated cells (Fig. 4–1d).

INITIATION OF SPERMATOGENESIS

In most urodeles studied to date, spermatogenesis is not initiated randomly throughout the year. Within a zone of the testis, initiation is highly synchronous, and within the testis as a whole it is restricted to a 2- to 3-month period. Consequently, the distribution of stages is relatively homogeneous among animals in the population, although more than one stage is present because of the spread in initiation. However, it is very different for animals examined at different seasons.

If, instead, entry into spermatogenesis were totally random, then the proportion of cysts found at a particular stage would depend only on the relative time spent in that stage. There would be no seasonal variations, and there would still be no significant variation between animals.

Neither *Xenopus* nor the axolotl appear to fit either of these two patterns. In *Xenopus,* Kalt (1976) found no indication of seasonality, although in the wild these frogs reportedly breed only in the early spring (Deuchar, 1975). However, he did find considerable variation in the distribution of cysts among the various stages and concluded that spermatogenesis occurs in continuous waves, with "clustered, non-random entry of stem cells into the secondary spermatogonial stage."

In the axolotl, we also found no obvious seasonal trends (Miltner and Armstrong, 1983). However, when individuals were compared, striking differences were seen in the proportions for different stages. For example, for animals examined in the same month, the proportion of primary spermatocytes ranged from 3 to 75% and of sperm from 0 to 100%. This is inconsistent with random entry, but consistent with spermatogenesis passing through the testes of each animal in a cyclic wave that depends on the animal's "biological clock," but not on the season. Consequently, two animals examined at the same time may be at widely varying places in their cycle. Presumably, wild animals are entrained by seasonal variation in temperature, precipitation, or some other cue missing in the constant laboratory environment. To our knowledge, the reproductive behavior of the wild stock in Lake

Xochimilco has not been reported. However, Brandon (1970) has studied the closely related neotenic species, *Ambystoma dumerilii,* from Lake Pátzcuaro, about 250 km (155 mi) to the west. This species appears to have a spring breeding season, and Brandon suggests that both temperature and rainfall may be important cues for synchronizing the maturation of gametes and spawning. Since the environmental parameters are similar for the Lake Xochimilco area, *A. mexicanum* probably responds to the same cues and normally breeds in the spring.

DURATION OF SPERMATOGENESIS

The time course for spermatogenesis has been examined in the red-backed salamander, *Plethodon cinereus* (Morgan, 1979), and the newt, *Triturus vulgaris* (Callan and Taylor, 1968). Both are temperate species and were collected from the wild in the spring, a time at which they might reasonably be expected to be initiating a new cycle of spermatogenesis. Callan and Taylor noted that in the fixed testes there was a "well defined progressive zonation of spermatogonial and spermatocyte follicles."

Kalt's study (1976) on the frog, *Xenopus laevis,* differed in that animals were obtained at intervals throughout the year. In our study of the axolotl (Miltner and Armstrong, 1983), animals were bred and raised in the laboratory. The time course was examined in both the spring and the fall. No difference was found.

The basic methodology employed in all these studies was essentially the same. Animals were injected with tritiated thymidine and were subsequently sacrificed at regular intervals. Squashes and/or sections of testes were examined by autoradiography and scored for the most advanced stage labeled to give the fastest rate of progression of cells through spermatogenesis (Fig. 4–2). In our study, spermatophores were also collected at weekly intervals to determine at what time label first appeared.

Fig. 4–2. Timing of spermatogenesis. The horizontal bars indicate the range of times for different animals over which radioactive label first appeared in the indicated stages. (Adapted from Miltner and Armstrong, 1983.)

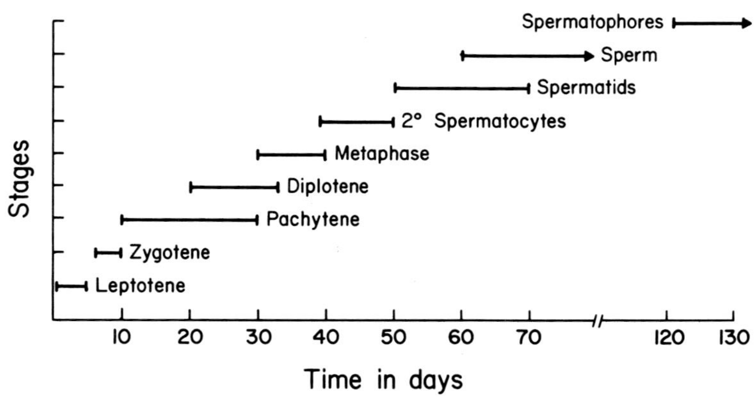

Table 4–1 Duration of Spermatogenesis: Time (in days) of First Appearance of Label in the Indicated Stage

Species	Stage[a]								Reference
	Z	P	D	M	2°SC	SPT	SPM	SPP	
Triturus vulgaris	6	14	18	20	21	22	—	—	Callan and Taylor (1968)
Xenopus laevis	5	10	22	23	23	24	37	—	Kalt (1976)
Plethodon cinereus	8	13	24	26	26	27	—	—	Morgan (1979)
Ambystoma mexicanum	5	10	25	32	40	50	65	121	This chapter

[a]Z, Zygotene; P, pachytene; D, diplotene; M, first metaphase; 2°SC, secondary spermatocyte; SPT, spermatid; SPM, sperm; SPP, spermatophore.

Previous workers have taken a fairly simple approach to interpreting their data. The duration of a stage has been calculated as the time from the first appearance of label in that stage to the first appearance of label in the next stage. However, one can reasonably expect some variation between animals. This will inevitably cause some overlaps in the data. For example, in our study, one animal showed labeled sperm at 60 days, whereas another showed labeled spermatids as late as 65 days. There seems to be no greater validity in defining the beginning of a stage by the first animal to show label in that stage than there would be in defining it by the last animal to show the label in the preceding stage. Therefore, we have used the median point between these two times to estimate the "average" duration of a stage. Our results are compared to those obtained for other amphibian species in Table 4–1. The duration of diplotene may be overestimated, because label in this stage alone was only observed twice; it was most often seen together with metaphase. If the animal showing diplotene labeled at 20 days is regarded as anomalous, the transition from pachytene to metaphase would be very rapid, the pachytene would be longer than indicated above, lasting from 10 to perhaps 31 days. Diplotene would be much shorter, probably lasting no more than 1 day (from 31 to 32 days). The other amphibian species also show rapid cycling through diplotene.

The time interval between labeling of spermatocytes and the appearance of that label in spermatophores was found to be 121 days. We have not directly determined the time spent as mature sperm in the testes and as stored sperm in the vas deferens. However, a rough estimate can be made from cyst counts based on DeFelice and Rasch's assumption (1969) that the relative frequency of cells in a given stage is proportional to the duration of that stage. Mature sperm cysts were found to make up, on the average, 58% of the postspermatogonial cysts. The 42% for the earlier stages would thus correspond to the 65 days taken to reach the mature sperm stage. On this basis, the duration of the mature sperm stage should be about 89 days, giving a total duration of 154 days. The numbers are certainly not in perfect agreement, but they suggest that, at least for these animals, sperm storage in the vas is minimal. In the wild, however, the situation could be quite different. In temperate urodeles, sperm matured in the summer may be stored for 6 months or more before the animal spawns the following spring (Lofts, 1974).

5

Regulation of Gene Expression During Early Development

JACQUES SIGNORET

All of the cells in an organism contain identical sets of genes, but these genes are differentially expressed. Development is essentially the result of a specific spatial and temporal regulation of gene expression. Germ cell genes must therefore be programmed to control two distinct processes of differentation: male and female gametogenesis. When associated in the egg cytoplasm, both sets of parental genes must then be reprogrammed in order to synthesize the gene products required for embryonic development. The new pattern of gene expression, in this particular case, results from complex interactions among (1) the zygotic nucleus, which has a specific set of regulated states for the genes of each parent; (2) the cytoplasmic environment, which changes from an oocyte to a developing egg through maturation and activation events; and (3) external influences, either physical or chemical, which are generally mediated through the cytoplasm. Later in development, different cell types express specific subsets of genes and exhibit appropriately specialized characteristics. This chapter will deal with the regulatory mechanisms of a particular set of genes expressed during early development.

NUCLEAR TRANSPLANTATION

Direct experimental analysis of the problem of early gene regulation began with nuclear transplantation. Originally, those experiments were designed to test whether a nucleus taken from an embryonic cell was able to play the same role as the nucleus of the zygote. The axolotl was first used for such experiments in 1962 by Signoret et al. Axolotls are very well suited to this type of experiment, particularly because of the many genetic markers that can be used to demonstrate the success of nuclear replacement.

It was soon discovered that normally fertile adult animals could be obtained from eggs containing nuclei taken from blastula stage embryos. Most nuclei from early developmental stages were able to replace the zygotic nucleus, and thus could be considered totipotent. However, such nuclei were rare on the second day of development (i.e., after gastrulation) and exceptional on the third day. Since injury due to experimental manipulation could not account for all the cases of abnormal or arrested development, it was assumed that the embryonic nuclei (or at least most of them) underwent progressive changes ("nuclear differentiation") that made them unable to participate in all aspects of normal ontogenesis (R. Briggs et al., 1964).

A review of the results of nuclear transplantation experiments suggests the need for two improvements in the system. First, we must recognize that the extreme "development or no development" result is an oversimplification of the subtle gradation of results actually observed. A more direct approach would be to focus on one gene, or on a system of a few genes, that could be introduced under different regulated states. The production of specific end product(s) would be an indication of gene expression. Second, the kinetics of cells during early development must be clearly understood. The kinetics of both the implanted nucleus and the host cytoplasm must then be taken into consideration.

THE CELL CYCLE, CELL KINETICS, AND GENE EXPRESSION

We have analyzed the cell cycle during the early development of the axolotl. Cleavage begins with a series of 11 relatively short, synchronous mitotic cycles (Signoret and Lefresne, 1971). During this time there is no G_1 phase, and G_2 is either absent or very short. S phase is a period of extremely rapid DNA synthesis, at over 10^4 replication forks, occurring at a rate of about 10^8 bp/sec per nucleus. Mitosis accounts for most of the cell cycle. Therefore, the chromatin is either condensed or actively replicating during synchronous cleavage, and very little time is left for gene expression. This is reflected by the fact that no newly synthesized RNA accumulates during this time. Discrete pulses of uridine incorporation have been reported, but these probably last no more than 5 minutes and vanish as soon as the nuclear envelope disappears (Signoret and Lefresne, 1975).

Following the eleventh cleavage, the midblastula transition occurs. This is marked by the desynchronization of cleavage and the beginning of expanded cell cycles with longer S phases and significantly longer G_1 and G_2 phases. Mitosis remains unmodified (Signoret, 1977). At this time de novo RNA synthesis can be detected and the expression of embryonic genes begins.

The one-cell stage deserves special consideration, quite apart from the kinetics of cleavage. This stage lasts about 7 hours and has a definite G_1 phase (>2 hours), a moderately short S phase (>1 hour), and a rather long G_2 phase (>3 hours). The chromatin of the parental pronuclei is diffuse, and there is no obvious reason to rule out gene expression during this stage. For technical reasons (due largely to the peculiarities of anurans) the one-cell stage has remained largely beyond the scope of molecular exploration. An investigation specifically designed to look for RNA synthesis in activated one-cell axolotl eggs proved that some synthesis does indeed occur (Thiebaud et al., 1983). Therefore, instead of looking at gene expression during progressive developmental stages, an interesting alternative is to limit observation to the one-cell stage.

When nuclear transplantation is performed, an unfertilized egg is preferred over a growing oocyte as the host cell, since it has a higher rate of protein synthesis and can be maintained for days. In addition, a 4°C treatment for 24 hours has proved very effective in allowing foreign substances to penetrate the egg. Nuclear transplantation, by itself, generally does not activate the egg when performed gently.

Artificially activated eggs may be kept for about 9 hours when used as host cells. Protein synthesis in activated eggs remains at the high level attained immediately after ovulation. As well, the events of the one-cell stage can be slowed down by lowering the temperature to 12°–14°C. It should be noted that these eggs often begin a process of fragmentation that mimics cleavage, but they never undergo gynogenetic development.

THE DNA LIGASE SYSTEM

DNA ligase was first identified as the enzyme able to join free ends of DNA molecules. This enzyme, required for replication, repair, and recombination, plays a key role in cell biology. In eukaryotic cells DNA ligase is present in two different forms. This molecular duality, first discovered in mammals, also occurs in chickens, newts, toads, trout, sea urchins, and crabs (Signoret and David, 1986a,b). When isolated tissues are studied they are generally found to contain either of two forms of the enzyme, and to operate at a single distinct level of activity. As a rule, nonproliferating cells use only DNA ligase II, the slow-sedimenting form of enzyme, whereas DNA ligase I, the fast-sedimenting form, is found in proliferating tissues. The early axolotl embryo offered us a unique system for the study of DNA ligase during development (J. C. David et al., 1979).

To distinguish between DNA ligases I and II, the DNA ligase activity of fractions separated by sucrose gradient centrifugation is measured. One unit of DNA ligase is defined as the amount that renders 1 nmol of 5′-^{32}P-labeled oligo(dT) resistant to alkaline phosphatase after 30 minutes at standard conditions. When the extracts from nonproliferating differentiated axolotl tissues such as brain, muscle, or liver are tested, single peaks of activity are obtained. These have a sedimentation coefficient of 6S and are defined as DNA ligase II. When developing embryos are tested in this manner, a single peak of activity is again observed. This, however, is due to a faster migrating (8S) form, DNA ligase I. This heavy form of DNA ligase is significantly more resistant to thermal denaturation than DNA ligase II, but is much more sensitive to inhibition by *N*-ethylmaleimide.

Rabbit antisera against both forms of DNA ligase have been obtained. Each enzyme is totally inhibited by its own antiserum, even when the antiserum is highly diluted. No cross-inhibition has ever been observed in amphibians. RNA has been extracted from cells possessing one or the other form of DNA ligase and translated in vitro. The translation products contained only the type of DNA ligase that was present in the original cells (Thiebaud et al., 1985). Taken together, these results suggest that DNA ligases I and II are different molecular entities that arise from distinct structural genes. In the axolotl, the DNA ligase genes are a model for study of gene expression and regulation during early development (Lefresne et al., 1983).

The replacement of DNA ligase II by DNA ligase I has been observed in both fertilized eggs and in eggs artificially activated by electric shock. This change occurs during the second half of the one-cell stage. It begins 4 hours after activation and is almost completed by 9 hours, at which time the egg is either cleaving or undergoing fragmentation. This change in DNA ligase can be prevented by treatments with agents that inhibit protein synthesis. Inhibitors of transcription, such as actinomycin D or α-amanitin also prevent the early switching of DNA ligase. Other metabolic and cytological activities appear to remain unaffected by the presence of these drugs (Brachet and Denis, 1963; Brachet et al., 1972). Similarly activated eggs that are mechanically enucleated or in which the chromatin is damaged by UV irradiation or by irradiation with 500 rad of γ rays from ^{60}Co do not switch ligases. DNA ligase I does not appear, and DNA ligase II remains as in the unfertilized egg.

Therefore, the appearance of DNA ligase I, and the concomitant disappearance of DNA ligase II, require the presence of the nucleus and involve both transcription and translation. This provides some proof of gene expression during the one-cell stage (Signoret et al., 1981). In an effort to be more precise, we have determined that a destructive dose of γ radiation inhibits the replacement of DNA ligase II by DNA ligase I only up to 3 hours postactivation. The same dose has no effect at 4 hours or later. Therefore, the nucleus is responsible for the appearance of DNA ligase I and the disappearance of DNA ligase II between 3 and 4 hours after activation. Other studies have determined that a significant amount of RNA synthesis occurs at about the same time. This causes the appearance of polysomes containing 24S poly(A)$^+$ mRNA, which codes for DNA ligase I (Thiebaud et al., 1983).

REGULATED STATES OF THE DNA LIGASE I GENE

The gene for DNA ligase I is not expressed in oocytes, and the unfertilized egg has no detectable levels of mRNA coding for this form of the enzyme. At 3.5 hours after activation, the gene begins a period of intense expression involving both transcription and translation. DNA ligase I activity becomes maximal 9 hours after activation. This pattern of DNA ligase activity is maintained throughout development, even though no significant de novo synthesis of mRNA can be detected during cleavage. Apparently both the enzyme and the corresponding mRNA are very stable.

The DNA ligase I gene begins to be highly expressed at about the same time as DNA synthesis occurs in the pronuclei. Moreover, this change in the regulated state of the gene is either prevented by inhibitors of DNA replication (such as aphidicolin) or delayed until the inhibition is relieved (as with cytosine arabinoside, and with late addition of deoxycytidine). Therefore, the gene is in an open state during the G_1 phase of the early one-cell stage. During this time the gene is not overtly expressed but is able to switch to an actively expressed state following the first round of DNA replication. This process may involve an interaction with a transient transcription factor present in the zygote.

When nuclei taken from early developmental stages are implanted into enucleated egg cytoplasm, they are able actively to express the gene for DNA ligase I.

This expression is very sensitive to α-amanitin and probably involves immediate transcription of the gene. As with the zygotic nucleus, the implanted nucleus must undergo DNA replication before the intense expression specific to the one-cell stage is seen. The DNA ligase I gene in the donor cell would thus appear to be in a regulated state similar to that of the early one-cell stage; in other words, it would be in an open state until DNA replication (in the specific environment of the egg cytoplasm) triggers the switch to an actively expressed state.

Older embryos (tailbud stages or young larvae) possess only DNA ligase I, just like those undergoing cleavage. The change to DNA ligase II occurs gradually several days later in the swimming larvae. However, studying the repression of the DNA ligase I gene, one must remember that the gene may be repressed even in cells with DNA ligase I activity, because of the relative stability of the enzyme previously produced. When implanted in the cytoplasm of the egg, a nucleus taken from any of these later stages is unable to express DNA ligase I. The gene is in a repressed state that seems to be stable and self-propagating, even after DNA replication and mitosis in the egg cytoplasm.

The change from the open state to the repressed state has been observed more precisely in notochord cell nuclei. It occurs during late-neurula stages and progresses cephalocaudally along the body axis. The DNA ligase I gene is repressed in both the somatic cells and the germ cells of older larvae. Therefore, a second change must occur during late gametogenesis to cause the gene to enter the derepressed (open) state found in unfertilized eggs. Maturation, which is known to cause large changes in gene expression (DiBernardino, 1980), could cause this particular change in the regulatory state of the DNA ligase I gene.

These results were obtained with normal diploid embryos. When unfertilized eggs are artificially activated, DNA ligase I replaces DNA ligase II under the control of the female pronucleus (as for the zygotic nucleus), even though development is never observed. When the female pronucleus is removed from a normally fertilized egg, haploid androgenetic embryos can be obtained. Although these embryos appear to cleave normally and undergo some development, DNA ligase I never appears. Androgenetic embryos at any stage of development possess only the DNA ligase II already present in unfertilized eggs. The gene for DNA ligase I appears to be repressed in the male pronucleus, and in any mitotic progeny produced by it. These haploid androgenetic nuclei, when transplanted into enucleated egg cytoplasm, behave identically to the original male pronucleus and never begin producing DNA ligase I.

This remarkable observation has led us to conclude that the gene for DNA ligase I is regulated differently in the male and female pronuclei, and in their respective mitotic progeny. Therefore, it can be assumed that the maternal and paternal copies of the gene for DNA ligase I in the zygote are inherited from gametes in different regulatory states. The maternal and paternal genes are open and repressed, respectively. These genes remain in the same states through self-replication and may, in this way, be differentially expressed during early development. The presence of an inactive copy of the DNA ligase I gene in the paternal genome has been experimentally confirmed by treatment with spermine. Spermine allows the gene to switch from the repressed state to the derepressed open state and thus allows it to be actively expressed following DNA replication in egg cytoplasm.

EXPRESSION OF THE DNA LIGASE II GENE

DNA ligase activity in differentiated cells of larvae is due to DNA ligase II. RNA extracted from differentiated axolotl tissues can produce DNA ligase II in cell-free systems and therefore must contain the corresponding RNA. During haploid androgenetic development, DNA ligase II is the only form of the enzyme produced at any time. Nuclei taken from cells producing DNA ligase II and implanted into enucleated egg cytoplasm produce only DNA ligase II. When the experiment is performed within the same species it remains inconclusive, however, since the egg cytoplasm already contains DNA ligase II. Heterospecific transplantation can circumvent this problem, since axolotl DNA ligase II can be clearly distinguished from the enzyme found in *Pleurodeles.* The two species-specific forms of DNA ligase II have different sedimentation coefficients and do not cross-react immunologically (Signoret et al., 1983). The appearance of axolotl DNA ligase II in *Pleurodeles* thus indicates that the gene is being expressed.

The paternal DNA ligase II gene is expressed when the paternal genes are alone in the egg cytoplasm. However, the paternal gene is undetectable when present with maternal genes in the same zygote. This remarkable interaction has been described as the "exclusion process" and will be discussed in greater detail. The exclusion process normally operates during early development. It is the result of an interaction between the maternal genes (which actively express DNA ligase I) and the paternal genes (in which the DNA ligase I gene is repressed), which causes the DNA ligase II gene to be repressed as well (Fig. 5–1). This may be imitated experimentally using nuclear transplantation. Neither androgenetic nuclei implanted in the cytoplasm of *nucleated* activated eggs nor those implanted in enucleated cytoplasm at the same time as diploid nuclei from blastulas (which express DNA ligase I) can express DNA ligase II. In both cases expression of DNA ligase I results in an inhib-

Fig. 5–1. Expression of genes for DNA ligase II (left) and DNA ligase I (right). The gene in a rectangle is under the repressed state. Stars represent transcription factor.

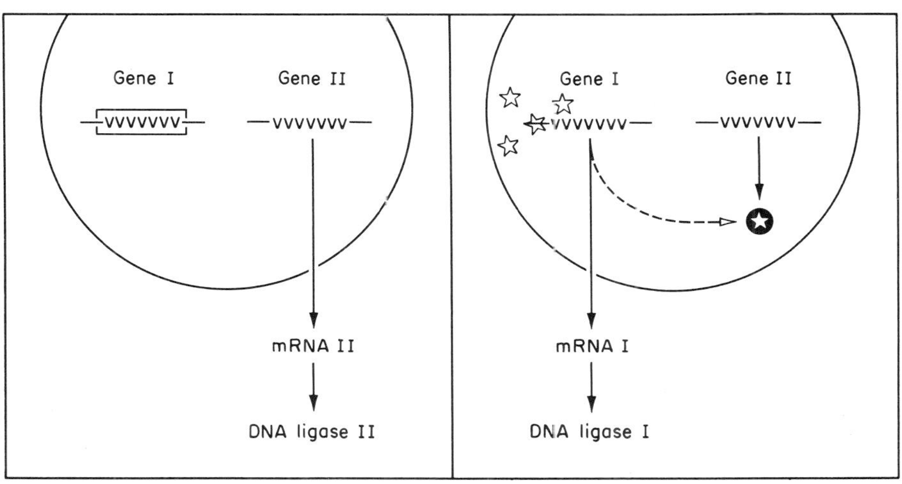

ition of the DNA ligase II gene. This is true whether the DNA ligase I is from the same set of genes or from another nucleus in the same cytoplasm. For these reasons, we believe that the gene for DNA ligase II is always ready to be expressed but that its expression depends on the absence of DNA ligase I.

We have noted that nuclei taken from early embryos and implanted in egg cytoplasm express only DNA ligase I. This expression of DNA ligase I is sensitive to inhibitors of transcription, such as α-amanitin. When treated with α-amanitin, these nuclei begin to produce significant levels of DNA ligase II. This must be due to posttranscriptional expression of DNA ligase II genes. Therefore, the DNA ligase II genes in early embryos must produce transcripts that can accumulate in the nuclei in a relatively stable form without being translated. This situation must then be maintained for as long as the DNA ligase I genes are actively expressed. When this expression ceases because of the α-amanitin, the inhibition stops and transcripts for DNA ligase II may be processed and translated. During later stages of development the DNA ligase I genes are repressed, after which the genes for DNA ligase II can be freely expressed. Since the interaction between the DNA ligase II transcripts and the inhibitor gene product fails to operate between species, it is probably very specific.

A CHRONOLOGICAL MODEL FOR DEVELOPMENTAL GENE REGULATION

A system of gene regulation operating during development should be considered chronologically. Genes undergo a series of regulatory changes as they form germ cells, fuse, and develop into differentiated tissues. The time required for the mRNA to become available, and for the enzyme to accumulate, causes a significant delay between any change in the regulatory state of a gene and the subsequent observable change in gene expression. These considerations, combined with our analysis of the early regulation of DNA ligase genes during development, suggests the model shown in Fig. 5–2.

The diploid oocyte has two copies of the DNA ligase I gene, both of which are repressed, and two copies of DNA ligase II gene, both of which are expressed. The cytoplasm contains mRNA coding for DNA ligase II, and all DNA ligase activity in the oocyte results from this form of the enzyme. Maturation of the oocyte involves important genetic reprogramming during which the genes for DNA ligase I are derepressed, even though they are still not significantly transcribed. At this point the unfertilized egg still contains only DNA ligase II mRNA and enzyme.

Fertilization triggers the completion of meiosis and yields a zygote containing a haploid female pronucleus containing a single derepressed copy of the DNA ligase I gene and a copy of the DNA ligase II gene. The zygote also contains a haploid male pronucleus containing a *repressed* copy of the DNA ligase I gene and another copy of the DNA ligase II gene. The one-cell stage of the fertilized egg begins with a G_1 phase of about 2.5 hours, during which the pronuclei remain separate. When DNA replication occurs in the pronuclei, the paternal genes retain their original regulatory state because the repressed status of the paternal copy of the DNA ligase I gene is self-propagating. The maternal copy of this gene, which is already derepressed, undergoes a specific activation at the time of DNA replication. A specific

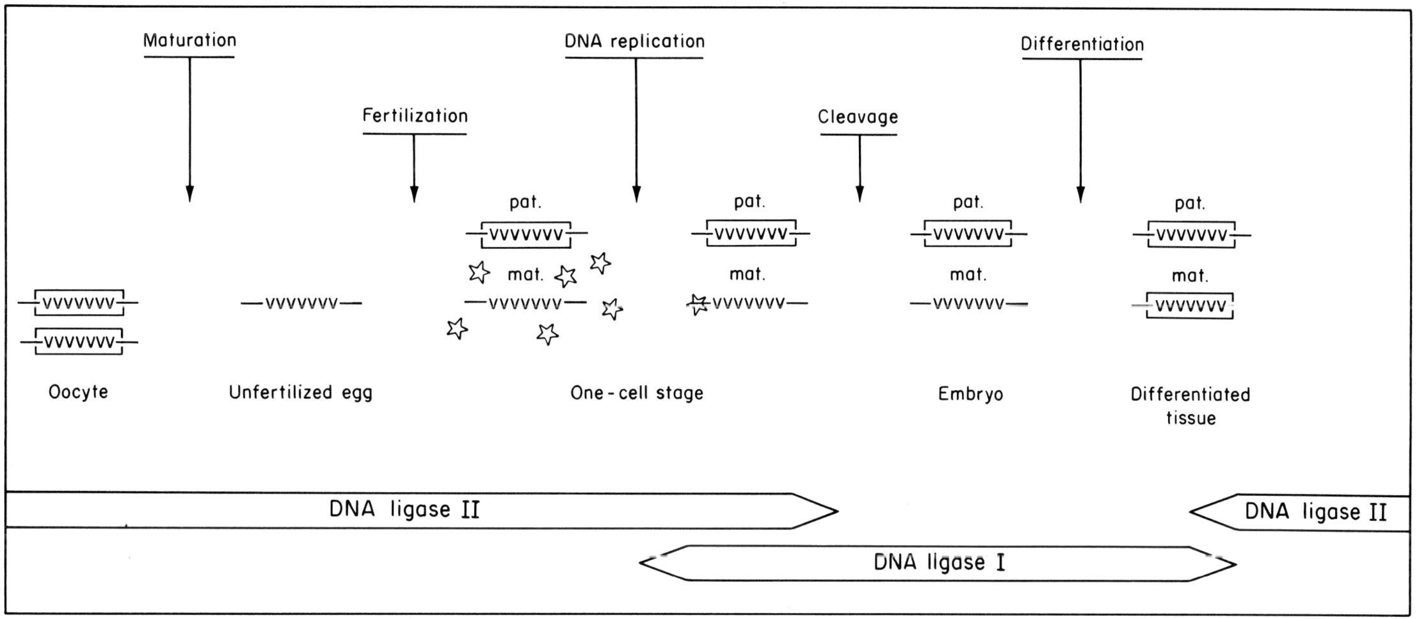

Fig. 5–2. Sequential events regulating the expression of DNA ligase molecular forms in development. Genes represented are for DNA ligase I, respectively, paternal (pat) and maternal (mat). Gene in a rectangle is under the repressed state. Stars represent transcription factor.

factor present during replication probably associates with the derepressed maternal gene in order to form an active transcription complex. Consequently, the replicated maternal copies of the DNA ligase I gene begin a period of intense expression.

At the one-cell state, G_2 is a critical time for DNA ligase gene expression. Newly synthesized DNA ligase I mRNA appears and accumulates, and the enzyme itself gradually becomes predominant. At the same time, mRNA for DNA ligase II gradually disappears and this enzyme vanishes. DNA ligase I therefore becomes responsible for all the ligase activity present in the egg by the time it begins cleavage.

During early cleavage the regulatory states of the DNA ligase I genes are self-propagating, but the specific transcription factor for the maternal gene gradually becomes less available. Therefore, of the two copies of the gene found in the nuclei of the blastula, the paternal copy remains repressed and the maternal copy remains derepressed but lacks the necessary transcription complex. The maternal copy of the gene is transcribed only at a low level, which is just sufficient to maintain a constant level of DNA ligase I mRNA and enzyme. Later in development the lengthening of the cell cycle, combined with the increased number of cells, favors gene expression in general.

Throughout this process both paternal and maternal DNA ligase II genes have been transcribed, but the "exclusion process" does not permit the transcripts to be completely processed and the corresponding mRNA never appears in the cytoplasm. The derepressed maternal copy of the DNA ligase I gene is expressed constantly, and thus DNA ligase II gene expression is inhibited. DNA ligase activity in the cell is due exclusively to DNA ligase I. The cytoplasm contains mRNA for DNA ligase I, but immature transcripts for DNA ligase II are found only in the nuclei.

Later in development, the different cell types begin to differentiate, and the mitotic rate gradually decreases. The maternal copy of the DNA ligase I gene is eventually repressed. This occurs at different times in different cell types. From this point on, the cell contains only two repressed copies of the DNA ligase I gene and transcription no longer occurs. The enzyme nevertheless persists for some time. DNA ligase I mRNA disappears gradually and, somewhat later, so does this form of the enzyme. Thus the exclusion process gradually becomes less important and expression of DNA ligase II begins. DNA ligase II replaces DNA ligase I in the differentiated cell. The stability of the DNA ligase I gene products (both mRNA and enzyme) causes a delay between the time the gene is switched off and the actual change in the enzyme. Germ cell precursors are just another specialized cell type containing two repressed copies of the DNA ligase I gene and using only DNA ligase II.

CONCLUSION

DNA ligase genes are a unique system for studying the regulation of gene expression during development. Though not completely understood, and certainly worthy of further investigation, this system has permitted us to formulate some interesting hypotheses concerning gene regulation. The DNA ligase I gene is transiently expressed during early development, whereas the gene for DNA ligase II is expressed only after differentiation. We have also established that the paternal and maternal copies of a gene are transmitted through the gametes, inherited by the zygote, and replicated in the embryo while maintaining different regulatory states. In addition, a remarkable interaction, which we have described as the exclusion process, allows one gene to control the expression of another (nonallelic) gene.

In the axolotl these studies should be continued at the molecular level, perhaps through the use of cDNA, specific antibodies, etc. In addition, other genes (such as oncogenes or heat shock genes) may be studied in the same way. DNA ligase can also be investigated in other developing systems, such as the sea urchin and the mouse. This study of early development in the axolotl has allowed us to cast some light on a few of the fundamental problems associated with the early control of gene expression.

6

Regional Specification in Early Development

JONATHAN M. W. SLACK

Amphibian embryos have long been the first choice for the experimental embryologist on account of their ease of availability, large size, and accessibility at all developmental stages. In the early days, embryos for study were collected from the wild, which meant that local species were used, for example, *Triturus vulgaris* and *Rana temporaria* in Europe and *Ambystoma punctatum* and *Rana pipiens* in North America. The axolotl was one of the first species to be domesticated and reared in the laboratory for embryological purposes, although it probably now ranks second in popularity to *Xenopus laevis*. In this chapter I review what is known of the mechanism of regional specification of the axolotl body plan and comment on those features that seem different from *Xenopus*. That these differences indeed exist—particularly for biochemical data—should serve as a caution against using only a single model species and the consequent unwarranted generalizations.

PRACTICAL CONSIDERATIONS

From the practical point of view, the main difference between axolotl and *Xenopus* embryos is that axolotls develop at one-third the speed of *Xenopus*. This has some advantages when several meticulous dissections must be performed on a short-lived stage, such as the early gastrula or the open neural plate neurula. It is also possible by juggling temperatures to arrange that blastulas, gastrulas, and neurulas should become available on successive days. It means, however, that explants from early embryos take 10–14 days to differentiate fully when cultured at the usual temperature of 18°C. Unfortunately, the polyspermic character of axolotls means that the dorsoventral polarity cannot be inferred from the position of the sperm entry point, as in *Xenopus*, and so before stage 10 one must rely on the gray crescent, which is clearly visible only on a minority of egg batches. On the other hand, the early neurula of the axolotl is a delight both to behold and to dissect, and operations can also easily be performed on individual organ rudiments such as the limb anlage until the time that the epidermis becomes elastic, about stage 36.

FATE MAP

The classical fate map for the stage 10 gastrula was prepared by Pasteels (1942), using the method of localized vital staining (Fig. 6–1). Like all vital-stain maps, it

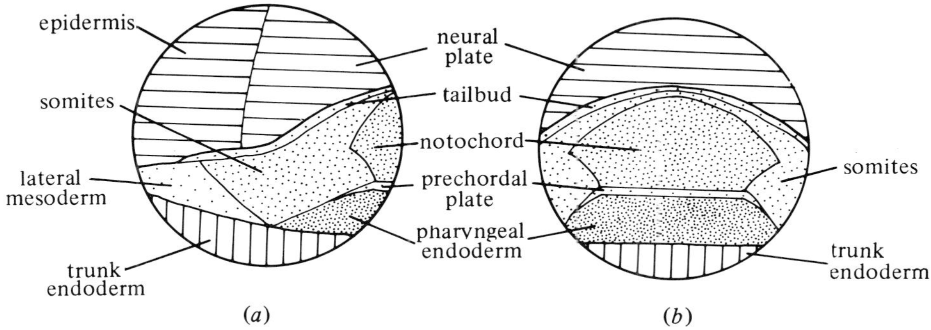

Fig. 6–1. Fate map of the early gastrula of the axolotl. (a) Side view; (b) dorsal view. (From J. Pasteels, J. Exp. Zool. *89*:255–281. Copyright © 1942, Alan R. Liss, Inc. Reprinted with permission.)

exaggerates the precision of the topographic projection from early to late stages. Later studies, using injected cell lineage labels, show that substantial cell mixing occurs during gastrulation, and we now think it is impossible to represent the positions of small structures such as single somites on a pregastrulation fate map (Dale and Slack, 1987a). However, the general form of Pasteels' fate map seems correct, and since no complete study has yet been performed with lineage labels, it remains still the initial point of departure. A study by Cleine and Slack (1985) (Fig. 6–2), using orthotopic grafts from embryos labeled with fluorescein dextran amine (FDA), showed an arrangement of prospective regions in the marginal zone similar to Pasteels' map, but suggested that the boundary between prospective epidermis and neural plate lies nearer the dorsal pole. This correction would bring the map closer to that for anurans. In another experiment, Smith and Malacinski (1983) labeled the external surface of early gastrulas with ^{125}I-labeled Bolton Hunter reagent and located the progeny cells at the early-tailbud stage. In *Xenopus* the outer surface ends up as the epidermis, the lining of the gut lumen, and the lining of the neural tube lumen. However, in the axolotl there is also extensive labeling of mesoderm, indicating that much of the prospective mesoderm lies on the blastula surface and also showing that there can be interchange of cells between superficial and deep layers. This study supports the impression given by dissection of embryos that gastrulation in the axolotl follows closely the course described for urodeles by Vogt (1929).

SPECIFICATION MAP

The development of small explants of tissue in isolation gives us an idea of the regional developmental commitment at the stage in question. Isolates from the animal pole region, conventionally regarded as ectoderm, were studied by Slack (1984b) using various techniques. Epidermis always differentiated around the outside, with a time course similar to intact embryos. The inner cells, however, seemed to remain undifferentiated for longer, to die, or sometimes to form structures

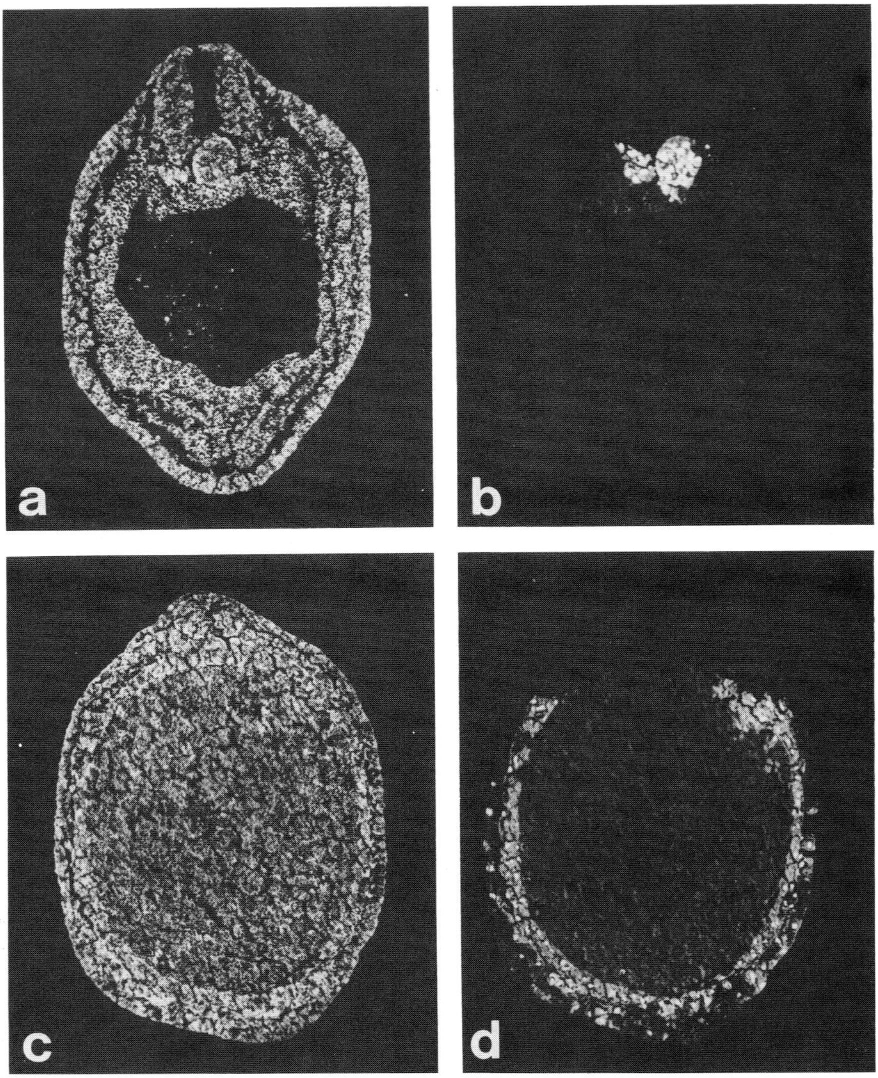

Fig. 6–2. Fate mapping of the axolotl by orthotopic grafts labeled with FDA. (a,b) Dark-field and fluorescent views after graft of dorsal marginal zone. (c,d) Dark-field and fluorescent views after graft of similar-size piece from ventral marginal zone. Note that the dorsal convergence movements during gastrulation have compressed the dorsal graft and extended the ventral graft. (Taken from Cleine and Slack, 1985.)

resembling neural tubes (autoneuralization, Fig. 6–3a,b). In *Xenopus*, ectoderm explants do not autoneuralize, suggesting that the choice between epidermal and neural development in the axolotl is more finely balanced and can be tipped by a variety of minor perturbations.

Explants from around the marginal zone of early gastrulas have been studied by Koebke (1977) and by Cleine and Slack (1985). The earlier study used conventional histology to asses the results, the later one the appearance of specific glycoprotein markers. Both concur in finding that notochord often develops from explants from the lateral as well as the dorsal marginal zone. This is similar to the behavior of other urodeles studied by Holtfreter (1938) but is not found for *Xenopus* where notochord develops only from a small sector of dorsal marginal zone coincident with its prospective region on the fate map (Slack and Forman, 1980; Dale and Slack, 1987b). Axolotl eggs probably undergo similar cytoplasmic movements after fertilization to those of *Xenopus,* with concomitant determination of the dorsoventral axis. However, the dorsal marginal zone shows only a weak bias toward notochord formation as late as stage 10, suggesting that there must be some intermediate steps between the cytoplasmic movements and regional specification within the mesoderm.

MESODERM INDUCTION

Studies on *Xenopus* have led us to the conclusion that there are three distinct inductive interactions in early development. Mesoderm induction involves the emission of signals from the vegetal hemisphere that cause a ring of tissue around the equator to become specified as mesoderm. There seem to be two distinct sig-

Fig. 6–3. Cell differentiation in the axolotl. (a) Isolated explant from the animal pole region of an early gastrula. (b) A similar explant after 10 days culture. It has formed an epidermal vesicle containing a neural tube-like structure and some undifferentiated cells. (c) Axial structures at the end of neurulation (stage 20). The neural tube, epidermis, notochord, and somites are all distinct tissue blocks, but cytodifferentiation has hardly started. (d) Axial structures at 10 days showing complete cytodifferentiation of the various tissues.

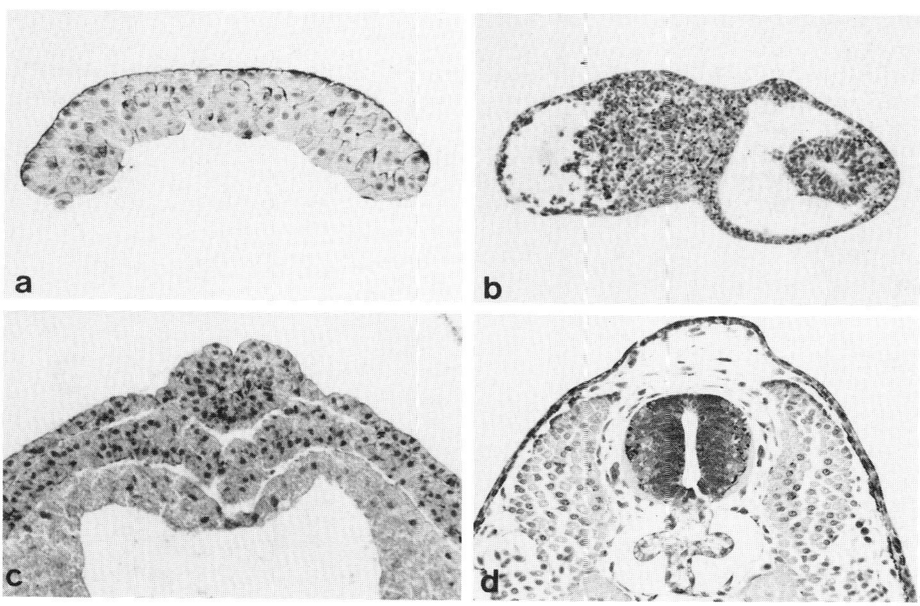

nals: a dorsal one confined to the extreme dorsal region that induces the organizer; and a ventral one that extends around the rest of the circumference and induces mesenchyme and blood cells. During gastrulation a third signal emitted by the organizer induces the intermediate structures such as somite and kidney from the mesodermal mantle as a function of distance (J. C. Smith et al., 1985; Dale and Slack, 1987b). This course of events is probably also followed in the axolotl.

The first experiments on mesoderm induction were, in fact, carried out using the axolotl (Nieuwkoop, 1969). The experimental design was derived from the classical studies of Hörstadius on the sea urchin; blastulas were divided into four zones along the animal–vegetal axis, and their self-differentiation and development in combination were studied (Fig. 6–4). Zones I and II formed only ectodermal structures in isolation, and zone IV formed only endoderm. Zone III, corresponding roughly to the marginal zone, formed a variety of mesodermal tissues and could on occasion develop into an approximation of a complete embryo. But extensive development could also be obtained by combinations of zones I + II + IV, suggesting that an inductive interaction was responsible for mesoderm formation. Although this work was carried out before monoclonal antibodies and cell lineage labels became available, the later studies strongly suggested that all the induced mesoderm came from the former animal hemisphere tissue, and that its regional character (dorsal or ventral) was a property of the inducer and not of the responding tissue (Nieuwkoop and Ubbels, 1972; Boterenbrood and Nieuwkoop, 1973). It has been shown that axolotl ectoderm, like that of other amphibian species, is competent to respond to the "vegetalizing factor" of Tiedemann (Minuth, 1978; Cleine and Slack, 1985).

In *Xenopus* it is well established that the dorsal inducing center is not present as a determinant in the egg but arises as a result of postfertilization cytoplasmic movements. These can be inhibited by prior irradiation of the vegetal hemisphere

Fig. 6–4. Nieuwkoop's mesoderm induction experiment. The axolotl blastula is divided into four zones along the egg axis. Mesoderm is formed either by the isolated zone III, or by a combination omitting zone III.

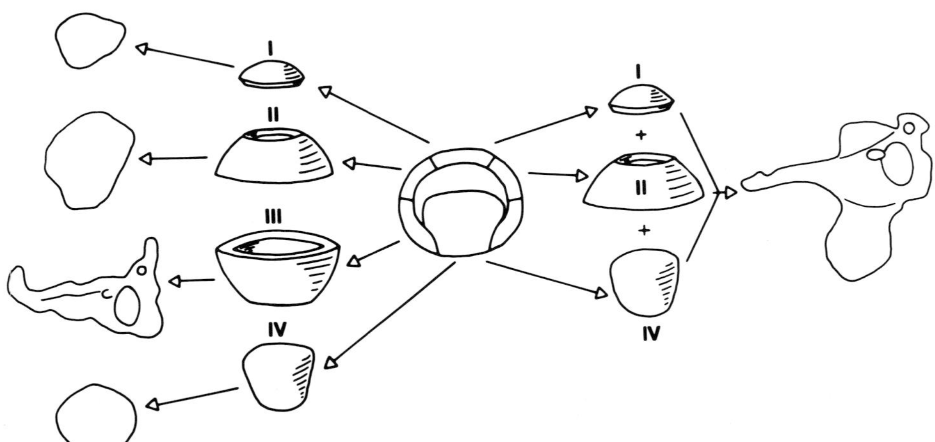

with UV light. Probably the same is true in the axolotl, since axis formation can also be inhibited by UV irradiation of the egg (Malacinski et al., 1975) and it has not been found possible to produce axis-containing inductions with grafts of dorsal egg cortex (Malacinski et al., 1980). Further indirect evidence for the importance of postfertilization cytoplasmic movements in the establishment of the dorsal inducing center comes from the study of 180°-rotated eggs, which arrest during gastrulation and whose blastopore lip does not have its usual inductive activity (Malacinski and Chung, 1981; Chung and Malacinski, 1982).

DORSALIZATION

In *Xenopus,* we believe that pattern formation within the mesoderm is due to a third inductive signal, emitted by the dorsal marginal zone, which partitions the rest of the mesoderm into differently specified territories in dorsoventral sequence. In the axolotl, we may infer that dorsalization occurs within the mesodermal mantle from two types of observation. First, when an organizer graft is performed, host cells from the ventral midline enter the somites and kidney of the secondary axis (J. M. W. Slack, unpublished observation). Second, dorsal marginal tissue from the axolotl can induce large muscle masses from isolated ventral marginal zones from *Xenopus* (Slack and Forman, 1980).

There is also a dorsalizing signal emitted by the notochord in the early neurula, and this may well be the same as the dorsalizing power of the organizer during gastrulation. In *Xenopus* such phenomena probably cannot be demonstrated, since terminal differentiation starts in most tissues at the end of gastrulation; in urodeles, however, some lability persists and was studied by Yamada (1937, 1940). Yamada found that prospective somite grafted to a ventral position usually differentiated into kidney unless it was accompanied by notochord. Conversely, prospective blood island tissue from the ventral midline could form muscle or kidney if grafted to a more dorsal position or if cultured together with notochord. Since Yamada's work had to rely on vital staining for tissue identification, some of his results may not be correct. However, Forman and Slack (1980) made a thorough study of the prospective somite from axolotl neurulas using triploidy as the cell marker and found that it could indeed form pronephric tubules or blood cells up to the end of neurulation if cultured in isolation or if removed from its normal paraxial position to a more ventral one.

NEURAL INDUCTION

The occurrence of autoneuralization in isolated ectoderm means that the axolotl is not the ideal embryo for the study of neural induction. However, when it comes to the determination of parts within the central nervous system, the axolotl embryo, with its large, open neural plate, is very suitable for grafts and electrophysiological experiments. An intriguing question here is whether there is any relationship between the electrochemical gradients that can be detected, and regional pattern formation within the neural plate.

Warner (1973, 1981) showed that the resting membrane potential in neural plate cells rose from about −35 to −70 mV in the course of neural tube closure. This change was due to an increased activity of the sodium pump; furthermore, blockage of the pump by strophanthidin could block the differentiation of neurons in culture (Messenger and Warner, 1979), suggesting that it is not simply an early marker of neuronal differentiation. C. O. Jacobson (1959) established the fate map for regions within the neural plate by vital staining. Studies of determination by transplantation or rotation of neural plate fragments indicate that some reprogramming is possible, especially for small grafts carried out at early-neurula stages (C. O. Jacobson, 1964; Model, 1982).

CYTODIFFERENTIATION

Terminal differentiation is the final stage of the series of developmental decisions responsible for pattern formation in an animal embryo. It consists of the synthesis of such quantities of specific molecules as to render cells visibly distinguishable from one another through the light microscope (Fig. 6–3c,d).

Those who work on early development generally agree that early markers of cell state are an essential precondition for accurate experimentation. Regrettably, in the early amphibian embryo the search for early markers has so far yielded only terminal differentiation products. However, this has had the effect of shortening the time base of experiments, since modern biochemical and immunological methods usually enable earlier observation of a substance than does classical histology.

A comprehensive study of regional differences in protein, glycoprotein, and glycolipid synthesis was attempted for the early axolotl neurula by Slack (1984a). This yielded a number of specific markers for epidermis, including some cytokeratins, two glycolipids, and a sugar-rich glycoprotein, epimucin. Also, two minor glycoproteins were found to be specific for the notochord, but other subdivisions of the mesoderm and the neural plate seemed devoid of specific markers at the level of resolution obtained. In *Xenopus* the pattern of glycoprotein synthesis is surprisingly different (J. C. Smith and F. M. Watt, 1985), and most tissues probably commence their terminal differentiation in late gastrula or early neurula (Dawid et al., 1985). In particular, muscle differentiation commences in the neurula, while in the axolotl the synthesis of α-actin and the appearance of the first thin-filament bundles do not occur until stage 26 (Mohun et al., 1980). In general, therefore, we may suspect that events occur later in the axolotl than in *Xenopus*—not only in absolute time, but also with reference to morphological stage. This may account for the apparently greater plasticity of the axolotl in various respects, including the wide area of marginal zone capable of becoming notochord in isolation, the tendency of isolated ectoderm to autoneuralize, and the continued lability of somite determination in the neurula.

LIMB DEVELOPMENT

By the end of neurulation, all the main parts of the body are in their final positions and are determined to become the appropriate organs. However, many later processes involving cell interactions act within each organ rudiment to produce the eventual pattern of cell differentiation. An example of such a "secondary field" is the limb rudiment, lying ventral to the pronephros at the level of somites 3–5 (Fig. 6–5). It is classically supposed to consist only of somatic mesoderm, but there may be a contribution from the splanchnic layer as well. At the level of somite 5 the rudiment thins sharply, and the flank mesoderm posterior to this level is not divided into two layers. Formation of the limb bud itself does not occur until about stage 38, but operations on the limb rudiment are most easily carried out about 2 to 3 days previous to this at stage 30–34.

Classically the work of Harrison (1921, reviewed by Swett, 1937) on *Ambystoma punctatum* had led to the concept of the limb rudiment as a homogeneous group of cells that autonomously became polarized first in the anteroposterior (AP)

Fig. 6–5. Horizontal section through a stage 30 axolotl embryo showing the location of the limb rudiment relative to somites, pronephros, and flank tissue.

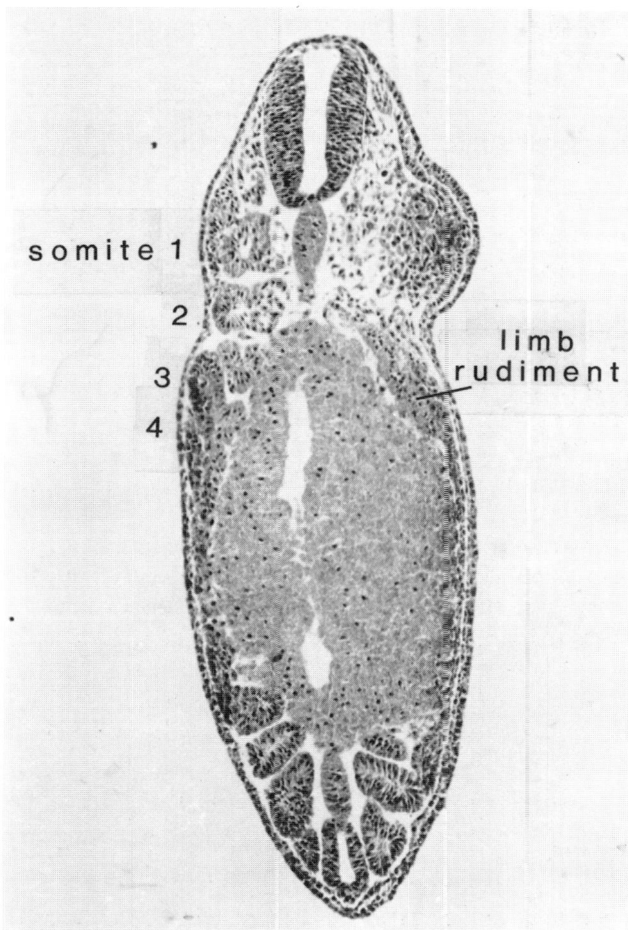

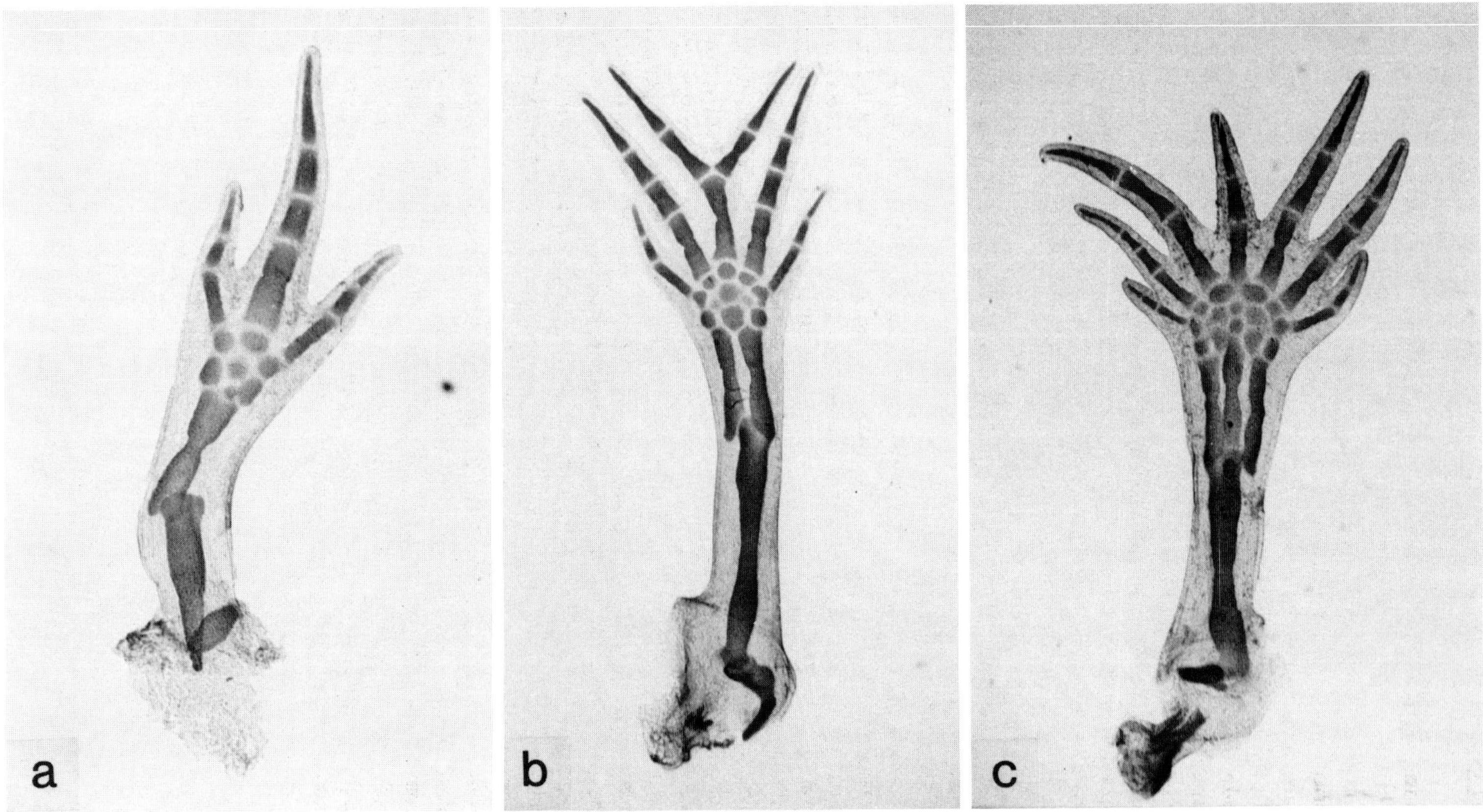

Fig. 6–6. Double posterior-duplicated limbs, each arising from a piece of limb rudiment sandwiched between two pieces of flank tissue. Whole mounts stained with alcian green.

axis, then in the dorsoventral (DV) axis, and finally in the proximodistal axis. Later work has suggested that Harrison's "limb disk" actually included surrounding tissues as well as the limb rudiment proper, and that interactions with the dorsal and the posterior regions are essential for bud outgrowth and for pattern formation (Stocum and Fallon, 1982).

Slack (1980) made a study of the development of half-limb rudiments grafted to the head of the embryo. This operation is the equivalent of a test of specification by explant culture for early embryos, since the host seems to have no morphogenetic effects on the graft. Anterior and ventral halves do not develop, while posterior and dorsal halves do. Moreover, anterior halves can be "rescued" and caused to develop to complete limbs by cografting a strip of tissue from the flank—that is, taken from far behind the normal limb rudiment. This suggests that *both* the posterior and the dorsal surrounding tissue are required for bud formation. But the posterior tissue (flank mesoderm) not only is necessary for growth, it also has a powerful polarizing effect like the ZPA in the chick wing bud (Tickle et al., 1975), and behaves as though it is a source for a morphogen gradient that specifies the pattern of structures in the AP axis.

The polarizing effect can be shown in two ways: Either a piece of flank tissue can be grafted anterior to the normal rudiment, or a normal rudiment can be grafted to midflank. In either case a high frequency of double posterior duplications is produced (Slack, 1976, 1977a,b; Fig. 6–6). These duplications have the structure expected if they were formed in response to an AP concentration gradient of a morphogen, with digit IV at both sides and anterior elements frequently missing from the midline. The muscle pattern as well as the skeleton is duplicated and symmetrical, while the pattern of nerves is somewhat less symmetrical but departs widely from normal. Since the operations are carried out some time before the appearance of the limb bud, a reasonable hypothesis would be that the flank tissue induces a ZPA at the point where it contacts the limb rudiment proper.

The axolotl is probably the best organism for the study of the prebud phase of limb development, and much work remains to be done. For example, we badly need an accurate fate map to clarify the role of cells lying in the splanchnic mesoderm and deep to the pronephros. This could be obtained using orthopic grafts from triploid animals. Triploid grafts could also be used to establish whether muscle cells come from the somites, as in the chick, and whether the polarizing tissue of the flank actually enters the limb bud. Furthermore, there is the issue of the relationship between larval limb development and limb regeneration, which is only now becoming accessible through the use of modern biochemical techniques (Slack, 1982; D. M. Fekete and J. P. Brockes, 1987).

7
Neurulation

RICHARD GORDON AND G. WAYNE BRODLAND

At the beginning of gastrulation, the axolotl embryo is a hollow sphere some 2 mm in diameter. During gastrulation, approximately half of the blastula invaginates through an indentation at its posterior end known as the blastopore. The topology of the embryo does not change during this process (Needham, 1936, p. 113).

The resulting gastrula is shown in Fig. 7–1. Its hollow, spherical outer layer is known as ectoderm. Neurulation involves differentiation of the ectoderm into epidermis and neuroepithelium—tissues that ultimately form the skin and central nervous system, respectively—and subsequent mechanical deformation of the neuroepithelium through which it forms the neural tube (the precursor of the spinal cord and the brain).

The inner and outer layers of tissue that form the dorsal lip of the blastopore adhere to each other. The mesodermal and ectodermal tissues involved in this dynamic adhesion at the end of gastrulation are known as the notochord and notoplate, respectively. The ectoderm thus has three parts: the neuroepithelial tissue (notoplate) to which the notochord adheres; the rest of the neuroepithelium (neural plate); and the remaining (ventral) hemisphere (epidermis). The part of the archenteron wall to the left and right and anterior to the notochord, we call the mesoplate. (The notochord plus mesoplate constitute the mesoderm.) The mesoplate contributes to the skeletal musculature, and the endoderm (Fig. 7–1) becomes the internal organs. Most evidence discounts any direct role of these tissues in neurulation.

The advantage of the nomenclature we have adopted is that it is hierarchical. In the following list, each tissue comes from some distinct portion of the tissue in the hierarchical level above, although corresponding levels of indenting do not necessarily correspond in time:

- Ectoderm
  - Epidermis
  - Neuroepithelium
    - Neural plate
    - Notoplate
- Archenteron wall
  - Endoderm

Mesoderm
  Mesoplate
  Notochord

The term "presumptive" is often applied to a portion of a tissue. "Presumptive epidermis," for example, means that in normal development the indicated portion of the ectoderm will later differentiate into epidermis. The above nomenclature

Fig. 7–1. Midsaggital section (A) and dorsal view (B) of a urodele embryo at induction. Neuroepithelium comprises notoplate plus neural plate. Ectoderm comprises neuroepithelium plus epidermis. Mesoderm comprises notochord plus mesoplate. Note that the blastocoel and archenteron are hollow, liquid-filled cavities. Since the tissue separating the two cavities is flexible, they sustain the same pressure. Part of the endoderm plugs the blastopore, permitting the maintenance of pressure inside. The direction of the end of the arrow indicating the archenteron in the midsaggital section is in the direction of the dorsal view. Involution movements of gastrulation are indicated by unlabeled arrows.

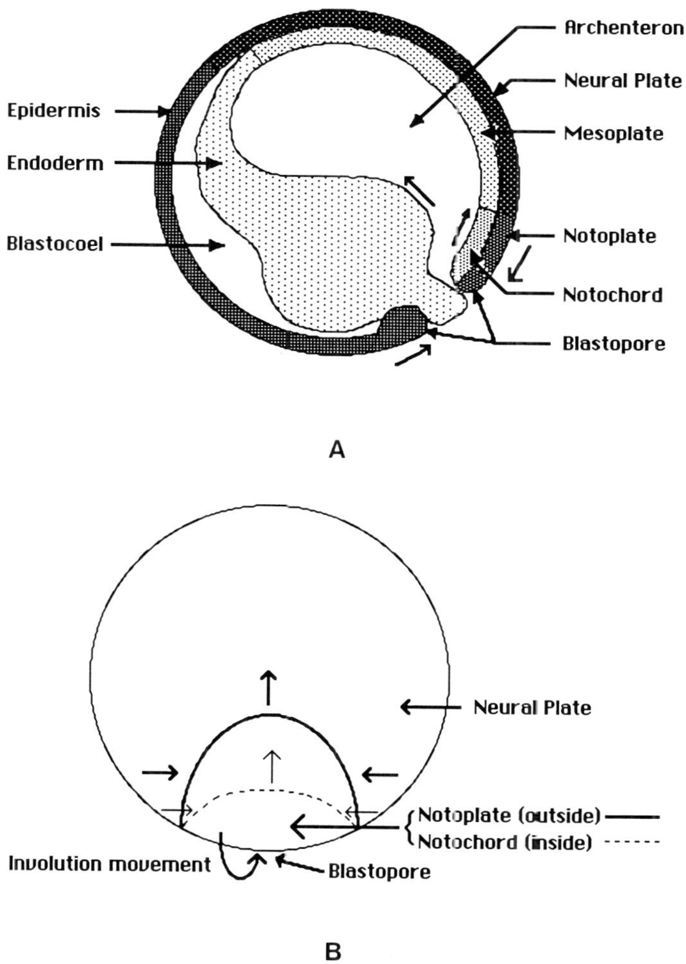

largely eliminates the need to use the adjective "presumptive," and thus makes tissue descriptions both more concise and more precise. Once a tissue is distinguished (by whatever means, including fate maps), the hierarchical label sticks with the cells, wherever they go, as they move through space. In the vocabulary of continuum mechanics, this means that the labels are Lagrangian (cf. A. G. Jacobson and R. Gordon, 1976).

WHY USE THE AXOLOTL?

In many ways, the axolotl is an ideal animal in which to study the process of neurulation. Axolotl embryos can be obtained year-round (Chapter 20) and are sufficiently hardy that they are amenable to experimental surgery. In addition, they have certain particularly useful physical features.

In frogs (order Anura) the neuroepithelium is covered by a superficial layer of epithelial cells. In salamanders, newts, and axolotls (order Urodela = Caudata, Duellman and Trueb, 1986), however, the neuroepithelium is not covered. Hence, the parts of the embryo that are involved in neurulation can be seen easily once the embryo has been dejellied. In addition, individual cells are large enough to be observed with a low-magnification microscope, and since each embryo cell contains its own yolk, groups of cells and even individual cells are easily cultured in a simple salt solution (Chapter 23). Because the natural pigmentation of most axolotls is not evenly distributed from cell to cell, light and dark patches result. This makes it possible to trace the motion of cell patches and even individual cells without the use of vital dyes (N. Burnside and A. G. Jacobson, 1968). This is a particularly useful feature, since significant cell motion is one of the most important things to happen during gastrulation and neurulation. In addition, pigmentation variants allow implanted tissue fragments to be discerned from host tissue (Chapter 22). This feature played an important role in the famous induction experiment of Spemann and Mangold (1924).

The ectoderm remains a monolayer of cells throughout the developmental stages discussed here. From the standpoint of analysis, this makes its mechanical analysis significantly more approachable than that of anurans, since one can, in a relatively straightforward manner, relate the subcellular mechanical properties of a cell to the mechanical properties of the tissue it forms. The tissues involved during neurulation in mammals and birds are also one cell layer thick. Thus, whatever can be learned about normal and abnormal neurulation in urodeles will almost certainly shed considerable light on the nature of neurulation-related defects such as the neural tube birth defects of spina bifida and anencephaly. These are widespread in humans, and their causes are unknown.

Much of what is known about neurulation is a collection of observations from species of different classes. Subtle and sometimes important differences limit cross-class inferences to being qualitative only, at best. It is thus regrettable that one species in one class has not been adopted as a standard subject for neurulation research, so that a quantitative data base could be established. The axolotl would be an ideal candidate, since it is easy to obtain and work with, and since many important discoveries have, in fact, been made using it.

OUTLINE OF NEURULATION

Neurulation involves at least 12 identifiable processes:

1. Global tension in the ectoderm builds up.
2. The notochord induces the notoplate.
3. The notoplate induces the neural plate.
4. The neural plate flattens from a hemisphere to a disk.
5. The neural plate changes shape from a disk to a keyhole form, while the notochord and notoplate lengthen in the anterior–posterior direction.
6. The rate of notochord and notoplate elongation increases suddenly by a factor of 10.
7. The neural plate rolls or folds up to form the neural tube.
8. The seam in the neural tube seals.
9. The rate of notochord and notoplate elongation returns to its previous value.
10. Different regions of the nascent brain become identifiable (Nieuwkoop et al., 1985).
11. Individual, motile, neural crest cells separate from the region of the seam.
12. Neurons and glial cells differentiate.

Some of these processes overlap in time. The first 10 steps are discussed in this chapter: the last two are discussed in Chapters 9 and 15, respectively.

Neurulation Step 1: Generation of a Global Tension

The whole ectoderm is under a global tension, as revealed by slitting experiments (Lewis, 1947; Karfunkel, 1974; A. G. Jacobson and R. Gordon, 1976). This tension is greater and more isotropic at earlier stages and is probably due to the tonus of the apical microfilaments, which are connected in a network via the belt desmosomes. Existence of this global tension implies that the blastocoel is under internal pressure and that the blastopore is mechanically sealed (by contact between the dorsal lip of the blastopore and the so-called yolk plug, which is the invaginating part of the endoderm). The tension is likely generated by the invagination movements of gastrulation, which stretch the cells that remain on the outside (approximetely one-half of the original outer layer) until they cover the entire embryo. Significant stretching must occur, as the surface area of the outside of the embryo does not change during gastrulation.

Neurulation Step 2: Induction of the Notoplate by the Notochord

Spemann and Mangold (1924; cf. Spemann, 1938; Hamburger, 1985) showed that a small patch of cells from the dorsal lip of the blastopore, corresponding roughly to the notochord, can change the course of development of presumptive epidermis, when implanted beneath or in the presumptive epidermis in a host embryo. This patch of cells not only induced presumptive epidermis to become neuroepithelium instead, but causes development of a whole new organism on the flank of the host embryo. A Siamese twin results (Fig. 7–2). Transplantation of the dorsal blastopore

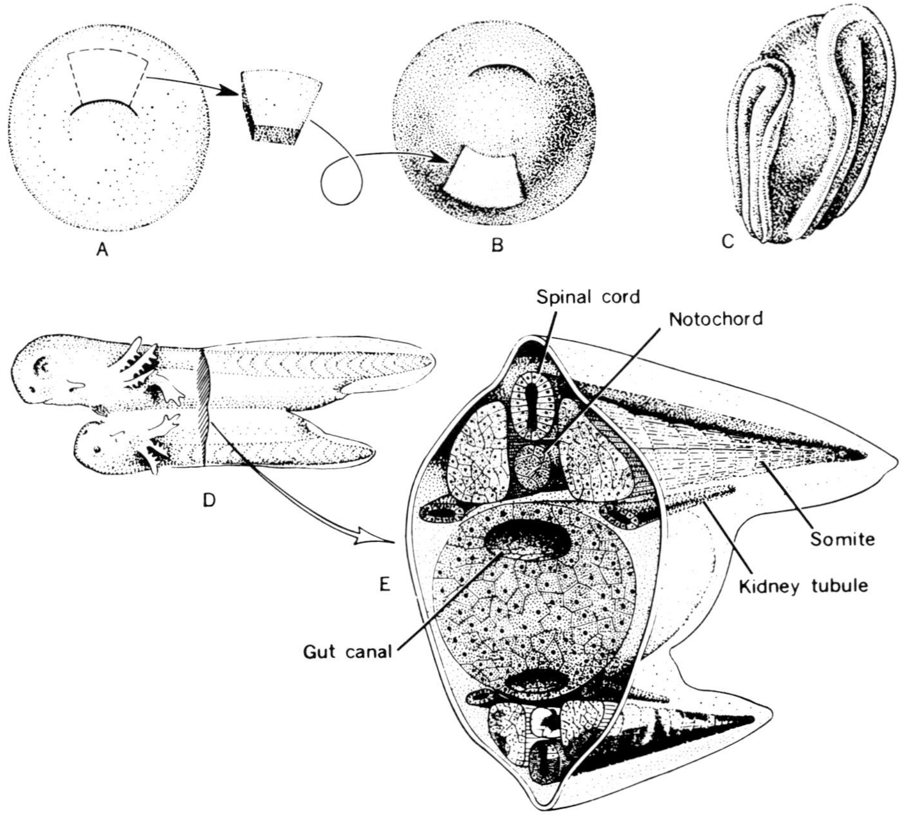

Fig. 7–2. The classical organizer experiment and its result. (From *Of Scientists and Salamanders,* by Victor Chandler Twitty. Copyright © 1966 W. H. Freeman and Company. Reprinted with permission.)

lip of a light-colored embryo into a dark-colored one showed that the implanted tissue actually induces host tissue to form the second twin. The piece of tissue from the dorsal lip of the blastopore thus became known as the organizer because of its power to induce or organize an otherwise plain portion of the epidermis to produce an often complete second embryo.

Once the potential of certain tissues to induce other tissues was established, a concerted effort was made to determine the mechanism of induction. The search for this mechanism was quickly—and prematurely—narrowed to a search for a particular chemical agent, the so-called inducer molecule, as a result of experiments in which the tissues were separated by a porous filter. Only casual reference was made to the possibility than an alternative, nonchemical mechanism might exist.

The long, unsuccessful search for the inducer molecule is documented in several books (Saxén and Toivonen, 1962; Nakamura and Toivonen, 1978; Nieuwkoop et al., 1985) and innumerable papers. The problem was not that an inducer could not be found, but rather that all sorts of strange substances and circumstances were found to cause induction. The situation, as summarized by Twitty (1966), is much the same today:

> Results followed thick and fast—and confusingly. In fact, many of them were so puzzling and seemingly paradoxical that Holtfreter was to remark that the analysis of organizer action was rapidly "bringing chaos out of order." It had been surprising enough to learn that a killed inducer could still induce, but it soon proved that other, ordinarily inert, parts of the embryo such as ectoderm became full-fledged neural inductors once they were coagulated by boiling. The roof really fell in when living or killed tissues of adult salamanders, and even from a variety of animals ranging from worms to man, were found to function in this same unexpected way.

Additional evidence (Gordon and Brodland, 1987) suggests that the mechanism of induction is actually mechanical. Inside each ectoderm cell is a cytoskeleton composed of apical and paraxial microtubules (B. Burnside, 1971, 1973), apical rings of microfilaments and belt desmosomes, and intermediate filaments. The apical microtubules and microfilament ring seem to act in an integrated manner. We call this coordinated apparatus, with its attachments to the apical belt desmosomes and probable linking to intermediate filaments, the "cell state splitter" (Fig. 7-3). The apical microtubules, randomly oriented in the plane of the microfilament ring, grow by polymerization due to the chemical potential of the microtubule monomer (tubulin) and thus generate an expansion force (Hill and Kirschner, 1982) at the apical end of the cell. Opposing this expansion is a contractile force produced by the microfilament ring. See, for example, Hiramoto (1979). At the time of primary induction, these forces are in balance with the global tension. The force equilibrium thus produced, however, is unstable. Hence, when a cell is subjected to an external contractile bias, the equilibrium is upset and the cell, driven by an increasingly unbalanced net apical force, continues to contract until the typical long, columnar, neuroepithelial cell shape is produced. A perturbation that expands the cell will lead to continued expansion, producing the characteristic flattened shape of epidermal cells.

Fig. 7-3. Idealized organization of the major cytoskeletal components in an ectodermal cell with a cell state splitter. The short lines connecting the microfilament ring to the belt desmosomes are only indicative of their attachment. However, they could represent the actin-membrane attachment protein (MAP) spectrin (Sadler et al., 1986).

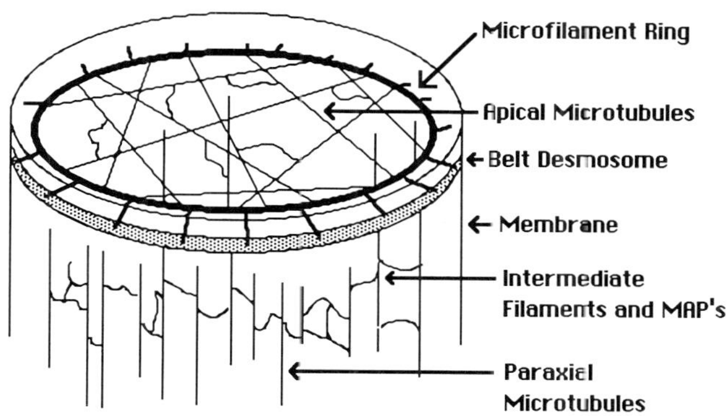

The global tension is the probable source of the biasing force. The reason that primary neural induction begins at the notoplate may be that the attachment of the notoplate to the underlying notochord approximately halves the load produced in each of these tissues by the global tension. Thus, contracted neuroepithelial cells are more likely to form. [An experiment by Youn and Malacinski (1981b), in which otherwise normal development occurred in anurans that developed no notochord as a result of UV irradiation, must be kept in mind. Perhaps the underlying mesoderm still attaches in an irradiated embryo, but fails to develop subsequently into a visible notochord. Notochordless embryos could not be obtained in urodeles following UV irradiation (G. M. Malacinski, personal communication).]

Ben-Ze'ev (1985) has shown that specific protein synthesis can occur within a cell as a result of altering its shape. It thus appears that differentiation may precede specific gene action. Gene expression may, then, result from differentiation rather than being its cause. The first visible sign of induction—thickening of the notoplate (Nieuwkoop et al., 1985)—is followed by the production of a variety of different proteins. Considerable changes occur, for example, in the mixture of intermediate filaments as neurulation progresses (reviewed in Gordon and Brodland, 1987). There is also evidence for an increase in intercellular adhesion at the time of primary neural induction (Suzuki et al., 1986): The production of "neural cell adhesion molecules" (NCAM) seems to start immediately after induction (M. Jacobson and U. Rutishauser, 1986).

Experiments conducted by Nieuwkoop et al. (1955) and Johnen (1956) suggest that primary neural induction takes only 5 minutes. The longer times of 0.5–12 hours, reviewed by Nieuwkoop et al. (1985), may be caused by delays in induction starting, rather than the time course of induction itself. This is perhaps due to a need to restore the global tension first.

Neurulation Step 3: Induction of the Neural Plate by the Notoplate

The notoplate induces the surrounding ectoderm (the presumptive neural plate) to become neuroepithelium as well. This process is sometimes referred to as "propagation of neural induction in the ectodermal layer" or homoiogenetic neural induction (Nieuwkoop et al., 1985). We think it may occur in the following manner (Gordon and Brodland, 1987). Once the notoplate cells have contracted, the adjacent ectoderm cells are stretched. They may respond with a stretch-induced contraction. (Wavelike contractions suggestive of stretch-induced contraction have occasionally been observed later in neurulation ["puckerings": A. G. Jacobson, time-lapse movies, personal communication]). These cells are thereby biased toward narrowing—that is, toward the neuroepithelial state. The transition to neuroepithelium thus propagates from cell to neighboring cell, and a large, confluent patch of neuroepithelium cells results. When exactly one hemisphere of the ectoderm has contracted, the high gradient in the meridional stress component (cf. Flügge, 1967) necessary for further propagation vanishes. A sharp boundary is thus formed between the neuroepithelium and the epidermis (the remaining ectoderm). This phenomenon thus explains compartmentalization and the binary nature of differentiation (Lawrence and Morata, 1979). The "positional information" (Wolpert, 1969) of a cell may just be its location relative to this sharp boundary.

One of the interesting questions that arose during early experiments on induction was the degree to which the subsequent develoment of embryonic tissue is dependent on its source and the degree to which it is dependent on the host into which it is implanted. A significant step toward answering this question was made by transplanting axolotl epidermis into the head region of a *Triturus,* where the latter species grows a larval organ known as a "balancer." The transplanted axolotl tissue did *not* produce a balancer (Mangold, 1931), but responded as it would have if it had not been transplanted. This and other similar experiments have emphasized the importance of the inherent properties of tissues.

Neurulation Step 4: Neural Plate Flattening

The mechanical transformation of the differentiated neuroepithelium into a neural tube has intrigued embryologists for more than a century, and has generated a legacy of theories that attempt to explain it. These dozen or so theories are reviewed in Gordon (1985), and an additional one appeared subsequently (A. G. Jacobson et al., 1986). Most of them were proposed and never either proved nor disproved, only forgotten.

The fairly recent discovery of the cell cytoskeleton and its force generation potential suggest that this transformation is driven by the cytoskeletal components shown in Fig. 7–3, although cell–cell interactions, possibly with differential adhesion (Steinberg, 1970), may also be involved (cf. Gordon, 1985).

The first step in reshaping of the neuroepithelium involves its flattening from a hemisphere to an almost flat disk. This major shape transformation, during which the apical area of the neural plate halves (A. G. Jacobson and R. Gordon, 1976), has yet to be investigated. Its exact relation to homoiogenetic propagation is unknown.

Neurulation Step 5: Shaping to a Keyhole

While remaining relatively flat, the neural plate undergoes a major reshaping. This reshaping was first observed in quantitative detail by Burnside and Jacobson (1968) using time-lapse movies. During this process, the apical areas of the neural plate cells shrink significantly and their heights increase, with the cells apparently retaining constant volume (except for cell divisons). A computer simulation of the kinematics of this transformation showed that both cell shrinkage and cell rearrangement (change of neighbors) are involved (A. G. Jacobson and R. Gordon, 1976; Gordon and Jacobson, 1978). The boundary between the neural plate and the epidermis becomes visible as the neural fold.

Neurulation Step 6: Acceleration of Elongation

Significant neighbor changes occur in the notoplate as it changes from a crescent shape to a rod shape (A. G. Jacobson and R. Gordon, 1976). (Similar cell rearrangements must be occurring in the attached notochord.) This axial elongation is responsible for much of the reshaping of the neural plate. The elongation continues after the keyhole shape is attained and is probably largely responsible for neural tube closure, perhaps by bringing about a mechanical buckling of this sheet of cells (Gordon, 1985). The notoplate's rate of lengthening increases suddenly by a factor of about 10 (A. G. Jacobson and R. Gordon, 1976), at which time neural tube closure commences. Until we have solved the mechanism of notoplate extension, which may involve gradients in cell–cell adhesion (Gordon, 1985; cf. Keller et al., 1985), we are not likely to understand the cause of this acceleration. An analogous acceleration has been artificially induced in *Drosophila* using the proteolytic enzyme trypsin (E. Fekete et al., 1975). One could imagine a natural short-term release of a similarly acting enzyme to be the source of the spurt in elongation of the notochord and/or notoplate.

Neurulation Step 7: Neural Tube Formation

At stage 15, the posterior approximately two-thirds of the neural plate begins to roll into a tube. Cells at this stage begin to take on a tapered "flask" or "bottle" shape as a result of further apical constriction not being accompanied by sufficient further elongation normal to the apical surface. Although this cell reshaping has often been implicated as the prime factor of neural tube closure, experimental evidence shows that excised neural tissue undergoes apical constriction but does not roll into a tube unless the notochord is left attached to it (A. G. Jacobson, 1978). (The epidermis is still under tension and would thus tend to impede neural tube rolling rather than assisting it.) Thus, the mechanical effects of the notoplate during its accelerated elongation seem to be critical to neural tube rolling. The viscoelastic behavior of the neural plate cells and their cytoskeletal components may account for the alternate cell forms in these two situations. Thus the suggestion that each cell has its own "height program" (A. G. Jacobson and R. Gordon, 1976) needs to be reexamined.

Neurulation Step 8: Sealing of the Neural Tube

When the edges of the neural plate have lifted and moved toward each other sufficiently, they meet and fuse. This process has been reviewed for all classes by Schoenwolf (1986), including possible roles for extracellular bridging materials. It is difficult to separate fusion from neural crest cell initiation (neurulation step 11), since both are occurring at the same time and place.

Neurulation Step 9: Decelerated Elongation

The rate of notoplate and notochord elongation suddenly returns to its previous value (A. G. Jacobson and R. Gordon, 1976). This sudden decrease is just as abrupt as its sudden decrease was at step 6. Its cause has yet to be investigated.

Neurulation Step 10: Regional Differentiation

Even before neural tube closure, local regions of the neural plate appear to be on their way to forming the optic vesicle and distentions of the brain (Nieuwkoop et al., 1985). While hydrostatic pressure may play some role in the morphogenesis of these distentions (cf. Desmond and Jacobson, 1977), their existence in embryos of the third amphibian order Caeciliidae *before* neural tube closure (Duellman and Trueb, 1986) suggests that pressure is not essential. The nonuniform pattern of apical shrinkage observed at the flat neural plate stage (A. G. Jacobson and R. Gordon, 1976) may be the forerunner of these features. Nothing is known about the origin of this pattern, which may be regarded as the first complex pattern to appear in the vertebrate embryo.

CONCLUSIONS

The above steps, together with subsequent neural crest cell motion and neuron and glial cell differentiation, constitute the process commonly known as neurulation. Differentiation and development of the brain and spinal column continue, however, as axons form interconnections between brain cells, and processes (axons) that grow outward from the neural tube (spinal cord) traverse the entire embryo to form the motor and sensory nerves, and thus complete the morphogenesis of the central nervous system.

Perhaps because we regard our brain as our most distinguishing feature, we entertain a certain fascination with its morphogenesis (cf. Chapter 15). The central nervous system is the first organ to develop in the vertebrate embryo, and thus, beyond its intrinsic interest, it provides a model system for all subsequent morphogenesis and organogenesis. Some aspects of neurulation have yielded to experimental, mathematical, and computer analysis, while others still evade us. We hope that this brief sketch, which could not mention many important investigations over the past century, will provide some impetus for future work. If that work could be concentrated on the axolotl, it would almost certainly proceed at a much greater pace.

8

Morphogenetic Waves During Elongation

JOHN B. ARMSTRONG

Following gastrulation, the axolotl embryo begins to elongate along its anteroposterior (AP) axis. The available data suggest that this growth is not uniform along the entire axis, but rather is due to the addition of new cells from an area actively undergoing mitosis at the caudal end (the tailbud). In this respect, elongation at least superficically resembles the growth of the limb bud from a terminal "progress zone," and some of the models have been adapted from ones proposed earlier (e.g., Summerbell and Lewis, 1975) to explain limb development.

During this period of elongation, somite segmentation and pronephric duct migration proceed in a cephalocaudal direction down the AP axis. In 1976, Cooke and Zeeman proposed a model (discussed below) suggesting that segmentation is related to the passage of a hypothetical wave of change through the mesodermal cells. Since then, various observations and experimental manipulations have revealed additional waves of change, apparently related to segmentation and duct migration, which pass down the AP axis. In this chapter, some of the evidence for these waves, and the interrelationship between them and elongation, will be examined.

SOMITE FORMATION

The most visible wave is that of somite segmentation (Fig. 8–1). In the axolotl, the paraxial mesoderm begins to segment during neurulation (at around stage 16), just caudal to what will become the third pharyngeal pouch. By stage 31 segmentation has reached the cloaca, which is normally located ventral to the border between somites 19 and 20, and then continues into the tail. The somites form at a constant rate (Armstrong and Graveson, 1988), and their total number (at least up to the cloaca) is unaffected by temperature.

The actual segmentation is the result of cellular rearrangements (Youn and Malacinski, 1981a). The two layers of cells, which are dorsoventrally elongated, become pyramidal and form a rosette (Fig. 8–2). This does not occur simultaneously throughout the rosette. Rather, the cells immediately posterior to the anterior segmentation line rearrange themselves first. Shortly afterward, the posterior cells

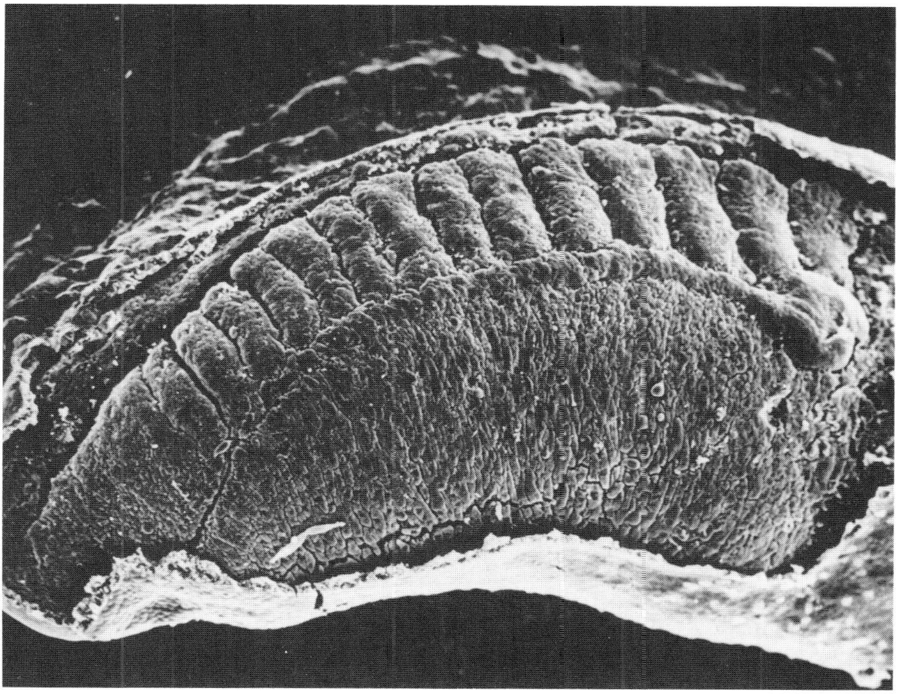

Fig. 8–1. Scanning electron micrograph showing the segmentation of somites and migration of the pronephric duct in a stage 29 axolotl embryo. ×54.

begin to change shape as the caudal segmentation line appears. Thus, while the formation of the segmentation lines is certainly discontinuous, the underlying cellular changes may progress as a continuous wave.

Prior to segmentation, cells of the paraxial mesoderm are electrically coupled (Blackshaw and Warner, 1976). Segmentation is accompanied by the breaking of electrical contacts between the cells forming the somite and the rest of the mesoderm. In the axolotl, each somite remains electrically insulated from its neighbors.

The forces driving rosette formation are not understood but are thought to involve changes in cell–cell adhesion (Bellairs et al., 1978). Youn and Malacinski (1981a) have suggested that, during rosette formation, cellular adhesiveness is not

Fig. 8–2. Schematic diagram of the changes in cell shape accompanying rosette formation at the time of somite segmentation in the axolotl. (Adapted from Youn and Malacinski, 1981a.)

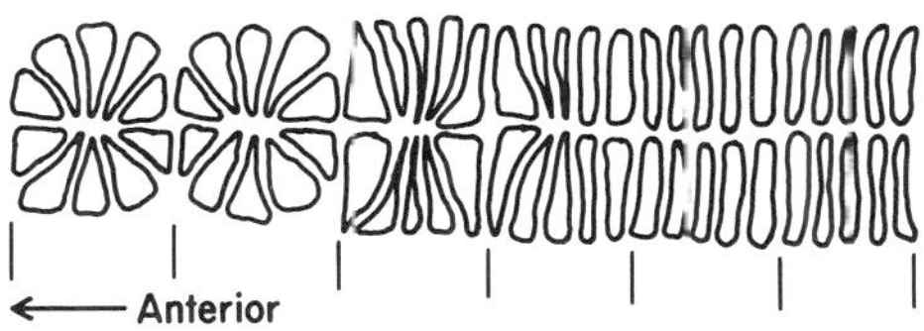

altered uniformly. Rather, the cells become more adhesive at one end. As the rosette forms, the cells arrange themselves radially around their highly adhesive apexes.

Following rosette formation, further changes, such as myoblast fusion, occur at definite time intervals. At 20°C, myoblast fusion begins about 25 hours after rosette formation and continues for a further 18 hours (Youn and Malacinski, 1981a). As these changes occur after the same interval for all somites, they also move caudally as waves.

The "Prior" Wave

Several years ago, Elsdale and his co-workers (Elsdale et al., 1976; Pearson and Elsdale, 1979) discovered that a 37°C heat shock dramatically affected the subsequent development of somites in *Rana* and *Xenopus.* Following heat shock, somitogenesis continues normally for three to five somites, then abruptly becomes abnormal. Posterior to the abnormal region, normal somite segmentation eventually resumes. The boundary of change from resistance to sensitivity defines a wavefront, termed the "prior" wave. Pearson and Elsdale (1979) suggest that heat shock disrupts the developmental coordination of the cells, and that the prior wave is a wave of somite determination establishing a prepattern that is resistant to the disruptive effects of the heat shock. The observations of A. G. Jacobson and S. Meier (1984) on *Taricha torosa* indicate that, in some species, a morphological prepattern can actually be seen in stereo scanning electron micrographs (SEM). The prepattern consists of squat, cylindrical domains of cells arranged in concentric circles. Meier (1979) has termed these units "somitomeres."

In the axolotl, even though somite abnormalities are observed following heat shock (Armstrong and Graveson, 1988), the boundary is not as clear-cut as that which defines Elsdale's prior wave. Following a 37°C heat shock at early-tailbud stages, a number of somites do form normally, but this number is reduced in 38.5°C heat shocks. These results have been interpreted to indicate that the region immediately posterior to the last-formed somite is more resistant to the disruptive effects of heat shock than is the presomitic mesoderm even further posterior. Armstrong and Graveson (1988) proposed that there is a gradual establishment of pattern (or prepattern), culminating in the formation of the actual segmentation line.

The abnormal zones, especially following a 38.5°C shock (Fig. 8–3), are symmetrical, with the greatest disturbance in the center. This can also be explained in terms of a model of progressive patterning. The most anterior cells are disturbed least as they are closest to having established a stable pattern. Cells lying further back in the column are disturbed more severely, as they are at earlier stages of pattern formation. Those even further posterior are less affected, because either patterning events have not yet begun or there is sufficient time to repair damage. The posterior boundary of the abnormal region coincides with a change in morphology of the presomitic mesoderm from a cohesive sheet to a noncohesive mesenchyme. Armstrong and Graveson (1988) believe that this change marks the beginning of the progress zone (see below).

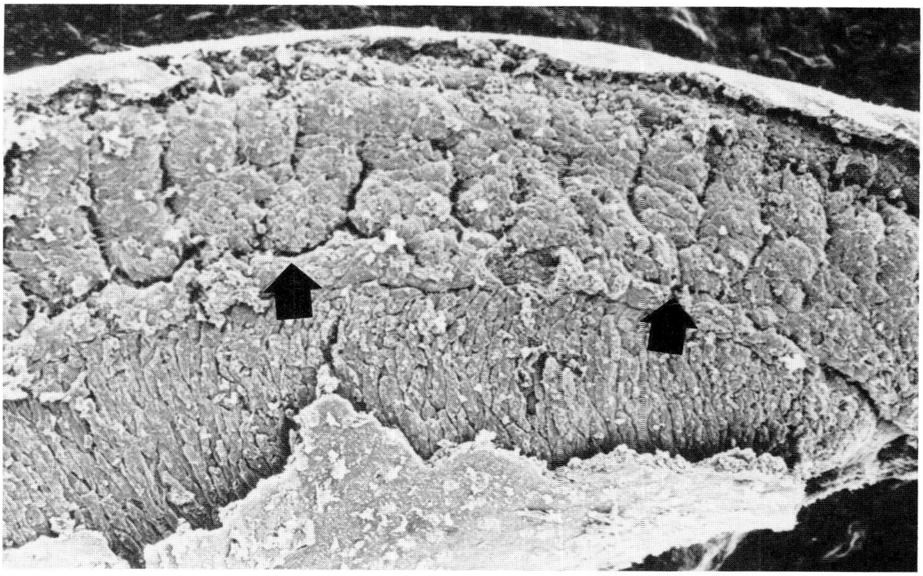

Fig. 8–3. Effect of a 12-minute, 38.5°C heat shock at the 6-somite stage on subsequent somite segmentation. The abnormal region includes ventral fusion of somites 9–11 and 15–17 (arrows), flanking a region of three somite-widths in which there are a number of furrows. Anterior is to the left. ×60. (From Armstrong and Graveson, 1988.)

What Process Is Sensitive to the Heat Shock?

Youn and Malacinski (1981a) propose that rosette formation depends on changes in cellular adhesiveness. Heat shock could therefore be inhibiting the synthesis of a cell adhesion molecule, or the molecule itself could be thermolabile. Previously, Gillespie et al. (1985) reported that a brief trypsin treatment of the mesoderm caused somite abnormalities similar to those produced by heat shock. The result is consistent with a trypsin-sensitive cell surface protein such as a cell adhesion molecule.

Models for Somite Formation

The simplest model for somite segmentation and elongation would be one in which segmentation partitions the paraxial mesoderm while the embryo elongates uniformly (Fig. 8–4a). The posterior, unsegmented region would initially be relatively large, but would shrink as segmentation progressed. The mitotic index would be relatively uniform along the length of the embryo, and somites would continue to widen after they had formed. However, there is no indication in the axolotl that the earliest segmented somites increase in width while the more posterior ones are segmenting.

As an alternative, one could imagine that narrow somite primordia exist in the

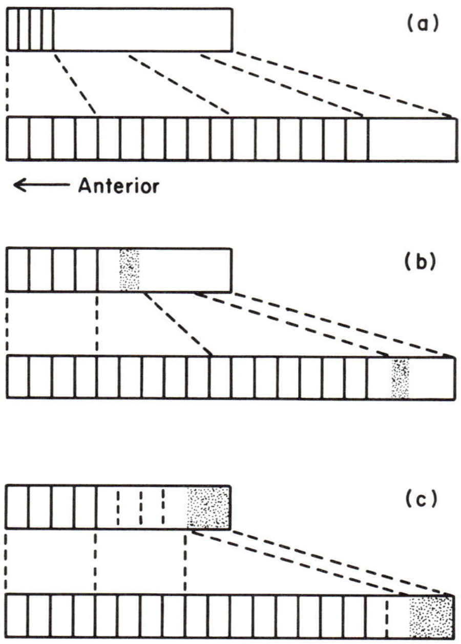

Fig. 8–4. Models of somite segmentation. In each case the upper bar represents the paraxial mesoderm at the 4-somite stage while the bottom bar represents the 16-somite stage. The solid vertical lines within each bar represent the boundaries of the somites, the dashed lines the boundaries of the somitomeres. The stippled area represents the mitotically active region in (b) and (c); mitosis is assumed to be uniform in (a). The lack of somitomeres in (a) and (b) is not intended to imply that they could not exist.

neurula. A wave of mitosis might then pass down the embryo expanding each primordium in turn. A short time later, each expanded primordium would segment (Fig. 8–4b).

In a progress zone model (Fig. 8–4c), we imagine that the mitotically active zone is confined to a small area in the tailbud. This area would be of relatively constant size, at least until segmentation and elongation were complete. Segmentation occurs in cells that have recently left the progress zone. Once formed, somites widen very slowly, and certainly not in proportion to the total elongation that is occurring. These last two models can be distinguished by determining the location of the zone that is mitotically active. As yet, such data have not been collected in the axolotl, although in *Xenopus,* Cooke (1979) has shown that from stage 20 onward mitosis is confined to a small posterior terminal region.

In the axolotl, Armstrong and Graveson (1988) fate-mapped presumptive somites and found that in the cohesive zone there are up to nine full-width somite primordia. The remaining primordia are densely packed within the noncohesive zone, which is only about as wide as two somites. This could be interpreted as a progress zone.

What Actually Controls Somite Segmentation?

One could also propose different models to account for the actual segmentation process. The simplest would be one in which a change in cellular adhesiveness proceeds down the AP axis, leading to rosette formation, as indicated above. The size of a somite would depend on the packing of the cells, which would in turn depend on their shape and considerations of free energy. Each somite would partition off a certain number of cells, and the total number of somites would be based on the total amount of (expanded) paraxial mesoderm. However, the number of cells per somite does not appear to be critical. In experiments with *Xenopus* where either the size of the embryo prior to gastrulation (Cooke, 1975) or the size of the somite-forming column (Cooke, 1977) was reduced, there were fewer cells per somite rather than fewer somites.

To account for these observations, Cooke and Zeeman (1976) proposed their clock and wavefront model. The clock is assumed to be a cellular oscillator for which the cells in the paraxial mesoderm are phase-linked. The wavefront may be regarded as a wave of intracellular change necessary but not sufficient for segmentation. A critical assumption is that the wave takes the same time to traverse the AP axis, regardless of its length.

The actual adhesive changes can only occur during one phase of the oscillation and after passage of the wavefront. Thus, when the front passes through the anterior portion of a prospective somite, adhesive changes do not occur because the "clock" is not at the correct "time." As it approaches the posterior border, the oscillator reaches the correct phase and the adhesive changes occur over the entire somite.

Wolpert and Stein (1984) have also used the idea of an oscillating chemical reaction, but suggest that oscillations only occur in a progress zone. When cells leave the zone, the concentration is frozen, generating a sine wave. If cells respond to a threshold of this morphogen, a repeating pattern is generated.

If the model is to have two components (i.e., the oscillator and the wavefront), they must almost inevitably be linked to a single underlying process if they are not to be differentially affected by such simple perturbations as changes in temperature. This underlying process could be the cell cycle. In Wolpert and Stein's model, the rate of exit from the progress zone would be determined by how rapidly cells are dividing (i.e., the length of the cell cycle). We might reasonably assume that the cell cycle also determines the rate of oscillation. Thus, lowering the temperature would simultaneously slow the oscillations and the rate of exit from the progress zone.

PRONEPHRIC DUCT MIGRATION

In the axolotl, the pronephros originates from the lateral margin of somites 2–4. The pronephric duct forms from a mass of loosely connected cells, apparently derived from the lateral mesoderm below somites 4 and 5 (Gillespie and Armstrong, 1985). By the eight-somite stage, these presumptive duct cells have migrated dorsally to the duct path and are found below somites 5–7. By the nine-somite stage they have begun to migrate caudally. The normal pathway of migration is very specific; it is the juncture between the somites and the lateral mesoderm (Fig. 8–1).

Duct migration is coordinated with somite segmentation. Consequently, at any stage up to the point where it reaches the cloaca, the duct tip can be found two somite-widths anterior to the posterior boundary of the last-formed somite. As the duct elongates, it recruits few, if any, new cells. Neither is there any significant cell division. Consequently, the duct becomes narrower as it becomes longer (Poole and Steinberg, 1981).

The Active Region

When a secondary duct is grafted below the primary duct, it is not able to migrate if its tip is placed anterior to that of the primary duct. Placed in a more posterior region, the secondary duct migrates dorsally to the normal duct path, then caudally to the cloaca (Poole and Steinberg, 1982). This experiment defines the "postmigration" wave that passes through the lateral mesoderm behind the migrating phonephric duct and renders it unable to support duct migration.

Secondary ducts transplanted more caudally than the primary duct reach the duct path ahead of the primary duct, indicating that the posterior lateral mesoderm is competent to support duct migration. However, this competent region appears not to include all of the posterior lateral mesoderm, as ducts transplanted more than two somites posterior to the normal position appear to be delayed in their initiation of migration (Gillespie and Armstrong, 1986). All end up only slightly ahead of the last-formed somite, or about two somites ahead of the normal position. Thus, the duct transplants delineate both the anterior and posterior boundaries of the active mesoderm, which is about two somites wide. The posterior boundary appears to lie just slightly posterior to the last-formed somite.

Zackson and Steinberg (1986) have demonstrated that other migratory cells, from the cranial neural crest, will also migrate along the duct pathway if transplanted to the posterior lateral mesoderm. However, unlike the duct, cranial neural crest cells are able to migrate when transplanted anterior to the proposed active region.

The Effects of Tunicamycin Treatment

To examine the possible involvement of cell surface proteins, embryos were treated with tunicamycin, a drug that specifically inhibits glycosylation of N-linked glycoproteins (Takatsui et al., 1975). Experiments were performed by microinjecting tunicamycin into the archenteron.

When embryos were injected with tunicamycin prior to stage 12, no effect on duct migration was observed. However, some embryos injected at stage 12 displayed abnormal duct migration. In these embryos, the pronephric duct was found to deviate from its normal path and proceed ventrocaudally over the lateral mesoderm (Fig. 8–5) for about two somite-widths before stopping.

Very few of the embryos injected at later stages displayed abnormalities, and

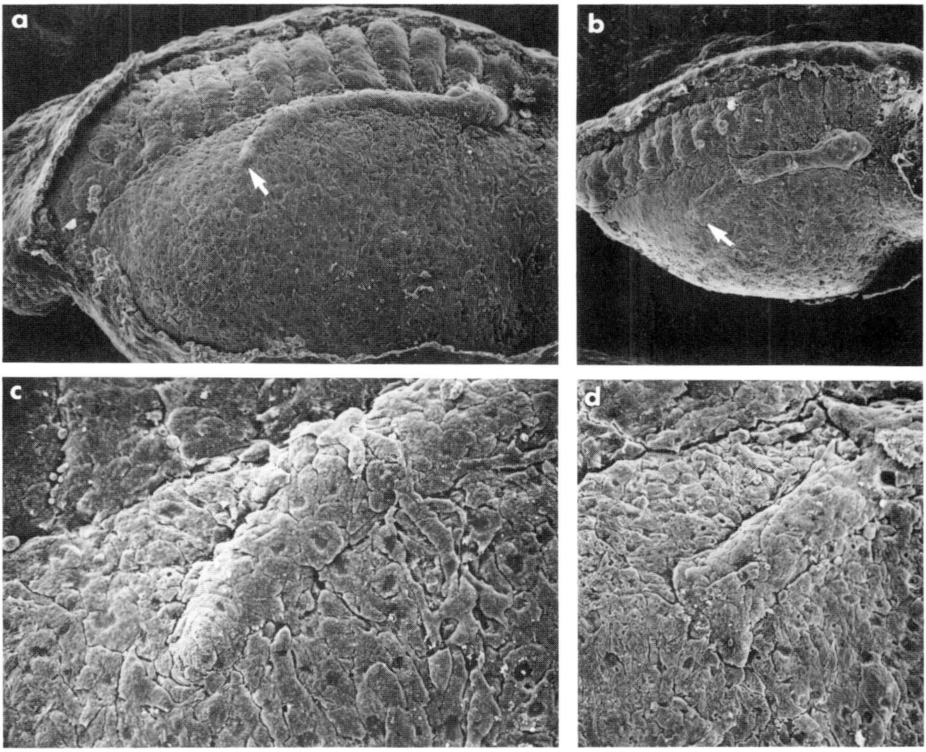

Fig. 8–5. Abnormal migration following tunicamycin treatment. Embryos were injected with 25 ng of tunicamycin at stage 12. (This was the highest dose soluble in a subteratogenic concentration of dymethyl sulfoxide.) The ducts (arrows) deviate ventrocaudally (a,b). Higher magnification of the duct tip (c,d) reveals its blunt appearance. Anterior is to the right in each micrograph. (a) ×66; (b) ×41; (c) ×241; (d) ×156.

none of these showed abnormal duct migration. Perhaps the mesoderm was no longer accessible to drugs injected into the archenteron.

In spite of the small number of affected embryos, some important information about duct migration can be deduced. The fact that the duct continues to migrate, albeit in an abnormal direction, suggests that the substratum rather than the duct is affected. If tunicamycin were interfering with the duct cells such that they could no longer recognize their path, one would expect the duct to migrate in a random fashion. However, in all cases, the duct migrated along the duct path to somite 8 or 9 and then ventrocaudally at an angle to the dorsoventral axis.

Since the various waves travel in a cephalocaudal direction, one might expect the synthesis of the presumptive tunicamycin-sensitive glycoprotein(s) to take place in the same sequence. Injection of tunicamycin at a specific time would affect only those cells that had not synthesized the important glycoproteins, leading to a sharp boundary between the posterior affected and anterior unaffected cells. When confronted with this boundary, one might expect the duct to migrate ventrally along the boundary because it cannot migrate posteriorly into the affected area. Migration would stop only after the duct was overtaken by the still-migrating anterior boundary of the active region.

The Wave of Extracellular Matrix (ECM) Formation

The ECM-filled space between the ectoderm and mesoderm is the space within which the pronephric duct migrates. The presence of matrix may well be necessary for duct migration, as an important role has been suggested for the ECM in several well-studied cases of cell migration (Toole, 1981).

When the distribution of the ECM between the mesoderm and epidermis was examined, Gillespie and Armstrong (1986) found that the posterior limit of the ECM was always posterior to the duct tip, but on the average, approximately one somite-width anterior to the last-formed somite at all stages. In the scanning electron microscope, the ECM appeared as a thick, matlike coating, which was resolved into a dense fibrillar network at higher magnification.

If the presence of ECM is a requisite for duct migration, then a reasonable supposition would be that its posterior boundary corresponds to the posterior boundary of the active region. The lack of an exact correspondence may indicate that the newly synthesized ECM at the posterior boundary is not as well preserved as the more anterior ECM or is not thick enough to be seen, or that the duct can migrate a short distance ahead of the ECM. These possibilities remain to be tested.

The Wave of Shape Change

The anterior boundary of the active region coincides with the primary duct tip and with a wave of shape change in the lateral mesoderm (Gillespie and Armstrong, 1986). During duct migration stages, the lateral mesoderm cells ahead of the duct tip (i.e., more posterior) are cuboidal and form a layer two cells thick (Fig. 8–6). In the region of the duct tip, the cells are columnar and the layer has been reduced in thickness to a single cell. Cells immediately behind the duct tip appear to be pyramidal, with the top and bottom cells interdigitating. Cells even further behind (in the most anterior position) are relatively flat, with one cell sitting on top of another to form a layer that, again, is two cells thick. These shape changes occur not just directly in the duct path, but along the entire dorsolateral axis, thus forming a wavefront. The position of this wavefront of shape change corresponds to that of the anterior boundary of the active region.

Fig. 8–6. Morphological changes in the cells of the lateral mesoderm accompanying pronephric duct migration. The duct tip is always found over the region where tall, columnar cells form a single layer.

The Forces That Drive Duct Migration

Steinberg and his colleagues (Poole and Steinberg, 1982; Zackson and Steinberg, 1986) have proposed that duct migration is driven by an adhesion gradient along the duct pathway. If their model is correct, then the anterior boundary of the active region must represent a relatively abrupt transition from an anterior region of relatively low and uniform adhesiveness to a posterior region of higher adhesiveness. This change may be associated with the wave of shape change. The change from a two-cell to a single-cell layer in the region of the duct tip must result in at least temporary exposure of the inner (medial) cells to the surface of the lateral mesoderm, and consequently, expose cell surface molecules that are quantitatively and/or qualitatively different from those on the outer (lateral) cells. Such differences would likely exist because the inner and outer cells have different microenvironments; the inner cells are in contact with the endodermal cells, while the outer cells are sandwiched between the inner cells and the epidermis.

If the duct tip does not adhere as well to the exposed inner cells, then as the cells of the two-cell layer interdigitate to form the columnar region, the cells of the duct tip should migrate to interact with the more adhesive outer cells in the region that still forms a two-cell layer. In this posterior region, there may still be a gradient of adhesiveness. This seems the simplest way to account for the migration of secondary ducts transplanted posterior to the primary duct.

RELATIONSHIP BETWEEN DUCT MIGRATION AND SOMITE SEGMENTATION

No firm causal relationship has yet been established between the events that control somite segmentation and those that control pronephric duct migration. However, the two are temporally coordinated to a degree that invites a unifying model.

The model I would like to propose is based on the idea of a tailbud progress zone. Such a progress zone would be a mitotically active but undifferentiated region occupying perhaps the posterior one-third of the region between the last-formed somite and the tailtip. This zone would be nearly constant in size, and would provide cells for the elongating AP axis. One might say that the cells are "kicked out" of the progress zone, and then begin a series of differentiative changes. If these changes occured at fixed times after the cells left the progress zone, then the changes would appear to move cephalocaudally in waves. The changes in the somitic and lateral mesoderm would not necessarily be the same, but would move coordinately.

PROSPECTS FOR EXTRAPOLATION TO OTHER VERTEBRATES

To my knowledge, the mechanism of pronephric duct extension has only been examined in the frog, *Xenopus,* and in the chick (Poole and Steinberg, 1984). In *Xenopus,* the duct grows by recruitment of new cells from the mesoderm, not by migration of existing cells. In the chick, the duct grows by proliferation of the tip cells and by active migration. In spite of these differences, the underlying morphogenetic processes may be regarded as progressing caudally in a wave.

Stern and Bellairs (1984) suggest that there may be other morphogenetic waves in the chick accompanying the obvious one of somite segmentation. They report a wave of heat shock sensitivity similar to that observed in *Rana* and *Xenopus*, and a wave of mitosis just posterior to the last-formed somite. The region of high mitotic activity is quite narrow, and is not interpreted as evidence for a progress zone. The authors suggest that it may be a component of some fundamental physiological wave, as yet undefined. It might, however, be taken as evidence for the model proposed in Fig. 8–4b.

In the mouse, Tam et al. (1982) have found that the region immediately posterior to the formed somites (the presomitic mesoderm) is organized into five or six segmental units (somitomeres) visible in stereo SEM. Postgastrulation, cells are added to the caudal end of the presomitic mesoderm from proliferating tailbud mesenchyme. Tam (1981) suggests that the specification of somite number and the control of embryonic axial growth may depend on the time cells spend in this proliferating zone, in a manner analogous to that proposed above.

ACKNOWLEDGMENTS

Much of the work carried out in my lab formed the basis for the doctoral thesis of Laura Gillespie, who was supported by a Natural Sciences and Engineering Research Council scholarship. Thanks are also due to the Kidney Foundation of Canada for research support, and to A.C. Graveson and D.C. Hoppe for technical assistance.

9

Neural Crest Cell Migration: A Pictorial Essay

JAN LÖFBERG, HANS H. EPPERLEIN,
ROBERTO PERRIS, AND MICHAEL STIGSON

The development of the neural crest (NC) is one of the best examples of embryonic morphogenesis based on directional cell movement (Trinkaus, 1984). The NC cells migrate from an original position along the dorsal neural tube to many regions in the embryo and give rise to a wide spectrum of cell types and body structures. Experiments on the amphibian embryo, including the axolotl and related species, have contributed considerably in the past to our understanding of the developmental capacity of NC cells (reviewed by Hörstadius, 1950; Weston, 1970). The NC derivatives include pigment cells, neurons and glial cells of the peripheral nervous system, connective and supportive tissues of the visceral skeleton, and a range of endocrine cell types.

The migratory capability and differentiation potential of the NC cells make them a good model system for investigations into some of the central questions of developmental biology. What mechanisms are responsible for the formation of precisely localized and specifically differentiated cell populations in the multicellular organism? How is onset of migration, guidance along specific pathways, and final localization regulated in directionally migrating cells? What control functions are intrinsic to the cells themselves and what external influences act on them? How is timing in morphogenesis and cell differentiation regulated? Accordingly, development of the neural crest is the subject of very active investigation (reviews by Hörstadius, 1950; Weston, 1970, 1982, 1983; Noden, 1978, 1984; Bronner-Fraser and Cohen, 1980; Le Douarin, 1980, 1982, 1986; Erickson, 1986; Newgreen and Erickson, 1986).

Since NC cells are difficult, if not impossible, to identify at late stages of migration, early studies on amphibian embryos were based on examination of serial sections of initial phases of migration, and on extirpation experiments. However, cellular markers were also used, involving xenoplastic transplantation of NC cells, the dispersion of which could be followed in the host embryo (e.g., Raven, 1931; Rosin, 1943). Introduction of the "quail-to-chick" cellular marker system (Le Douarin and Teillet, 1974; Noden, 1975) made it possible to localize NC cells at all stages of their development. This technique opened up a new era of detailed investigations centered around the chick embryo (e.g., Le Douarin, 1982, 1986; Noden, 1978,

1984). It is now hoped that the introduction of a new tracer technique, based on intracellular injection of a fluorescent marker (Gimlich and Braun, 1985), will revitalize the interest in using amphibian embryos, particularly the axolotl embryo, as material for research. As we shall see, the large size of the axolotl embryo permits micromanipulation in situ with a precision that would be impossible to attain in the chick embryo.

A regulating function of the local embryonic environment in determining the pattern of NC dispersion and differentiation is well documented in the chick (Le Douarin and Teillet, 1974; Le Douarin, 1982; Noden, 1975, 1978) as well as in the axolotl and related species (Detwiler, 1934; Dalton, 1950b; Keller et al., 1982; Löfberg et al., 1985, 1987; Perris and Löfberg, 1986; Perris et al., 1987; Perris, 1987; Perris and Johansson, 1987). On the other hand, there is also evidence from experiments on these embryos that the dispersing NC cells are not a homogeneous population of uncommitted cells on which environmental stimuli act. Rather, it seems that developmental restrictions are already imposed on some NC cells when they disperse, and their migratory capability may also be heterogeneous (Hörstadius and Sellman, 1946; Ciment and Weston, 1982; Weston et al., 1984a,b; Epperlein and Löfberg, 1984; Le Douarin, 1986).

We are studying NC cell migration in the axolotl embryo to find out whether the extracellular matrix (ECM) has a regulating function for the migration and phenotypic expression of NC cells. We also investigate pigment cell patterning in larvae of normal and white axolotl larvae (S.K. Frost and G.M. Malacinski, 1980; S.K. Frost et al., 1984a), in order to elucidate the role of cell–cell and cell–ECM interactions during this process. The following (Figs. 9–1 to 9–27) is a pictorial essay, or a laboratory record of some observations made and experimental results obtained during this work—with particular emphasis on pigment cells.

The embryos are staged according to Bordzilovskaya et al. (Chapter 19). If not stated otherwise, the anterior part of the embryos is on the left in the micrographs. Abbreviations used: B, branchial ectomesenchyme; EP, epidermis; ECM, extracellular matrix; F, ECM fibrils; MC, "membrane microcarrier"; NC, neural crest; NT, neural tube; P, pronephros; G, proteoglycan "granules"; SEM, scanning electron microscopy; S, somite; TEM, transmission electron microscopy.

Fig. 9–1. At stage 25 the NC cells have segregated from the epidermis and occupy a medial position along the dorsal neural tube. From about the fourth trunk segment and posteriorly, the premigratory NC cells form a cordlike structure. Here, in the trunk, migration has not yet started but more anteriorly, dispersion of ectomesenchymal NC cells is well under way (Fig. 9–5). ×100.

Fig. 9–2. The onset of NC cell migration travels like a wave along the body axis in an anterior to posterior direction. At stage 30, many NC cells have started spreading laterally on the neural tube. Some cells have even advanced out on the somites. The majority of these early-dispersing cells are probably melanophores, because these cells are the first ones to migrate out into the subepidermal space (Keller et al., 1982; Epperlein and Löfberg, 1984). ×120.

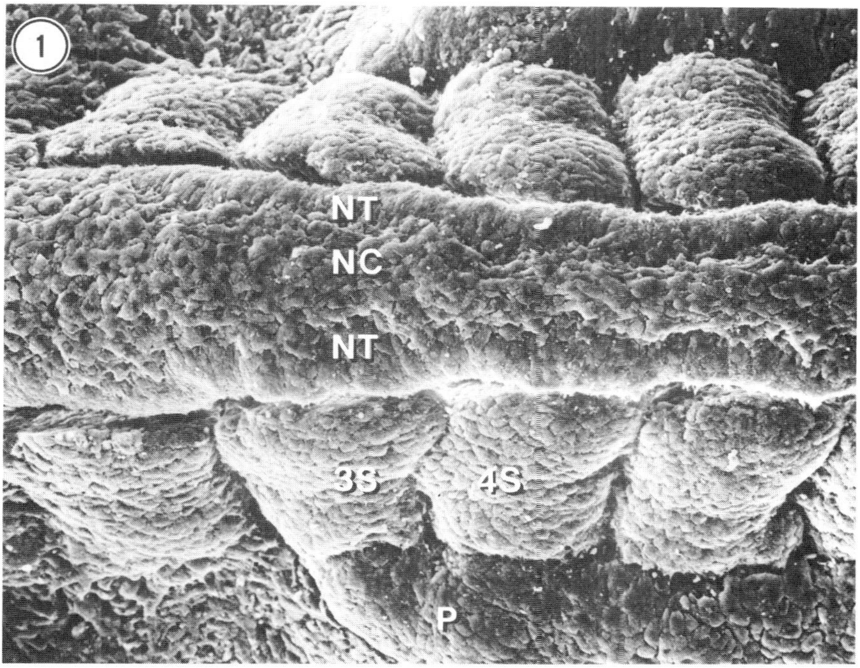
1
NT
NC
NT
3S
4S
P

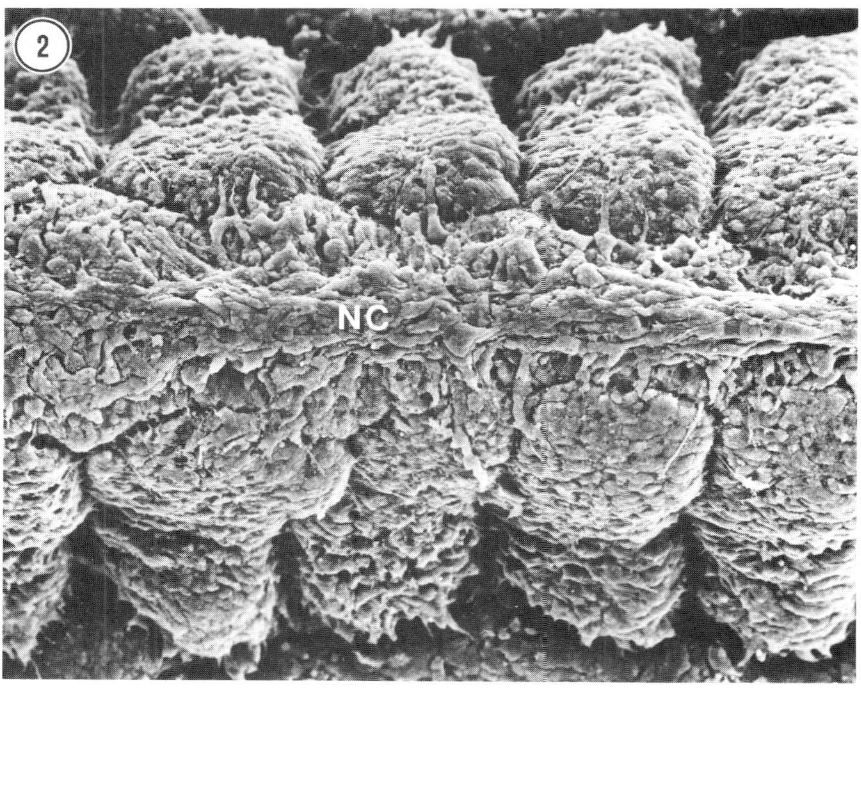
2
NC

Fig. 9–3. The leading tips of the migrating NC cells send out fine filopodia (arrows) by which they seem to establish new contacts and attachments to their substrate—here, ECM fibrils on the neural tube. An increase of such fibrils in the subepidermal space around the original NC cell population (i.e., on the neural tube, under the epidermis, and between the two) was found to be temporally and regionally correlated with onset of migration (Löfberg and Ahlfors, 1978; Löfberg et al., 1980). This suggests that ECM deposition might be a controlling factor for the onset of the migration by providing a suitable substrate for cell locomotion (see below). ×4500.

Fig. 9–4. From the head and posteriorly to about the fifth trunk segment, the NC cells flatten out when spreading on the neural tube. Further posteriorly, the medial NC cord persists in the premigratory position for several stages onward (see Figs. 9–6, inset, and 9–18). Stage 30, anterior is up. ×60.

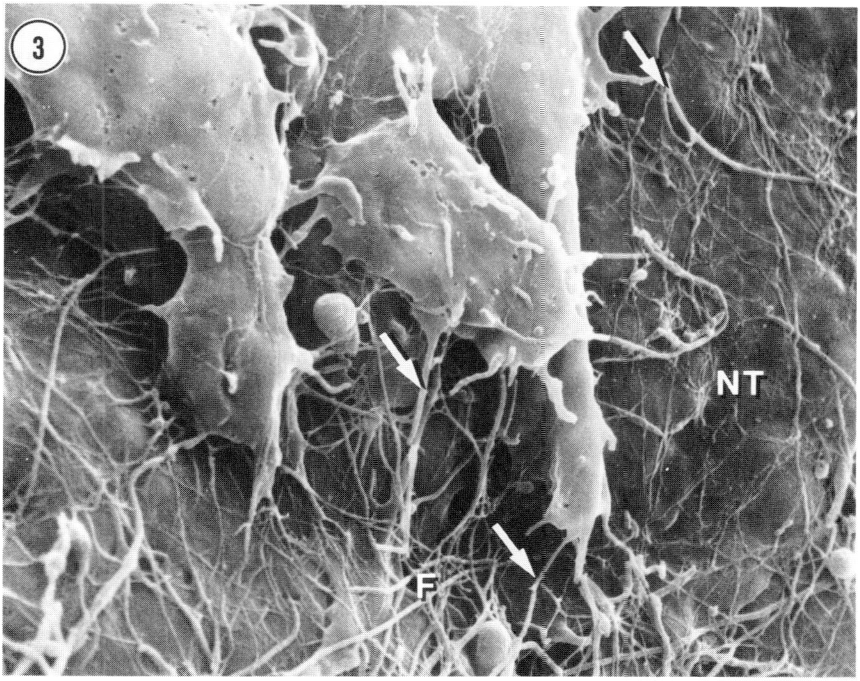
3
NT
F

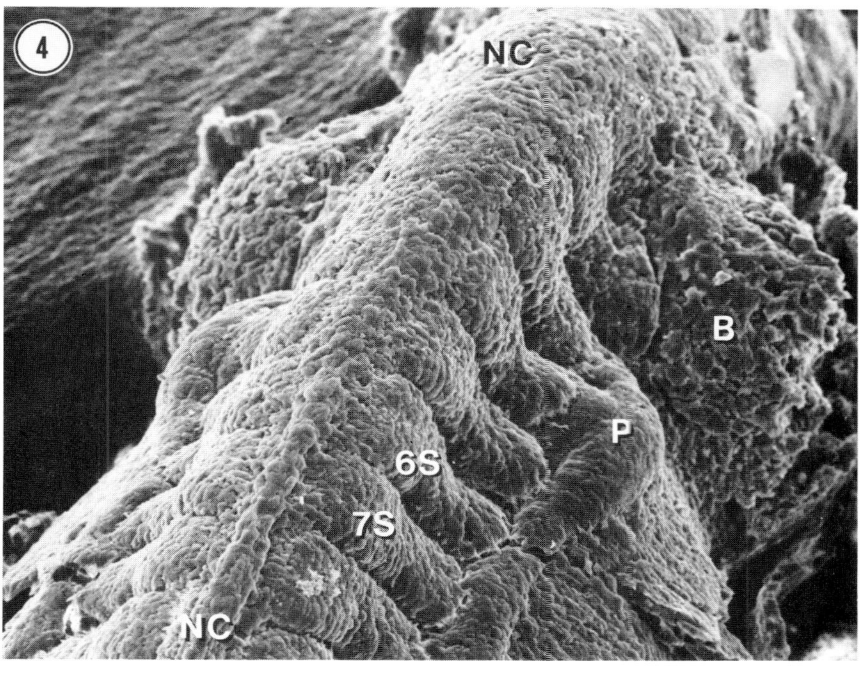
4
NC
B
P
6S
7S
NC

Fig. 9–5. In the branchial region, the ectomesenchymal NC cells seem to disperse ventrally on the branchial endoderm as coherent cell streams (arrows) on their way to form gill arches. Periotic mesenchyme was removed for clarity (asterisk). Stage 25. Compare with Fig. 9–6. ×120.

Fig. 9–6. In the anterior trunk region of stage 32 embryos, the NC cells spread on the neural tube like a coherent sheet with a leading edge. However, when reaching the somites the cells separate from each other and continue to proceed individually in the subepidermal space (i.e., along the dorsolateral pathway) to become pigment cells in the skin. This contrasts with their progression as a cell stream on the branchial endoderm (Fig. 9–5). More posteriorly, where the medial NC cord persists, the cells seem to separate directly from the cord as individuals always from its ventral part (inset). NC cells proceeding into the gap between the neural tube and the somites (i.e., along the medioventral pathway) to form elements of the peripheral nervous system, are mostly concealed under the dorsolaterally migrating cells. However, in embryos of the white axolotl mutant, medioventrally migrating cells are readily seen because subepidermal migration of pigment cells is inhibited (see Fig. 9–26). ×80; inset, ×300.

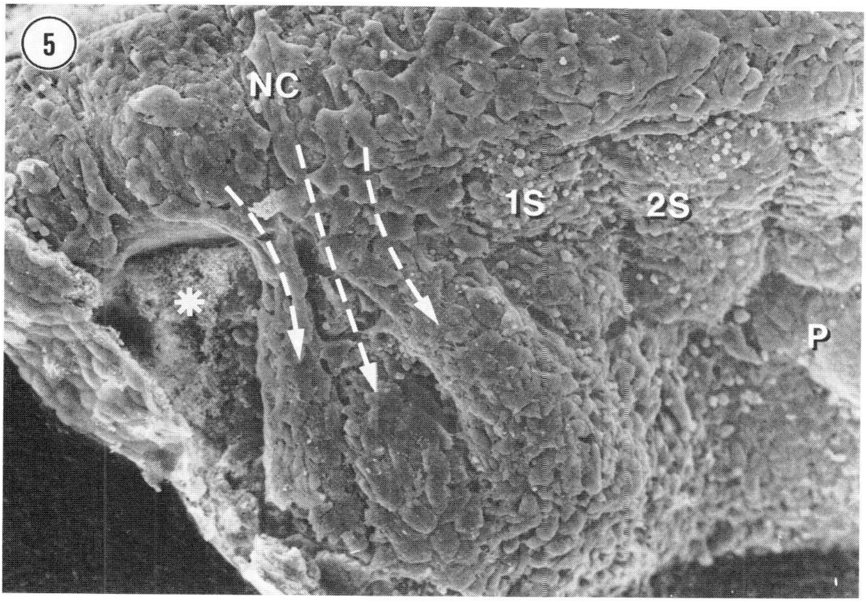
5
NC
1S
2S
P

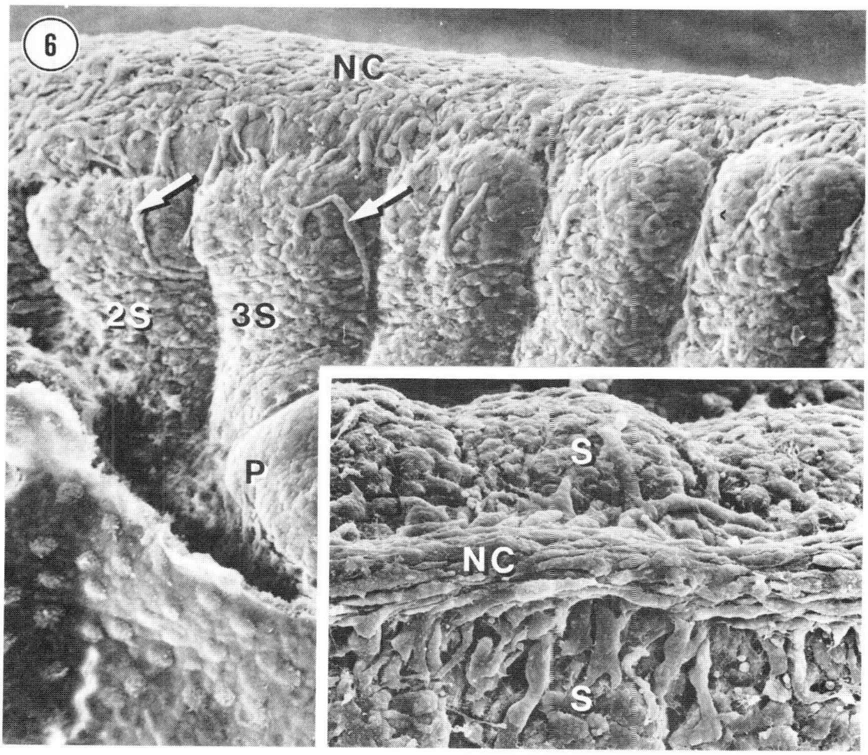
6
NC
2S
3S
P
S
NC
S

Figs. 9–7 and 9–8. In stage 34 embryos cryofractionated in frozen ethanol, an intricate network of ECM fibrils is revealed along the migratory pathways of the NC cells. The ECM of the subepidermal space is shown here. The fibrils seem to be associated with the basal laminae of the epidermis, the neural tube, and the somites (Löfberg and Ahlfors, 1978; Löfberg et al., 1980, 1985). The NC cells proceed in close association with the fibrils and often appear quite entangled by them. Hence, it seems that subepidermal NC cell migration largely occurs *within* the fibrous ECM network, which is spanned out between the epidermis and the somites, rather than on the somite surfaces as might be judged by examining embryos with epidermis removed (see above). Fig. 9–7, ×1500; Fig. 9–8, ×9000.

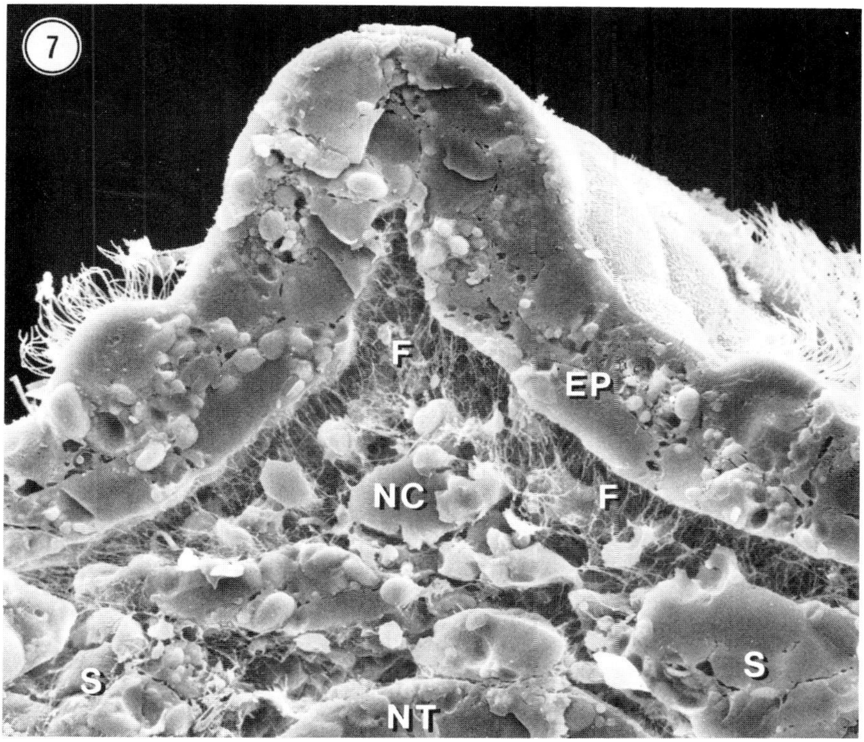
7
F
EP
NC
F
S
S
NT

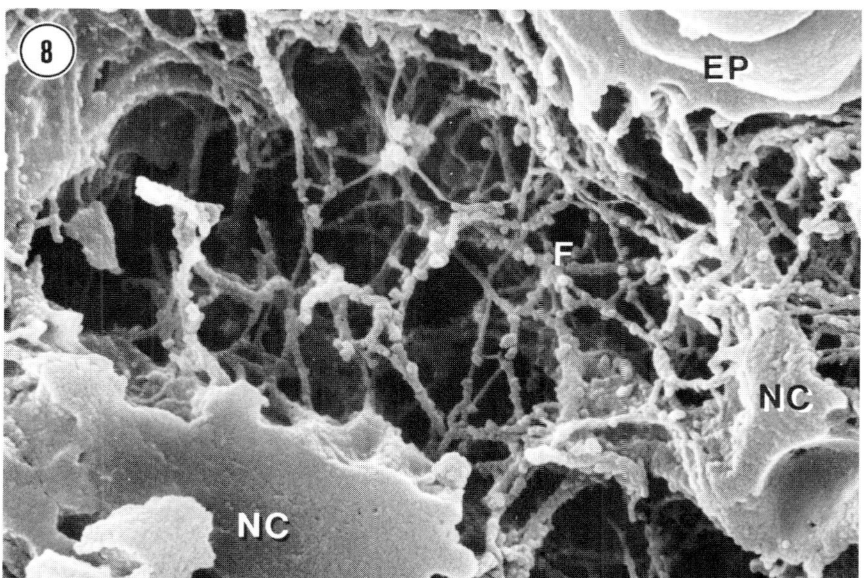
8
EP
F
NC
NC

Figs. 9–9 to 9–13. Ruthenium red precipitates glycosaminoglycans (Luft, 1971a,b). If this stain is included in the fixatives, granular precipitates (G) and filaments about 3 nm thick (arrows) become preserved in the ECM of both SEM (Fig. 9–9) and TEM (Fig. 9–10) specimens. The granules probably represent proteoglycans and the filaments hyaluronate (Turley et al., 1985). ECM fibrils, which show the characteristic cross-striation indicating collagen (Fig. 9–11, inset), as well as surface coats of NC cells are decorated with both granules and filaments (Figs. 9–9 and 9–10). At points of contact between NC cells and ECM fibrils, the cells have electron-dense plasma membrane-associated plaques indicating specialized adhesions between the cells and the fibrils (Fig. 9–11). Immunohistochemistry, using the indirect immunofluorescence technique, indicates the presence of fibronectin (Fig. 9–12) and the chondroitin 6-sulfate determinants of proteoglycans (Couchman et al., 1984) in ECM along the migratory pathways of NC cells. The latter component is detected mainly along the medioventral pathway between the neural tube and the somites (Fig. 9–13; the immunopositive, apical layer of the epidermis is an artifact produced by unspecific binding of the primary antibody to the mucous layer). These findings (Figs. 9–7 to 9–13) indicate that the NC cells use ECM structures as substrate for locomotion. The major component of the ECM is a network of collagenous fibrils with which proteoglycans, fibronectin, and hyaluronate are associated. Fig. 9–9, ×30,000; Fig. 9–10, ×56,000; Fig. 9–11, ×56,000; inset, ×200,000; Fig. 9–12, ×300; Fig. 9–13, ×400.

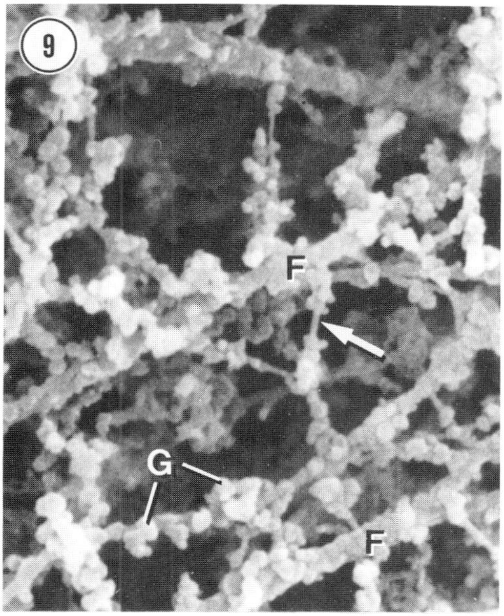
9
F
G
F

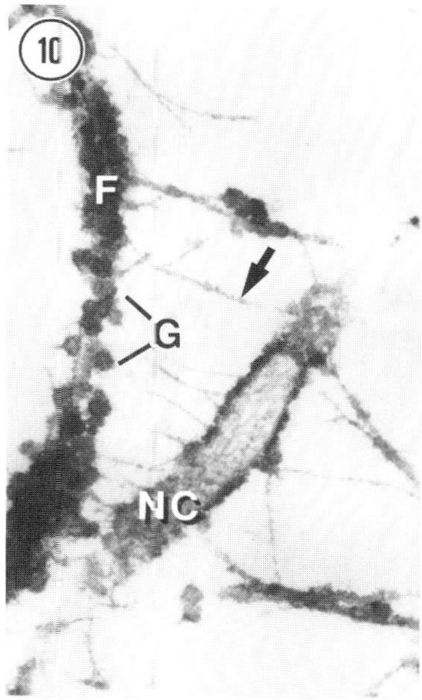
10
F
G
NC

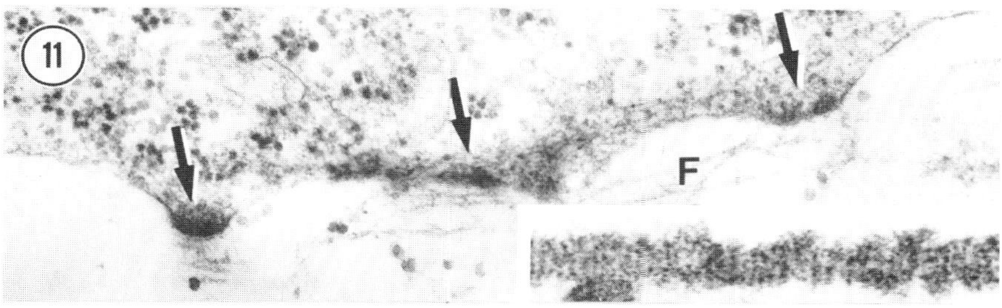
11
F

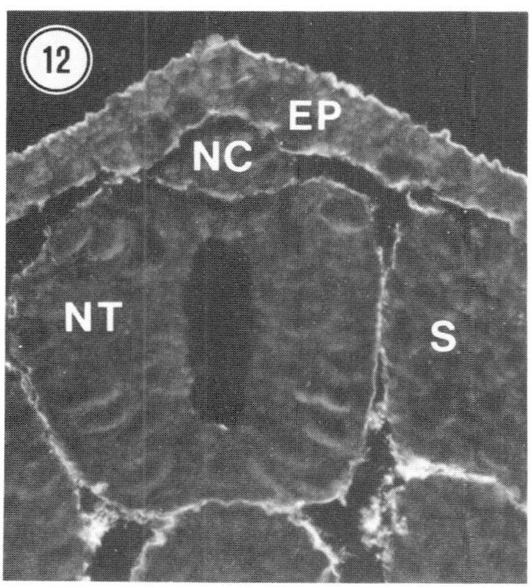
12
EP
NC
NT
S

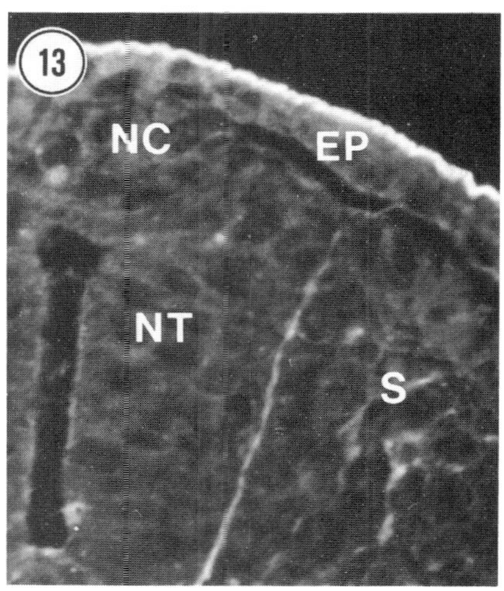
13
NC
EP
NT
S

Figs. 9–14 to 9–17. A series of experiments was made to test if the onset of NC cell migration in the embryonic axolotl trunk is stimulated by surrounding tissues and their associated ECM (Löfberg et al., 1985). Tissue grafts, or embryonic ECM adsorbed in vivo onto inert "microcarriers" prepared from Nucleopore filters, were placed close to the premigratory NC cells in stage 25 embryos. These were then incubated to stage 30, at which stage the experiments were terminated and subsequently evaluated with light microscopy, SEM, and TEM. It was found that grafts of dorsal epidermis were particularly effective in locally stimulating initial NC cell migration in the region under the graft. For instance, if dorsal epidermis was orthotopically and unilaterally transplanted from older to younger embryos, precocious onset of migration was triggered under the graft (Fig. 9–15). The figure shows a lateral aspect of an embryo that received a graft of older epidermis at the fourth trunk segment. A tongue of NC cells protrudes into the subepidermal space ahead of other cells on the neural tube. If grafts of dorsal epidermis were transplanted from anterior to posterior segments in stage 25 embryos, a similar migratory response was obtained. Figure 9–16 shows such a case: An epidermal graft was taken from the third trunk segment of a stage 25 donor and transferred to the eighth segment of a stage 25 host. A tongue of NC cells extends out over the somite where the graft was placed, demonstrating local activation of NC cell migration elicited by the graft. Figure 9–14 shows a dorsal aspect of a stage 30 control embryo (anterior is up), which received an isochronic-orthotopic epidermal graft at stage 25 over the left, fourth segment. Normal NC cell spreading, which is symmetrical on both sides, is shown at this segment; that is, no migratory stimulation has occurred. Figure 9–17 shows a cross section through a stage 30 embryo, into which a membrane microcarrier (MC), onto which subepidermal ECM had been adsorbed in vivo, was implanted under the epidermis at stage 25. A stream of NC cells has been triggered to migrate out (arrow) by ECM on the carrier. The contralateral, control side shows NC cell spreading normal for this region. In control embryos, receiving plain carriers, no migration was induced. These results demonstrate that subepidermal ECM alone can elicit NC cell migration, implying that the ECM attached to the underside of the epidermal grafts was the stimulating factor. To be effective, the grafts and the microcarriers had to be in contact with the premigratory cells. These experiments (Figs. 9–14 to 9–17) indicate that the premigratory NC cells along the trunk have migratory capability but that they need an activating stimulus from the environment—presumably from the surrounding ECM—to start migration. We propose that the subepidermal ECM, which acts as a substrate for cell locomotion, is a controlling factor for the onset of NC cell migration in the axolotl embryo. Fig. 9–14, ×80; Fig. 9–15, ×120; Fig. 9–16, ×140; Fig. 9–17, ×350.

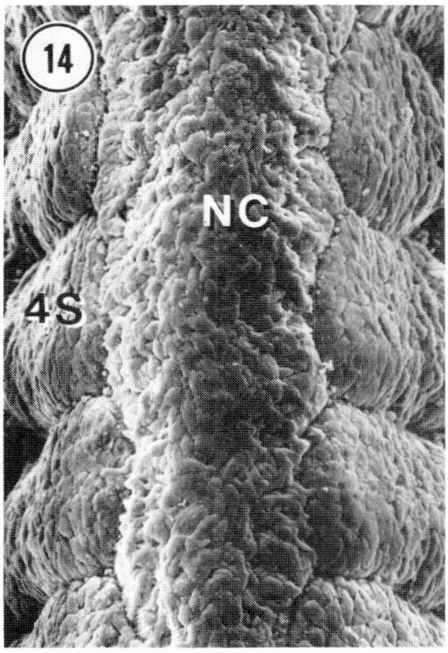
14
NC
4S

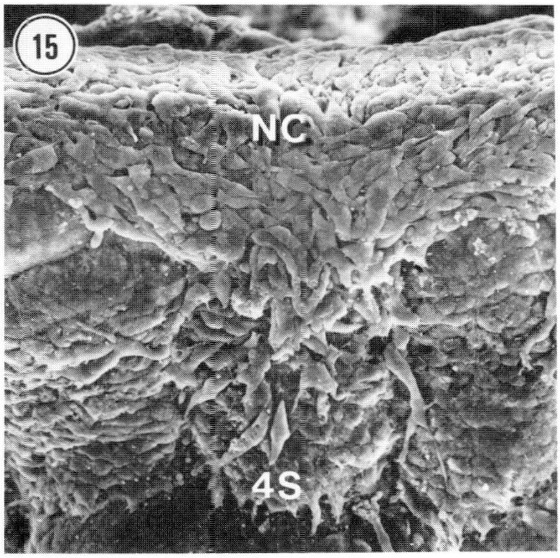
15
NC
4S

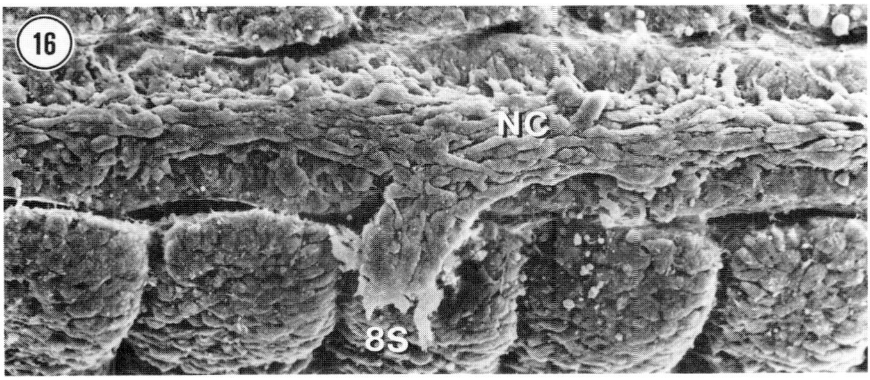
16
NC
8S

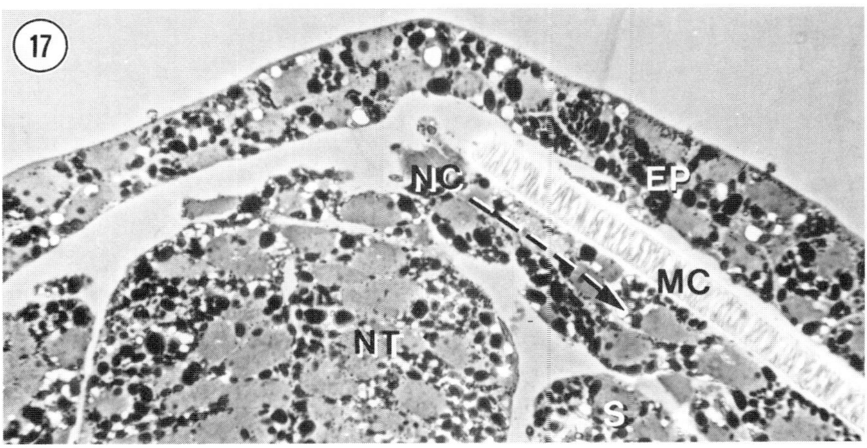
17
NC
EP
MC
NT
S

Figs. 9–18 to 9–20. In stage 34–36 larvae, prominent swellings or humps (Fig. 9–18, arrows), separated by thinner cell strands, occur along the original NC cord, which is still present in a premigratory position on the neural tube. It was shown (Epperlein and Löfberg, 1984) that these humps are mixed aggregates of melanophores, as revealed by the presence of tyrosinase, and xanthophores, visualized by fluorescence of their pterins (Fig. 9–19, arrows). Migrating melanophores could be identified by approximately stage 31. These cells have dispersed extensively over the flank by the time that xanthophores could first be identified at stage 36 (Fig. 9–19). The xanthophores now start dispersing from the chromatophore aggregates, which dissolve. In regions of xanthophore disperson (Fig. 9–20, arrows), melanophores migrate toward the nearest melanophore region (as revealed by video time-lapse recordings, unpublished). The typical barred pigment pattern of the axolotl larva (Fig. 9–20) is thus established by the formation of alternating clusters of "like" cells (Lehman, 1957; Epperlein and Löfberg, 1984). These observations indicate that the xanthophores have a higher homotypic cell–cell adhesion than the melanophores. This may account for the delayed spreading of xanthophores from the NC cord as well as for their aggregation, subsequent segregation from melanophores within the aggregates, and for their lack of mixing with melanophores during pigment pattern formation. Of principal importance is the finding that phenotype expression does occur and that an inherent patterning capability is present already in the premigratory NC cells. This contrasts to other findings, which underline the importance of the embryonic environment for processes of cell differentiation. Fig. 9–18, ×40; Fig. 9–19, ×30; Fig. 9–20, ×25.

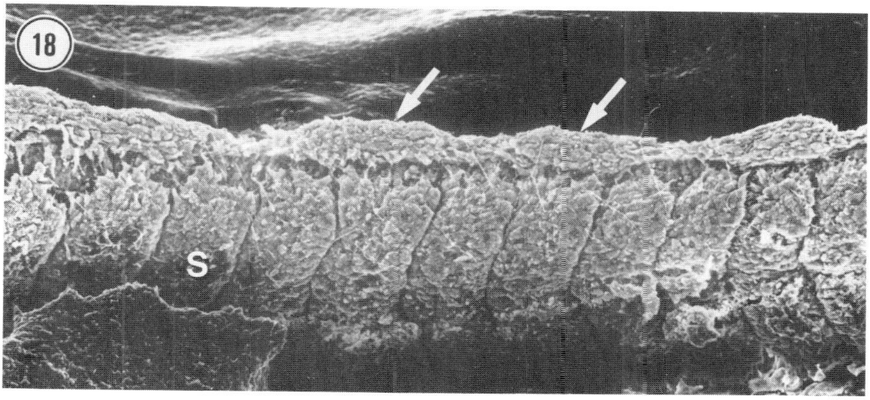
18
S

19

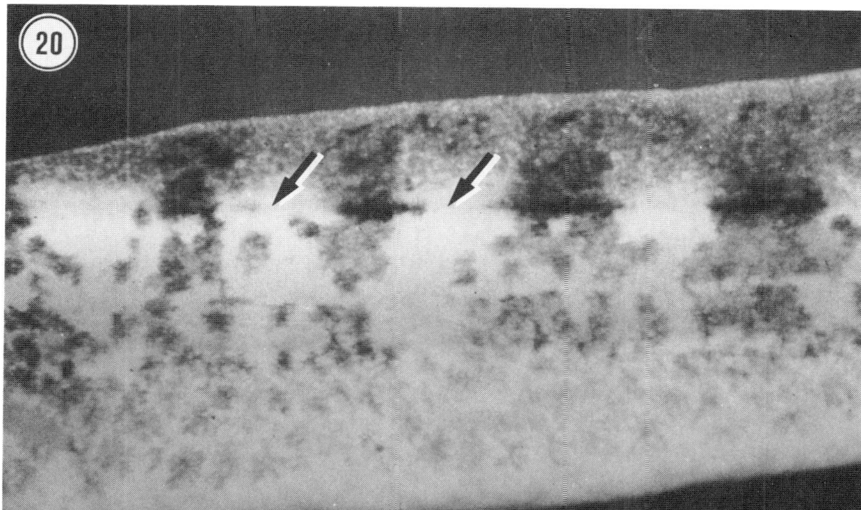
20

Figs. 9–21 to 9–24. In larvae of the white *(dd)* axolotl mutant, trunk pigmentation is restricted because the epidermis is unable to support pigment cell migration from the neural crest (Dalton, 1950b; Keller et al., 1982; Keller and Spieth, 1984; Spieth and Keller, 1984). Chromatophore aggregates (humps) do form, however, in the white embryo (Epperlein et al., 1986), but these are maintained during larval life close to their site of origin (Fig. 9–21, arrows). The pigment cells of the mutant can be stimulated to migrate down on the flank if a piece of its epidermis is replaced by an epidermal graft from a dark *(D/−)* donor (Fig. 9–22). Conversely, migration in the dark embryo can be inhibited locally by a graft of epidermis from a white donor (Keller et al., 1982). We wanted to find out if the subepidermal ECM is the defective component that inhibits pigment cell migration in the white embryo (Löfberg et al., 1988). Therefore, we transplanted subepidermal ECM, adsorbed in vivo onto membrane microcarriers, between white and dark embryos. ECM from the medioventral NC cell pathway was similarly transplanted. Carriers with ECM attached were implanted subepidermally in host embryos at a premigratory NC stage. Plain control carriers and subepidermal ECM from white donors did not affect NC cell migration either in dark (Fig. 9–23) or in white embryos. Carriers with subepidermal ECM from dark donors, however, triggered profuse NC cell migration in the subepidermal space of white embryos—where migration otherwise is blocked (Fig. 9–24, arrows indicate NC cell dispersion). Figs. 9–21 and 9–22, × 15; Figs. 9–23 and 9–24, × 350.

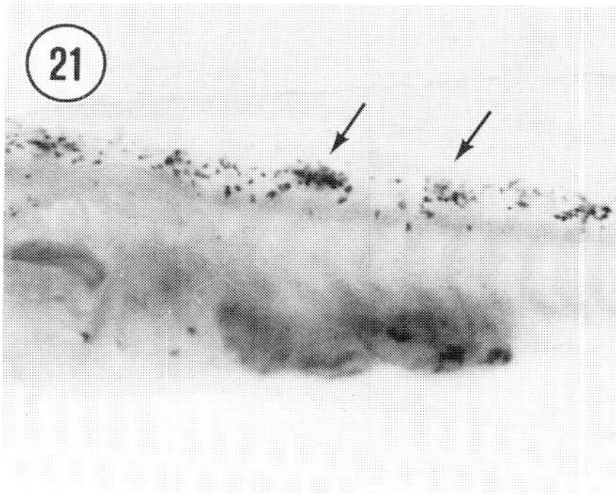
21

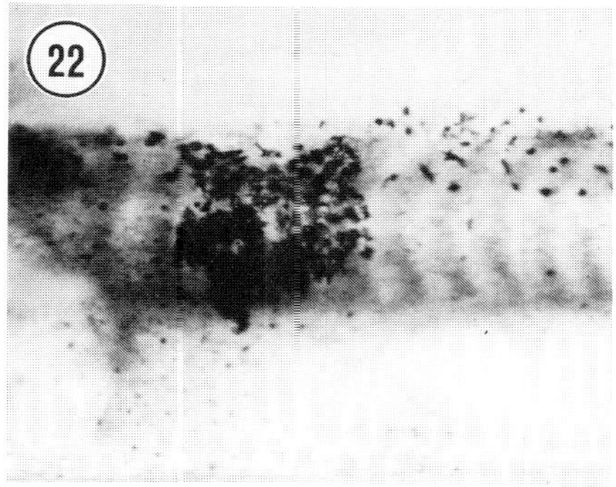
22

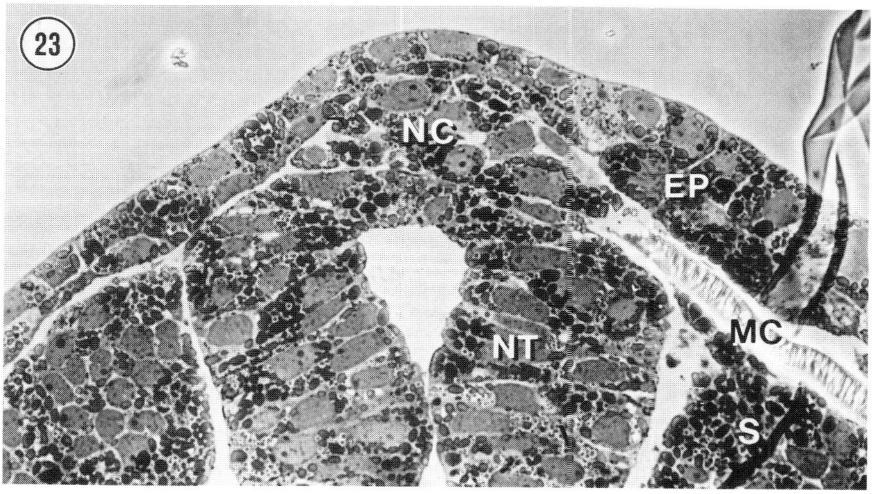
23
NC
EP
NT
MC
S

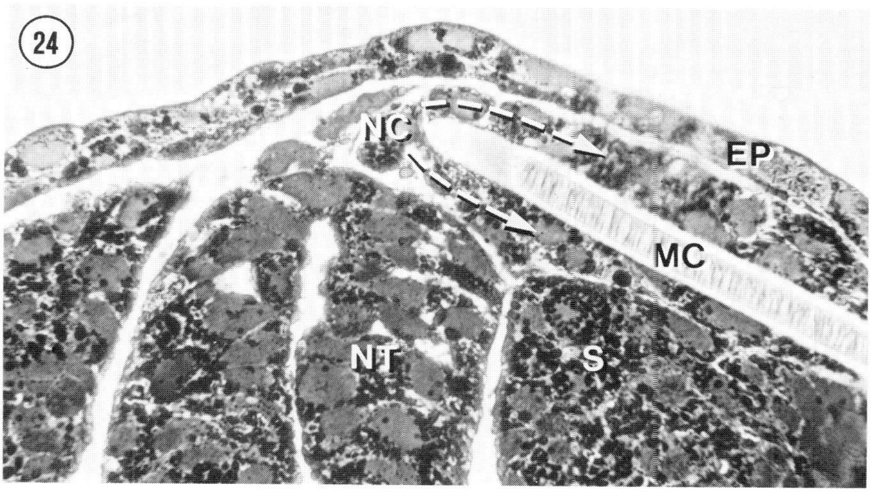
24
NC
EP
MC
NT
S

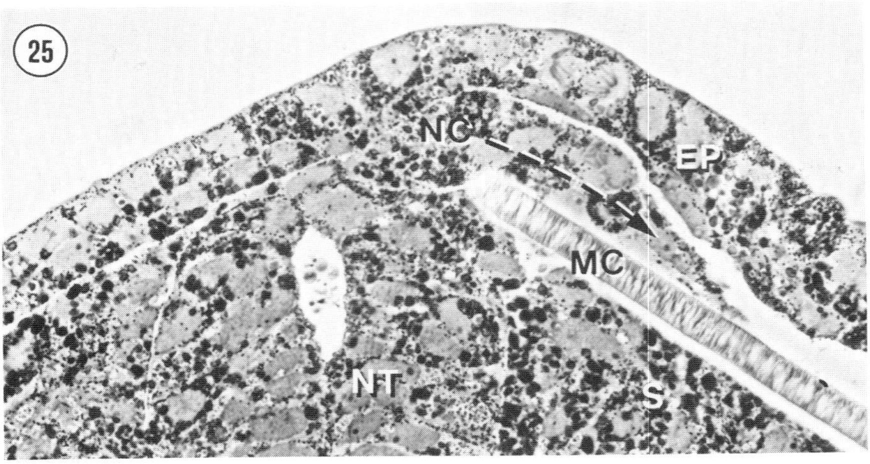
25
NC
EP
MC
NT
S

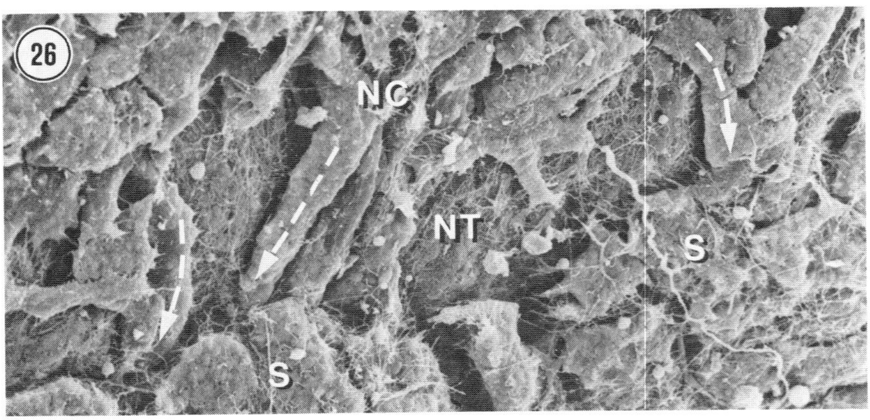
26
NC
NT
S
S

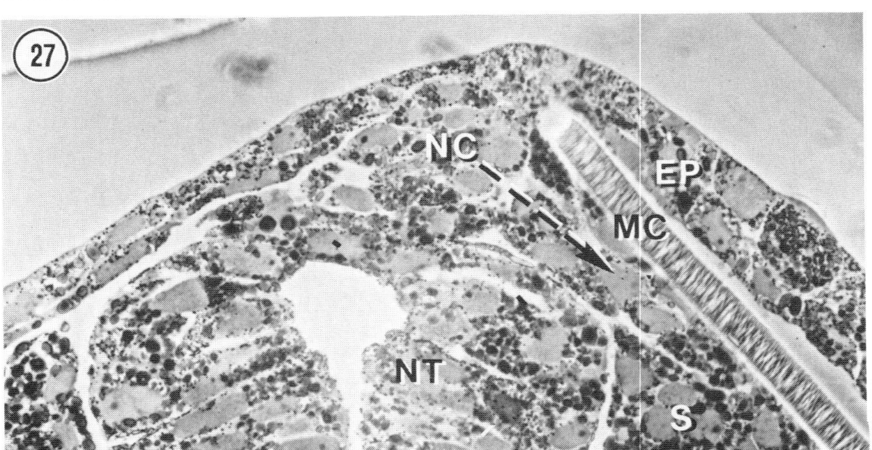
27
NC
EP
MC
NT
S

Taken together (figures and captions), these findings underline the importance of cell–cell as well as cell–ECM interactions during early neural crest cell migration in the axolotl embryo. Evidence was obtained that indicated phenotypic expression and inherent patterning capability in pigment cells already in the premigratory position along the neural tube. However, realization of the barred pigment pattern of the dark axolotl larva depends on chromatophore migration. This process, in turn, is under control of an environmental factor, the subepidermal ECM, which apparently furnishes a suitable substrate for chromatophore migration. If defective, as indicated in embryos of the white mutant, the ECM cannot support migration and no barred pattern develops.

The transplantation experiments between white mutants and normal dark embryos (Figs. 9–22 to 9–27) suggest that the effect of the *d* gene is expressed regionally through the subepidermal ECM during a limited period of development. Hence, the action of the *d* gene seems to bring ECM maturation out of phase with migratory capability of pigment cells. We propose that such a shift in relative timing of the developmental phenomena involved inhibits pigment cell dispersion in embryos of the white axolotl mutant. If correct, this interpretation emphasizes the necessity of precise timing in cell–ECM interactions for proper pigment cell patterning, and suggests that the restricted pigmentation of the mutant larva results from heterochrony (Gould, 1977; Raff and Kaufman, 1983; B.K. Hall, 1984). To find out if these assumptions are valid is a challenging task for future investigation.

ACKNOWLEDGMENTS

We wish to thank Dr. B. Caterson, Biochemistry Department, West Virginia University, Medical Center, Morgantown, West Virginia, for generously providing antibodies used in this study, and Dr. R.P. Tucker, Friedrich-Miescher-Institut, Basel, Switzerland, for carefully reading the manuscript. The study has been financially supported by the Swedish Natural Science Research Council (grant B-BU 3810-104 to J. Löfberg) and Deutsche Forschungsgemeinschaft (grant Ep 8/4-1 to H.H. Epperlein).

Figs. 9–25 to 9–27. Surprisingly, subepidermal ECM from white donors older than the donors usually used, did the same (Fig. 9–25). Migration of NC cells along the medioventral pathway seems normal in white embryos (Fig. 9–26, arrows). As might be expected, medioventral ECM from white donors also elicited migration in white (Fig. 9–27) as well as dark hosts. In all cases, the migratory response, a protruding tongue of NC cells, was localized to the region of the carrier. This had to be placed in contact with the premigratory NC cells in order to trigger migration. Pigment cells occurred among those NC cells that were stimulated to migrate. From these results we conclude that the subepidermal ECM of the white embryo is transiently defective as a substrate for pigment cell migration. The "maturation" of the ECM seems to be retarded beyond the stages during which pigment cells normally respond to it with migration. In contrast, the medial ECM of the white embryo appears to mature normally.
Figs. 9–25 and 9–27, ×350; Fig. 9–26, ×600.

10
Developmental Genetics

GEORGE M. MALACINSKI

As a system for studying developmental genetics, amphibians have much to offer. The large size of their eggs and the external development of the typical amphibian embryo are themselves features that experimental embryologists have exploited for over a century. In addition, however, the fact that the tissue and organ morphogenesis patterns displayed by amphibians are similar to those of higher vertebrates (e.g., mammals) increases the importance of amphibians as model systems.

It was natural, therefore, that once genetic approaches to the analysis of embryogenesis were acknowlegded as being legitimate, several researchers turned their attention to amphibians. Most probably, its was the axolotl's neotenic character that initially generated interest in it (see Chapter 1). In 1863 the first shipment (34 animals) from Lake Xochimilco (near Mexico City) was received in Paris at the Jardin des Plantes (A.H.A. Duméril, 1870b). Included in early shipments to Paris was a white variant (gene *d*??), which was originally described by Duméril (see Newth, 1960), and later by Häcker (1907).

Initially, the axolotl was used sporadically as experimental material by various European embryologists. Systematic attempts to understand its developmental genetics began, however, around 1935, when Humphrey started to study sex determination in axolotls. He employed *Ambystoma mexicanum* tailbud-stage embryos as hosts for gonad primordia transplants from *Ambystoma punctatum* donors. Sex reversal of the donor gonads was observed (Humphrey, 1935b). A few years later, Humphrey analyzed the compatibility of skin grafts among various *Ambystoma* species, including *A. punctatum, A. tigrinum,* and *A. mexicanum* (Humphrey and Burns, 1939). Then, together with Fankhauser, he studied the induction by cold treatment of triploids and haploids in axolotl eggs (Fankhauser and Humphrey, 1942).

During those early years of amphibian developmental genetics, wild-caught ambystomatid salamanders, especially *A. tigrinum* and *A. punctatum,* were used routinely. A switch from those species to *A. mexicanum* was prompted by a need to carry out breeding experiments in the laboratory. As Humphrey pointed out, *A. tigrinum* proved to be poorly adapted to breeding experiments under laboratory conditions. It was necessary to duplicate—in the laboratory—hibernation conditions in order to prompt breeding behavior in terrestrial species (Humphrey, 1942).

Around that time, most of the axolotls in American laboratories came from various European collections (e.g., Poland, Switzerland, Ukraine; see Humphrey, 1966), although wild stock imported from the Mexico City area was occasionally obtained (e.g., see Humphrey and Bagnara, 1967). Animals of laboratory origin proved then, as well as now, to be more healthy in captivity than freshly imported wild stock.

Serious genetic studies began with Fankhauser and Humphrey's characterization in 1942 of nucleolar number (2) and chromosome count (28) in tailtip squashes from diploid axolotl larvae. They even prepared polyploid larvae and counted 70 chromosomes in pentaploids. Later they mapped the centromeres (Lindsley et al., 1956) of a few of the axolotl chromosomes.

As mentioned above, the first mutant, a white variant, was described very early. The second mutant, the "bulldog dwarf" gene that generated disproportionate dwarfism, was reported by Lipsett in 1941. In 1935 Humphrey obtained a spawning of white animals and raised them to sexual maturity. When mated, some spawnings segregated into what was eventually identified as the gene *f* (fluid imbalance) (Humphrey, 1948b).

An intense inbreeding program with a small colony was initiated by Humphrey. It yielded the gene *v* (vasodilation), then microphthalmia (*mi,*) *g* (gill lethal), and so forth. It was inbreeding that generated virtually all of the early mutants, as well as each of the more recent ones.

Although at the time the first half-dozen axolotl mutants were recognized (ca. 1940–1960), several other developmental genetic systems were more advanced (e.g., silk moth—*Bombyx mori;* fruit fly—*Drosophila melanogaster*), the axolotl was clearly the most studied and best defined of the amphibian systems.

Although the axolotl was the choice of early amphibian geneticists, other amphibians are also presently being employed with great success. These include, most notably, the South African clawed toad, *Xenopus laevis,* and various Japanese *Rana* species (recently reviewed by Malacinski, 1988).

Compared with those other species, however, the axolotl exhibits several advantages. The eggs are relatively large (2.0 mm diameter versus 1.2 mm for the *Xenopus* egg) and develop slowly; both traits are useful for the developmental geneticist wishing to perform microsurgical manipulations in order to understand a mutant phenotype. The early embryo responds favorably to temperature regulation and can be speeded up at 25°C, or slowed down almost to a stop, at 5°C. This feature is especially valuable for the developmental geneticist. A portion of a spawning believed to harbor a mutant phenotype can be raised at 25°C, to speed up development and shorten the time required to recognize a mutant phenotype. The remainder of the spawning can be held back at 5°C, and, once the mutant search has yielded a positive identification, employed in an appropriate experimental protocol. Several genes—including *e* (eyeless), *c* (cardiac), and *o* (ova-deficient)—have been studied extensively using that temperature regulation protocol.

In addition, the axolotl displays appendage regeneration, a phenomenon never fully documented in the laboratory anurans such as *Xenopus, Rana,* or *Bombina.* Although no information on the developmental genetics of appendage regeneration has yet been collected, it is likely that initiatives in that direction will soon be forthcoming.

Despite the fact that the axolotl appears to be well suited for developmental genetics, several technical limitations have intervened to retard progress. At best, five spawnings per year can be obtained from a sexually mature female that is mated at approximately 2- to 3-month intervals. Artificial insemination protocols have been developed, but unlike those for laboratory anurans, no opportunity for stripping eggs by massaging the female's abdomen exists. Rather than inseminating several hundred eggs simultaneously, small batches of 12–24 eggs are harvested as the female sheds them, and then they are bathed in a sperm suspension (Chapter 21).

Progress in cell culture technology has also been retarded for urodeles such as the axolotl. Although techniques for primary cell culture are available (see Chapter 23), no established axolotl cell lines are available. Practical cell culture methods would be very advantageous for analyzing the large assortment of autonomous cell-lethal genes (see below). Finally, one technical limitation that may eventually hinder molecular genetics studies is the repetitive DNA content of the axolotl. Like that of most urodeles, the axolotl genome contains substantial amounts of surplus, repetitive DNA (Straus, 1971). It remains to be seen whether that excess DNA will hinder attempts to generate genomic DNA libraries for isolating mutant DNA sequences.

Despite these limitations, the axolotl has emerged as a remarkably versatile and diverse system for developmental genetics, as will be verified by reviewing the axolotl mutant genes.

MUTANT GENE COLLECTION

Several dozen mutant genes have been recognized, at one time or another. Some of the genes listed in Table 10–1, especially the maternal effects (e.g., genes *o* and *cl*), may in fact be extinct. Nevertheless, that list (and the diagram in Fig. 10–1) indicates which sorts of mutant phenotypes have so far been recognized, and therefore which types of mutants are most likely to be discovered in the future. For example, a disproportionately large number of autonomous cell-lethal mutants (see Chapter 11) has been collected, suggesting that future mutant searches will probably yield even more.

Virtually all of those mutants have been identified by carefully screening, with the aid of a dissecting microscope, spawnings of laboratory-inbred, or recently collected (from Lake Xochimilco, Mexico) animals. By keeping careful records and systematically mating siblings or backcrossing progeny to parents, a surprisingly large and diverse collection of mutants has been obtained.

As the illustration in Fig. 10–1 indicates, the gene collection represents a broad range of phenotypes. However, it is worth noting that most of those mutant phenotypes affect external, morphological features—a result of the manual screening method mentioned above. No doubt an even larger number of mutant genes has escaped detection because they affect more subtle, internal developmental events (e.g., genes associated with specific metabolic defects). If key developmental events, such as pattern specification, are driven by subtle processes such as the pathways of intermediary metabolism (e.g., DeLotto and Spierer, 1986), then a major cate-

Table 10–1 Axolotl Mutant Genes

Name and Developmental Stage Usually Associated with Gene's Action	Stages of Expression of Phenotype	Major Phenotypic Trait
Oogenesis (maternal effects)		
No cleavage (*nc*)	2	Eggs fail to cleave
Cleavage (*cl*)	2–4	Early cleavage defects
Ova-deficient (*o*)	11	Arrest at gastrulation
Temperature sensitive (*ts*)	11	Gastrular arrest; also paternal effect
Fluid imbalance (*f*)	9	Fluid accumulation at blastula
Vasodilation (*v*)	9	Arrest at blastula
Early development (e.g., cleavage)		
Nuclear variants (n^1; n^2; n^3; n^5)	10	Small nucleolus
Reduced rDNA (n^4)	10	Reduced copy number of rDNA genes
Organogenesis		
Cardiac (*c*)	38 to Hatching	No heartbeat
Cardiac-2 (*c-2*)	38 to Hatching	No heartbeat
Eyeless (*e*)	21 to adult	No eyes, sterility, enhanced pigmentation
Pinhead (*pi*)	12? to Hatching	Craniofacial defects; interocular distance decrease
Short toes (*s*)	Posthatching	Abnormal limb development
Spastic (*sp*)	Hatching	Abnormal swimming movements
Spastic lethal (*sp-l*)	Hatching	Abnormal swimming movements
Anemic (*an*)	Posthatching	Temporary anemia
Phocomelia (*ph*)	Posthatching	Retarded limb development
Ascites (*as*)	Hatching	Lack outlet for Wolffian duct
Microphthalmia (*micro*)	?	Small eyes; semilethal
Growth (e.g., cell and tissue function)		
Premature death (*p*)	37	Possible neural crest defect[a]
Stasis (*st*)	Hatching	Circulatory failure
Quivering lethal (*q*)	Hatching	Limited movement
Twisted gills (*t*)	Pre- to posthatching	Abnormal gills; do not feed
Utrecht lethal (*ut*)	Posthatching	Slow growth; abnormal gills
Microphthalmia (*mi*)	Hatching	Small eyes; lethal
Gill lethal (*g*)	37	Abnormal gills
London lethal (*l*)	Hatching	Small eyes, head; reduced growth rate
Lethal *r* (*r*)	Posthatching	Abnormal gills
Lethal *x* (*x*)	Posthatching	Abnormal gills
Lethal *y* (*y*)	Posthatching	Abnormal gills
Hand lethal (*h*)	Posthatching	Foot malformation
Brandon lethal (*b*)	Posthatching	Retarded limb development
Sexual maturity (e.g., pigmentation)		
White (*d*)	38 to Adult	Defective neural crest cell migration
Albino (*a*)	38 to Adult	No melanin synthesis
Melanoid (*m*)	38 to Adult	Increased melanophores
Axanthic (*ax*)	38 to Adult	No yellow pigments synthesized

Table 10–1 Axolotl Mutant Genes (*continued*)

Name and Developmental Stage Usually Associated with Gene's Action	Stages of Expression of Phenotype	Major Phenotypic Trait
Peg-leg (*pl*)	Posthatching	Abnormal limb development
Matilda (*mt*)	Posthatching to 3 months	Swims in circles; abnormal gills
Isozyme variants		
Soluble esterase (*est-3*)	?	Absence of enzyme activity
Phosphoglucomutase (*pgm*)	?	Altered electrophoretic mobility
Peptidase (*pepB*)	?	Altered electrophoretic mobility

[a]A. C. Graveson and J. B. Armstrong, personal communication.

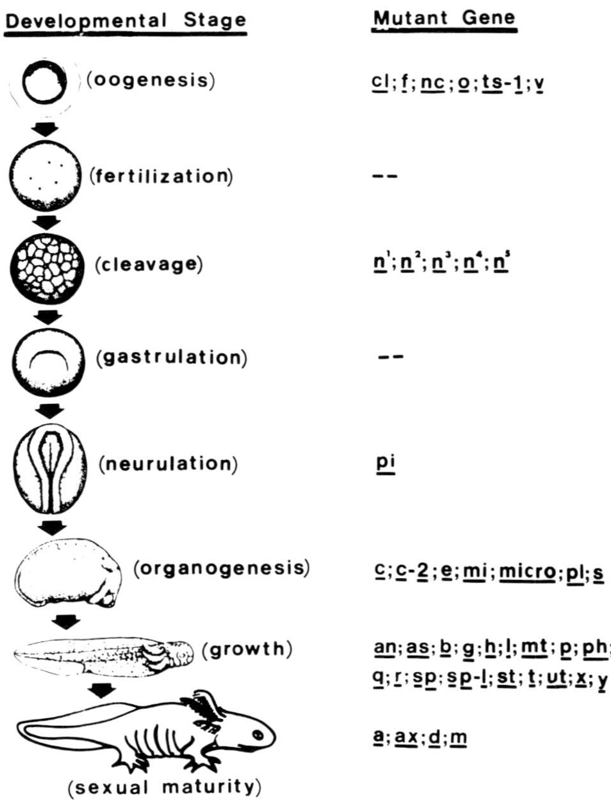

Fig. 10–1. Assignment of each of the mutant genes to a developmental stage at which the individual phenotype is most striking. Although some of the most interesting mutants display maternal effects (oogenesis), the largest number of mutants appears to affect the function and viability of *all* cells of the organism. These latter mutants are therefore referred to as "autonomous cell lethals" (see Chapter 11), and are assigned to the "growth" phase in this illustration.

gory of developmental mutants is being missed by the bias toward manual morphological screens.

Nevertheless, screens for isozyme markers have been partially successful. Isozymes of malate dehydrogenase, lactate dehydrogenase, esterase, and phosphoglucomutase have been identified by Cooper et al. (1985). Isozyme markers of that type may eventually be useful in genetic mapping studies and as cell lineage tracers.

Being limited, for the most part, to recognizing morphological developmental abnormalities has both advantages and disadvantages. Clearly those mutant genes that alter major developmental cascades and terminate in spectacular defects, such as eyelessness (gene *e*) (see Chapter 13), are worthy of study and almost cry out for serious research approaches. Ironically, however, it is those genes that affect developmental cascades and generate catastrophic effects that will probably be the most difficult to understand fully. That is because they most likely generate a wide range of secondary, or pleiotropic, effects, which will require an exhausting amount of effort to sort out.

Although it is labor-intensive, manual screening of spawnings of related animals is straightforward and has generated all of the mutants listed in Table 10–1. A side benefit of that method is the fact that inbred stocks, which are very useful for controlling experimental variation, result from the procedure. Mutagenesis and gynogenesis have both been attempted, but without notable success. Survival among treated embryos and adults is unfortunately relatively low (reviewed by Malacinski, 1988).

The culmination of this discussion is the sobering thought that classical approaches to axolotl developmental genetics (i.e., inbreeding, scanning spawnings, etc.) limit researchers to being satisfied with those mutant genes that are collectible. Those collectible genes (e.g., cell lethals) may—in reality—not generate phenotypes that are the most desirable for contemporary research. For example, limb regeneration mutants or egg polarity mutants may not be obtainable using classical approaches. Modern molecular approaches (see below) to identifying useful mutant genes will probably be necessary. Nevertheless, there is good reason for believing that mutations are continually occurring in laboratory strains (Humphrey, 1975), so as the search for new mutants goes on, room for optimism exists.

MAPPING MUTANT GENES

Conventional breeding exercises have provided linkage data for several of the genes listed in Table 10–1. For example, the genes *f* (fluid imbalance) and *g* (gill lethal) are closely linked, as reported first by Humphrey (1960). They are within approximately 2 map units of each other. Such tight linkage actually caused some early confusion; the two phenotypes were originally assigned to a single mutant gene (*f*). The white gene (*d*) and the nucleolar organizer genes (*n*) are also closely linked (Humphrey, 1975). The *y* (lethal) and *s* (short toes) genes are probably within 20 map units of each other (Humphrey, 1973), while the *e* (eyeless), *x* (fragile gills), and *r* (renal insufficiency) genes are probably also separated by 20–25 map units (Epp, 1978).

Employing the more direct approach of gynogenesis, several additional genes have also been mapped. For example, Lindsley et al. (1956) observed that the *zw* (sex locus), *d* (white), and *f* (fluid imbalance) genes are remotely located from their centromeres. Several other genes have also been mapped at the gene–centromere level. These genes, together with their respective map unit distances from the centromere, are listed in Table 21–2. Since the axolotl genome is relatively large (approximately 2600 map units), it is probable that many of the mutant genes are located too far from a centromere to provide reliable mapping data with this approach.

Accordingly, a multistep approach that relies on cytogenetics has been started. Giemsa banding patterns for related *Ambystoma* species (e.g., *A. mexicanum* and *A. tigrinum*) are first established. Second, the fertile hybrids that these species are capable of producing are generated. Third, the chromosome combinations that segregate in mutant progeny of the F_2 hybrid are tracked. In several instances, where important mutant genes with distinctive phenotypes are located on a chromosome with a very distinctive banding pattern, it should be possible to provide reliable chromosome assignments. An initial attempt in that direction has been provided by Cuny and Malacinski (1985). They have succeeded in accomplishing the first two steps in that three-part procedure.

It will be interesting to see whether molecular genetics (see below) will provide a short-cut alternative to the conventional mapping techniques just described.

FUTURE PROSPECTS

Several opportunities exist for further exploiting the axolotl as a model system for developmental genetics. Included are both short-term and long-term approaches. In the *short term,* perhaps the focus should be on utilizing more fully the mutant genes that are presently available in abundant supply. Those include the genes *e, c, pi,* and various pigment genes. However, exploiting the full potential of those genes will require preparing either gonad grafts (e.g., Humphrey, 1929b) or head grafts (e.g., Van Deusen, 1973) on embryos. By transplanting gonad primordia from darkly pigmented, homozygous mutant embryos into the gonad-forming area of lightly pigmented (e.g., albino) hosts (in place of the host primordia, which are removed), sexually mature adults that shed homozygous recessive gametes can be obtained. With those gametes, the researcher can design experiments on relatively early-stage embryos. Otherwise, it is usually necessary either to wait until the mutant phenotype is recognizable (by which time secondary as well as primary effects of the mutant are being exhibited), or to perform "blind" experiments on the progeny of mated heterozygotes (and wait until the mutant phenotype is exhibited in 25% of the progeny, before scoring the results).

Over the *medium term,* an investment of time into gene cloning is suggested. Attempts should be made to identify defective gene products associated with various mutant genes. Recombinant-gene technology is sufficiently well advanced so that differential screening of tissue-specific mutant RNAs can be carried out (e.g., Hedrick et al., 1984). An alternative strategy also is available. Genes that have been isolated (and in some cases, sequenced) in other systems (e.g., *Drosophila*) can be employed as probes in Southern blot analyses to search for homologous genes. In perhaps the most famous instance to date involving an amphibian, a *Drosophila* homeobox gene probe was employed to probe a *Xenopus* genomic Southern blot. Homologous genes were found (Carrasco et al., 1984), and the cDNA probes that eventually were generated were employed to study expression of the homeobox genes during *Xenopus* embryogenesis (Carrasco and Malacinski, 1987). Using that success with the homeobox genes as a guide, further attempts employing heterologous probes may provide medium-term insights into the organization and function of the axolotl genome, which are not readily obtainable using the methods of conventional transmission genetics.

Long-term goals should include attempts to develop a gene transformation system. To date, attempts to transform amphibians (mainly *Xenopus laevis*) by gene injection have met with limited success, particularly with genes transcribed by RNA polymerase II. However, certain of these genes, namely a cardiac α-actin gene from *Xenopus borealis* (Wilson et al., 1986) and a gastrula-specific gene *DG17* (Kreig and Melton, 1985), have been correctly expressed. The injection of cloned DNA into *X. laevis* appears to produce animals that are mosaics (L. D. Etkin and B. Pearman, 1987), probably because integration of such sequences does not occur before first cleavage. The axolotl may offer an advantage for such studies in that the interval before first cleavage is long (up to 20 hours at 10°C), potentially allowing sufficient time for integration at the single-cell stage. In addition, the regenerative capabilities of salamanders provide a convenient method for following the fate of injected DNA.

ACKNOWLEDGMENTS

The activities of the author associated with axolotl developmental genetics are supported by NSF BSR 84-21177. Susan Duhon assisted in the preparation of Table 10–1.

11

Cell-Lethal Genes

ANTON W. NEFF

Since the first demonstration of recessive lethal genes in the mouse (Cuenot, 1905), lethal genes have been utilized in such diverse areas as population genetics, radiation genetics, developmental genetics, cell differentiation, and cell metabolism (reviewed by Hadorn, 1961; T.R.F. Wright, 1970; McLaren, 1976b). Lethal mutations provide unique tools for those studies because they can be employed to generate very precise, reproducible experiments that can seldom be duplicated by other laboratory methods. Recessive lethal genes (genes that produce effects severe enough to cause the death of the individual organism before it is able to reproduce) differ from other mutant genes only in the severity of their effects. They can cause primary lesions that lead to abnormalities at any biological level (cell, tissue, organ) or during any stage of ontogeny. Cell-lethal genes are those that cause abnormalities in every cell in an organism. They represent, therefore, a subcategory of recessive lethal genes. Operationally, they are defined either by the inability of mutant tissue to survive when transplanted to a normal wild-type host, or by decreased survival when cultured in vitro. If the effect of such mutations cannot be rescued by genetic or nongenetic factors, they are considered to be autonomous (Gluecksohn-Waelsch, 1963). Transplantation experiments and cell, tissue, and organ culture experiments have shown that not all cell-lethal mutations are, however, autonomous (Murphy, 1974). This review will discuss the criteria employed to place mutant genes in the cell-lethal category, list the axolotl cell-lethal genes, and discuss several examples.

CELL-LETHAL CLASSIFICATION SYSTEM

Of the axolotl mutants, the largest category is the cell-lethal genes (see Chapter 10). This is a heterogeneous group of genes with a wide range of phenotypes (e.g., short toes, *s*; small eyes, *mi*). As well, they display a broad range in time of development when the lethal phenotype can be recognized, and when the embryos die. Some die before hatching (e.g., gene *p*), shortly after hatching (e.g., gene *g*), and months after hatching (e.g., gene *h*) (see Table 11–1 and Armstrong, 1985, for a recent review). The main criteria for including genes in this category have been two embryonic manipulation experiments: (1) parabiosis and (2) tissue transplants. Embryos homozygous for cell-lethal genes do not survive parabiosis with a wild-type embryo, and gill and limb primordia from homozygous cell-lethal genes typically fail to develop on a wild-type host (Malacinski, 1978b). Table 11–1 shows that not all of the putative cell-lethal genes have been fully tested for cell lethality, including genes *q* and *ph*. Many of the genes that have been tested show no long-term rescue by parabiosis and no rescue of homozygous mutant limb and gill primordia on wild-type hosts. They therefore apparently fit easily into the cell-lethal category (e.g., genes *g, mi, x, l, ut,* and *h*). Other genes appear not to be cell lethal, in that although parabiosis does not rescue the mutant embryo, tissue transplants are indeed rescued by growth on the wild-type host. The genes *t* and *st* are examples. Another set of genes appears to show incomplete cell-lethal characteristics such as incomplete rescue of transplants (gene *y*), limited or long-term survival of parabionts (genes *y* and *r*), and selective rescue of tissue transplants (gene *s*).

There are, however, some inherent shortcomings to the cell-lethality tests that have so far been used on axolotl mutants. First, the classification of a mutant gene as a cell-lethal gene is based solely on negative evidence—that is, failure of survival. Therefore, genes can only with certainty be eliminated from this category, rather than added. Second, there is an inherent underestimate built into the cell-lethal tests. Parabiosis of whole embryos and selected tissue transplant tests are used to classify mutants with respect to cell lethality. Individual cells that might be capable of being rescued are easily overlooked.

One way of circumventing those shortcomings is to examine more tissues and to analyze cells directly under culture conditions. Since Ephrussi (1935) cultured Brachyury lethal mutant tissues in explant cultures, researchers have used tissue and organ culture as tools to understand homozygous cell-lethal genes (Anderson and Ede, 1985). Neff et al. (1980) have since proposed utilizing the cultivation of mutant axolotl embryo explants as a model system to study cell lethality. Finally, Armstrong (1985) has proposed the explant culture scheme included in Table 11–2 for classifying cell-lethal genes. Using these cell–organ culture procedures, mutant cells that fail to survive are classified as *cell autonomous.*

More specialized categories of cell-lethal genes can be developed, since the results of the double tests (tissue transplantation–cell, organ explant) provide four possible outcomes: transplant rescue and explant rescue; transplant nonrescue and explant nonrescue; transplant rescue and explant nonrescue; transplant nonrescue and explant rescue. Even more categories are possible if explant culture results determine that some cell types are able to survive in culture while others are not.

CASE STUDIES

In order to determine whether in vitro-explant cultivation of axolotl embryonic tissues is feasible, and to establish if the categories of cell-lethal genes outlined by Armstrong (1985) would be useful for the eventual classification of virtually all axolotl cell-lethal genes, preliminary data on tests for cell lethality utilizing in vitro-explant cultivation were collected.

Table 11–1 Characteristics of Cell-Lethal Mutant Phenotypes

Gene Symbol	Structures, Functions Affected	Onset	Death	Test for Cell Lethality: Parabiosis	Test for Cell Lethality: Transplantation	Reference
g	Gills; pigmentation, yolk metabolism	Prehatching	1–2 weeks posthatching	Not rescued	Gill, limb, gonads not rescued	Tompkins (1970)
mi	Eyes	Feeding stage	1–2 weeks postfeeding	Not rescued	Gill, limb not rescued	Humphrey and Chung (1977)
x	Gills	Hatching	Postfeeding	Not rescued	Limbs, branchial arch not rescued	Humphrey and Chung (1978)
y	Limbs, mesonephros, lungs	Hatching	5–6 weeks posthatching	No long-term rescue (survive 6 months)	Gonads, mesonephros incomplete rescue	Humphrey (1973)
l	Head, eyes, gills, limbs; growth	Hatching	1–2 weeks posthatching	Not rescued	Limb, gill, pronephros, gonads, head not rescued	Chung and Briggs (1975)
ut	Gills, limbs; growth	Hatching	2–3 weeks posthatching	Not rescued	Limb, gill not rescued	Humphrey et al. (1978)
h	Limbs, digits	1–2 months posthatching	2 months posthatching	Not rescued	Gills, limb not rescued	Humphrey and Chung (1977)

r	Gills, limbs; feeding	1 week postfeeding	3 weeks postfeeding	Not rescued, limited rescue	Gill, limb not rescued	Humphrey (1964); Humphrey and Chung (1978)
t	Gills, limb, cartilage, rectus cervicus muscle	Prehatching	2–3 weeks posthatching	Not rescued	Gill, limb rescued	Armstrong et al. (1983)
st	Liver; blood circulation	Hatching	Hatching—4 months	Not rescued	Gill, limb rescued	Humphrey and Chung (1977)
s	Limbs, digits, ribs, Müllerian duct	Postfeeding	1 year	Not rescued	Limbs not rescued; pronephros rescued	Humphrey (1967b)
q	Gills; swimming, feeding	Prehatching	Postfeeding	Not rescued	—	Humphrey (1975)
p	Gills, branchial pouches, gut, heart	Prehatching	Prehatching	Not rescued	Limb, eye, skin rescued; gill not rescued	Trottier and Armstrong (1977); Mes-Hartree and Armstrong (1980)
ph	Limbs, long bones	Hatching	Feeding—3 months	—	Limbs not rescued	Humphrey (1977b)

Table 11–2 Subdivisions of Cell-Lethal Genes

Group	Parabiosis	Transplants	Explants
Nonrescuable cell lethal	Fails	Fail	Die[a]
Rescuable cell lethal	Fails	Some survive	Die[b]
Pseudo-cell lethal	Fails	Some survive	Survive

[a]Classically, this category has been classified as autonomous.
[b]Survive if essential culture medium is provided.

Gene *r*

Embryos from a spawning of heterozygous (*R*/*r*) parents were collected at early stages, before the phenotype was expressed, and placed in explant culture. Their growth was compared to control cultures (Fig. 11–1). Homozygous mutant cells survived and proliferated. Either *r* is not a cell-lethal gene, or it is a rescuable cell-lethal gene. The rich (20% fetal calf serum) medium might provide an essential nutrient that facilitated growth. This observation is consistent with Humphrey's original hypothesis (Humphrey, 1964) that the *r* mutation affects tissue-specific metabolism. This hypothesis could be tested with in vitro cultivation of specific mutant cells, tissue, and/or organs in a defined culture medium.

Genes *t* and *st*

Cells taken from embryos homozygous for the gene *t* (at stages after the phenotype could be identified) that were grown in explant culture showed cell autonomy. They did not proliferate in culture and degenerated by 36 days in culture (Fig. 11–2). This is a surprising result, since parabiosis and transplant experiments classified the *t* gene as *not* being in the autonomous cell-lethal category. Rather, it was considered to be a rescuable cell-lethal gene (i.e., it was rescued by transplantation). The nonsurvival in cell culture could indicate that the medium is not providing an essential component. Another possibility is that by the time the *t* phenotype can be recognized, irreversible pleiotropic events have already adversely affected the cells. To test the possibility that once the lethal phenotype can be recognized cells are

Table 11–3 Preliminary Results of Embryonic Explant Cultures of *mi*/*mi*, *st*/*st*, and *g*/*g* Embryos

Mutant	Number of prehatching embryo explant cultures showing autonomous (no mitosis) and nonautonomous (mitosis) culture behavior		
	No Mitosis	Mitosis	Days Cultured
mi/*mi*	1	2	64
st/*st*	1	2	64
g/*g*	4[a]	1	67, 76
Wild type	3	8	64, 67, 76

[a]Morphology similar to cultured *t*/*t* cells.

already committed to death, another lethal gene, *st,* was tested in vitro. This lethal gene is similar to the *t* gene in that parabionts are not rescued, while transplants are. Preliminary results revealed that cells from prehatching-stage *st/st* embryos survive and proliferate in culture like controls (Table 11–3). The unique behavior of cultured *t* cells may provide an experimental approach using the cultivation of specific cells under defined culture conditions, at different stages of development, for understanding the physiological basis of the mutant phenotype. Explants cultured from embryos before the phenotype can be recognized would serve to identify the stage at which the cells become committed to death. Experimentally, pieces of embryos, such as limb or gill primordia, could be surgically removed from an embryo. The embryo would heal and survive, and later indicate the genotype. The defined medium could be enriched with various supplements in an attempt to impart growth on the *t/t* cells, and therefore identify the physiological block in mutant cells.

Genes *g* and *mi*

These mutants fit the classical definition of autonomous cell-lethal genes (no rescue of either parabionts and transplants). Embryos that were identified by their phenotype as being homozygous for either the gene *mi* or *g* were explant-cultured. Preliminary data showed two different results. The *mi/mi* cells proliferated and grew in culture like controls, while the *g/g* cells showed no proliferation and eventual degeneration under similar culture conditions (Table 11–3). Those paradoxical results call out for explanation and suggest that the simple category, *cell-lethal,* may require more sophistication.

POSSIBLE EXPLANATIONS FOR CELL-LETHAL PHENOTYPES

For the sake of discussion, the view that cell-lethal genes will not be particularly useful for developmental biologists should be considered. First, most—if not all—of the cell-lethal genes may affect housekeeping functions and therefore would tell us nothing about major developmental events. Second, when working with recessive alleles, it is difficult to get spawnings yielding 100% homozygous embryos for difference analyses (see Chapter 10). Third, the late onset of most of the cell-lethal phenotypes complicates the study of the primary gene defect; it is easy to get bogged down with pleiotropic effects.

One can easily envision possible primary defects in cell-lethal genes affecting either common metabolic pathways or regulatory mechanisms operating during development. In the axolotl collection, some mutants may affect intermediary metabolism, others early cell interactions involving inductions; yet others might affect growth factor elaboration or the ability to respond to growth factors. Additionally, programmed cell death regulation could possibly be expressed in all cells rather than a selected cell population, as is the case for wild-type embryos. Some may even represent a sort of maternal defect like the *nu* mutation in *Xenopus.* That is, cells run out of maternal stores at around hatching; a new zygotic component replacing the maternal component is not replaced correctly or adequately, and the

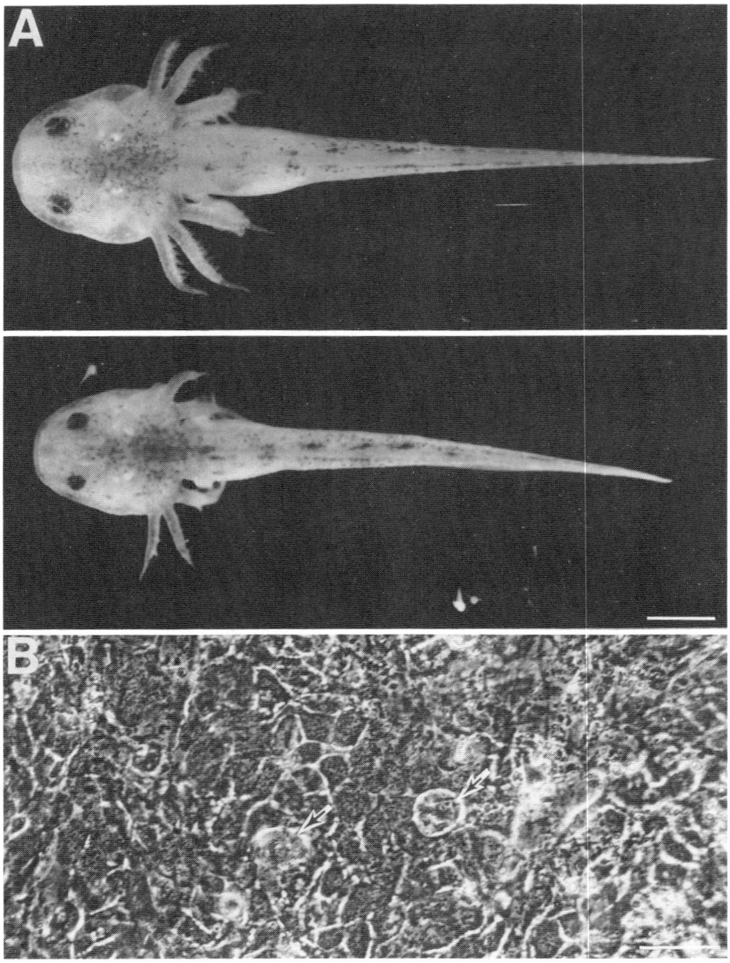

Fig. 11–1. Results of gene *r* "renal insufficiency" explant culture experiments. (A, top), Normal 1-week postfeeding larva. (bottom) Same-age sibling showing *r/r* phenotype. Note fragile, irregular gills. All *r/r* larvae from this spawning died within 3 weeks postfeeding. Bar = 1 mm. (B) Phase micrograph of explant-cultured cells from an *r/r* posthatching embryo at 43 days in culture showing area of active cell proliferation. Also, 22 of 24 explant cultures of prehatching embryos with a 25% chance of expressing *r/r* (cultured up to 107 days) showed active proliferation. Active proliferation occurred in 18 of 19 control wild-type explant cultures under identical culture conditions. Clearly, *r/r* embryo explants do *not* show autonomous cell lethality in culture. Approximately 1-mm^3 pieces of single, whole embryos were plated on collagen-coated plastic dishes in culture medium [55% L-15 minimal essential medium (MEM); 20% fetal calf serum; 23% distilled H_2O; 1% kanamycin solution; 1% fungizone solution; 0.02 units/ml insulin; 1 μg/ml thyroxine, pH 7.4] and incubated at room temperature with 3-day feedings. Under these conditions, 80% of the explants displayed cell outgrowths by 2 days in culture. Cell proliferation occurs by day 8 in both epithelial and fibroblastic cells, and ciliated cells and heart explants retain their beat up to 2 months in culture. Arrow indicates mitosis. Bar = 100 μm.

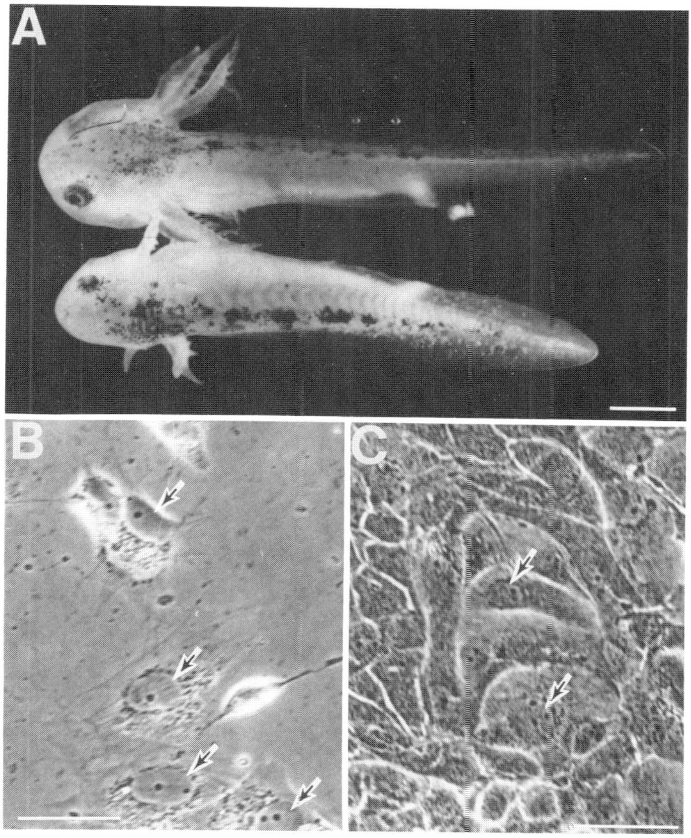

Fig. 11–2. Explant culture behavior of *t* "twisted-gill" mutant embryonic cells. (A, top) Normal prehatching embryo. (bottom) Same-age sibling showing *t/t* phenotype. Note reduced growth, twisted gills, and increased xanthophores (yellow) pigmentation. All *t/t* embryos died within 2 weeks of hatching. Bar = 1 mm. (B) Phase micrograph of explanted cells from a prehatching-stage *t/t* embryo cultured for 18 days. Note the reduced heterochromatin in the *t/t* cell nuclei (arrows) compared to the nuclei (arrows) of normal control cells cultured for 18 days (C). These nuclear differences were observed within 12 hours culture. Seven positively identified prehatching *t/t* embryos from two separate *t/t* spawnings were cultured for up to 36 days. The *t/t* cultured explants attached, and cells grew out of the explants in a manner similar to the controls. However, they showed more yolky cells, more xanthophores, less culture viability with respect to axon outgrowth, heartbeat maintenance, and cilia beating. The *t/t* cells did not proliferate and completely degenerated by 36 days in culture. No mitosis was observed in *t/t* cultures, whereas *all* control cultures showed active proliferation by 7 days. *t/t* embryo explants clearly display autonomous cell-lethal behavior in culture. (B, C) Bars = 100 μm.

cells die. Considering the shortage of mutants in these categories in vertebrates, compared to *Drosophila* and *Caenorhabditis elegans*, for example, the axolotl cell-lethal mutants should be maintained and a systematic search for the primary gene action undertaken. First, classical embryological approaches should be employed, then cell biological approaches, and eventually, as axolotl cDNA libraries become available, molecular approaches should be used.

FUTURE DIRECTIONS

The technology for explant culture of axolotl embryos is sufficiently advanced to obtain long-term cell function (heartbeat, cilia differentiation, etc.) and cell proliferation. It is possible to achieve in vitro survival of mutant cells past the stage of onset of death of the noncultured siblings. Similar culture conditions have been utilized by the author to maintain a proliferating axolotl epithelial cell line for up to 10 months (Neff, 1984). Clearly, the in vitro behavior of cells of embryos homozygous for cell-lethal genes cannot be predicted from classical parabiosis or transplanation studies alone. For example, of the two cell-lethal genes tested (*mi* and *g*), only *g* was autonomous. Also, the gene *t*, which is not autonomous (i.e., it is rescuable by transplantation), nevertheless displayed autonomous cell lethality when explanted. Although the classifications for cell-lethal genes proposed by Armstrong (1985) are valid, there is evidence that calls for new classifications, such as "nonrescue of parabiosis," "rescue of transplants," and "nonrescue of cultured explants." The data in this report are preliminary, so it is premature to prepare a completely new classification scheme for cell-lethal genes. However, there is an indication that under specific culture conditions certain cell-lethal cells can survive and proliferate, while cells with other cell-lethal genes do not. This disparate behavior hints that cellular metabolic pathways are involved in some of the pathologies. Armstrong's proposed classification (1985) of cell-lethal genes, which incorporated explant cultivation of embryonic tissues, is very appropriate. This author feels that a detailed and systematic analysis of the collection of axolotl cell-lethal mutants using in vitro cultivation of cells should be undertaken. Several additional techniques, however, need to be developed. For example, culture conditions for embryonic axolotl cells that utilize defined media need to be established. A method for assaying the in vitro behavior of mutant cells at an early developmental stage—close to the time of primary gene action (earlier than when the phenotype can be recognized)—is required. One way to accomplish this is to culture parts of early embryos, and allow the surviving donor to determine the genotype later in development. Another long-term approach would involve the identification of temperature-sensitive cell-lethal mutants. Temperature-sensitive mutants could potentially provide homozygous parents from embryos reared at permissive temperatures. Those parents could then be used to provide spawnings that are 100% mutant. It is possible that temperature-sensitive genes might be forthcoming. Although axolotl mutants have not been systematically screened, two temperature-sensitive mutants have recently been identified (*ts* by Malacinski and Barone, 1985; *pi* by Neff et al., 1987).

Molecular approaches such as those suggested by Malacinski (Chapter 10) may be inhibited somewhat by the size of the axolotl genome. Nevertheless, amphibian embryos clearly offer many potential advantages for this type of research in that they are large (thereby facilitating microinjection), yield many progeny, and can be easily analyzed by standard biochemical techniques.

ACKNOWLEDGMENT

The author expresses appreciation to the Indiana University axolotl colony for assistance with the work described in this report.

12
Pigmentation and Color Variants

SALLY K. FROST

Pigmentation and the pigment mutants of the axolotl, in particular, have been variously exploited over the years for many experimental purposes. Genes that alter pigment phenotype are relatively common in the axolotl and have, on occasion, provided useful visible markers for cell and developmental biology experiments (for reviews see Frost and Malacinski, 1980; Frost et al., 1984a). Pigment genes are in themselves interesting to developmental biologists, because in general such genes alter basic pattern formation by causing changes in the types of pigment cells that differentiate.

THE PIGMENTARY SYSTEM OF THE AXOLOTL

Recently, researchers have renewed attempts to understand the basic development of pigment cells and pigment patterns in wild-type and mutant axolotls (Frost et al., 1984b,c, 1986b,c). Like most amphibians, axolotls differentiate three pigment cell types—melanophores, xanthophores, and iridophores (Fig. 12–1)—all of which originate from the embryonic neural crest. Melanophores (black, melanin-containing cells) differentiate first in prehatch (stage 35; Chapter 19) axolotls, and xanthophores (yellow, pteridine-containing cells) differentiate shortly thereafter. From stage 41 onward, a "barred pattern" is visible (Fig. 12–2), consisting of populations of melanophores interspersed with populations of xanthophores forming a black/yellow pattern of longitudinally oriented stripes along the flanks of young axolotls (Epperlein and Löfberg, 1984). The barred pattern persists for a relatively short period of time in the life span of an axolotl and is really only visible in young pre- and posthatch larvae under 3 cm in length. Thereafter, the densities of chromatophores are such that stripes are no longer apparent in axolotl skin.

Iridophores (purine-containing reflecting cells) also differentiate in axolotl skin, but they do so relatively late in development—usually 3–4 months posthatch, in axolotls longer than 10 cm. In wild-type axolotls, xanthophores and melanophores appear to be distributed in fairly equal numbers throughout early development. Iridophores, when they differentiate, appear to be relatively few in number and sparsely distributed on the head, flanks, and gills of axolotls (Frost et al.,

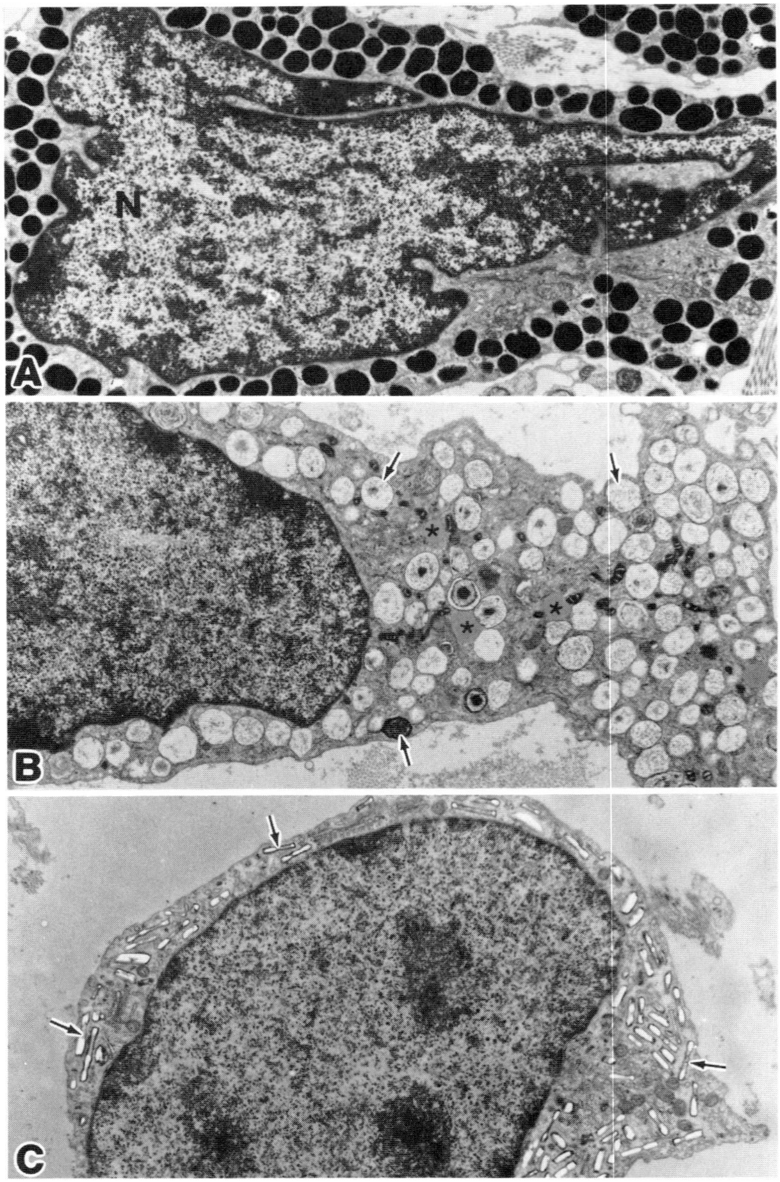

Fig. 12–1. Transmission electron micrographs of pigment cells from axolotl skin. (A) Melanophore. Note the electron-dense black pigment granules (melanosomes) in the cytoplasm of this cell. Long processes filled with pigment granules extend out from the nucleus (N) of the cell. (B) Xanthophore. Two types of pigment granules are present in this cell: pterinosomes (arrows) containing electron-dense fibrous material (pigment), and occasional carotenoid vesicles (*). (C) Iridophore. Reflecting platelets (arrows) are randomly scattered throughout the cytoplasm. (A–C) ×5940. (Micrographs courtesy of S. J. Robinson.)

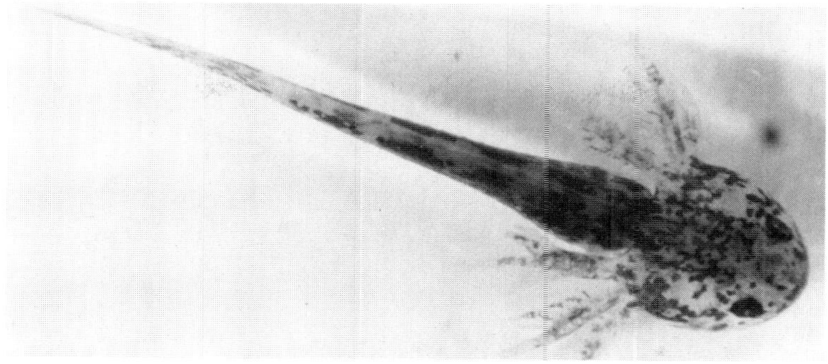

Fig. 12–2. Newly hatched (1.5 cm length) wild-type axolotl larva. Note the prominent barred pattern on the flanks. The black stripes are clusters of melanophores; the lighter stripes contain xanthophores and are yellow in color.

1984b). Older axolotls (sexually mature; several years of age) often darken substantially with age until there is little evidence of background coloration other than black. Why skin darkening occurs is not known, but it is likely due to some combination of genetic and/or environmental factors, some of which are currently being examined (Thorsteinsdottir and Frost, 1986; Frost et al., 1988). Nevertheless, the "typical" adult wild-type pigment pattern can vary from a dark mottled green to a muddy yellow-brown color.

In amphibians, pigment patterns are known to change dramatically during metamorphosis (Smith-Gill and Carver, 1981). Axolotls, which normally do not metamorphose, nevertheless exhibit changes in pigments and pigment patterns associated with growth and development (Frost et al., 1984b). Such changes are much more subtle in axolotls than, for example, other salamanders and frogs, in which an adult pigment pattern is established independent of the pattern that existed in the larval form. During typical amphibian metamorphosis, the dermis of the skin thickens, chromatophores proliferate, and the various chromatophore types migrate to specific regions in the dermis close to the basement membrane that separates dermis from epidermis. Often the chromatophores are stacked into dermal chromatophore units with xanthophores on the top (closest to the epidermis), melanophores on the bottom, and iridophores sandwiched in between. This basic type of chromatophore organization is typical of "green" amphibian skin, and variations in the pigments present in the chromatophores (xanthophores in particular) or variations in the organization of the chromatophore units give rise to different colors and patterns. Black spots, for example, are composed solely of melanophores; red skin often contains only carotenoid-laden xanthophores (also referred to as erythrophores) (Bagnara and Hadley, 1973; Frost and Robinson, 1984).

These types of metamorphosis-associated changes in the skin and pigment patterns do not occur in axolotls. However, some clear-cut developmental changes do occur, such as a gradual thickening of the dermis (primarily the collagen layer) with age and pigment cell proliferation, including the late appearance of iridophores.

Chromatophores are never organized into stacked units in the axolotl. However, these large cells (often several hundred microns in total length), with their extensive pigment granule-laden processes, overlap enough to produce the mottled green/yellow-brown phenotype. In general, chromatophores in axolotl skin are aligned side by side directly beneath the epidermal basement membrane (Fig. 12–1). Cell bodies do not overlap; however, there are also pigment cells interspersed in the epidermis and, in older animals, pigment cells may be found deep in the dermis (Frost et al., 1984b). Thus, the gradual thickening of the skin accompanied by overall increases in pigment cell numbers and distribution result in a darkly pigmented adult axolotl.

PIGMENT MUTANTS

For the purposes of this discussion the pigment mutants found in the axolotl include only the three genes, melanoid *(m)*, axanthic *(ax)*, and albino *(a)*. Other genes affect the phenotypic color of axolotls, thus producing color variants, and are discussed in a subsequent section of this chapter. The basic feature shared by the pigment mutants is that each of these genes is a simple recessive that affects how pigment cells differentiate from the neural crest (Frost and Malacinski, 1980). Thus, in each case, if mutant neural crest cells are exchanged with wild-type crest cells (Fig. 12–3), the resulting wild-type embryo will eventually exhibit an area of skin with the mutant phenotype, and vice versa. This type of genetic defect is thereby intrinsic to the neural crest. The origin of color variants, on the other hand, is either not known or clearly does not reside within the pigment cells (neural crest cells) themselves.

The discovery of and early experiments on the pigment mutants in the axolotl have been reviewed by Frost and Malacinski (1980) and Frost et al. (1984a). Thus, the intent of this discussion is to bring the reader up to date on subsequent descriptive and experimental studies that have involved these mutant genes.

Table 12–1 summarizes several years of intensive descriptive investigations that aimed at determining how these genes influence pigment cell differentiation and thus pattern formation. Phenotypically, each of the pigment genes produces a

Fig. 12–3. An example of the type of experiment that can be done to determine whether a suspected pigment mutant is due to a defect intrinsic to the neural crest-derived pigment cells. In this case if the crest cells are defective, then the wild-type embryo will eventually exhibit an area of skin with the mutant phenotype, and vice versa. (From Frost and Malacinski, 1980.)

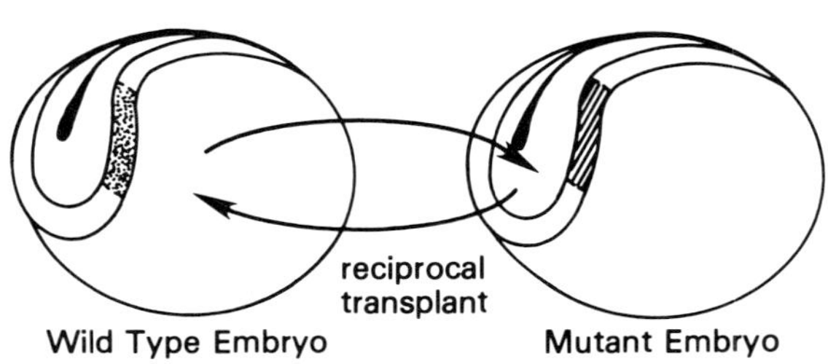

Table 12–1 A Summary of Pigment Cell Differentiation in Wild-type and Mutant Axolotls

	Pigment Cell Type		
Phenotype	Melanophores	Xanthophores	Iridophores
Wild type (Frost et al., 1984b)	Appear early in development (by stage 35) and are present in "normal" numbers thereafter	Appear early in development (shortly after stage 35) and are present in "normal" numbers thereafter	Appear late in development (in animals > 10 cm in length; several months old) and never in great numbers; present in eyes throughout development
Melanoid (Frost et al., 1984c)	Appear early in development (by stage 35) but hyperproliferate thereafter	Appear early in development (shortly after stage 35) but gradually disappear during development	Never appear in either the skin or eyes
Albino (Frost et al., 1986b)	Present in the skin early in development (as usual) in more or less normal numbers but without melanin	Appear early in development (as usual) and persist in a well-pigmented (bright yellow) state throughout development thereafter	Appear earlier than normal (by at least several weeks) and persist in greater than normal numbers thereafter
Axanthic (Frost et al., 1986c)	Appear early in development (as usual) and are present in "normal" numbers thereafter	Never appear, although an occasional cell that may be an unpigmented xanthophore has been observed	Never appear in either the skin or the eyes

variation from the wild-type pattern such that adult melanoid axolotls are a velvety-black color, adult axanthics are charcoal gray, and albinos are a shiny golden-yellow color.

One obvious conclusion from these descriptive studies is that even though each of these genes may only control a single metabolic event, the result in all cases is that *all three* types of pigment cells are affected. It is also clear that xanthophore and iridophore differentiation are metabolically more closely linked to each other than either is to melanophore differentiation. For example, in melanoid axolotls melanophore differentiation is somehow stimulated and melanophores proliferate in greater than normal numbers. Xanthophores appear when they normally should, but instead of continuing to proliferate they gradually disappear from the skin until, in the adult, there is no visible, chemical, or microscopic evidence of yellow xanthophores whatsoever. Iridophores in melanoid axolotls are apparently doomed from the beginning. Not only do they never appear in the skin; they also never appear in the eyes (not even in embryonic eyes), suggesting that early in development something prevents this cell type from being expressed.

In distinct contrast to melanoids, albino axolotls lack all evidence of melanin pigmentation, but from a very early age acquire an intensely yellow phenotype, which, as development progresses, becomes a shiny golden-yellow color as iridophores gradually appear in the skin. The intensity of the yellow color of albino skin is clearly due to an increase in the amount of yellow pteridine (xanthophore) pigment biosynthesis and may be due as well to greater than normal numbers of xanthophores (Frost et al., 1986b). The shiny quality of albino skin is acquired as iridophores become obvious in the skin, and this happens at an earlier developmental stage than normal. Thus, in albinos, the simple suppression of melanization seems to supply a stimulus for both xanthophore and iridophore differentiation, further emphasizing that interrelationships exist among these cells that can ultimately influence patterns of differentiation.

Axanthic is a much more extreme case than either melanoid or albino. Melanophores and melanin pigmentation appear to be normal in axanthic axolotls, but xanthophores and iridophores are altogether absent throughout development. Electron-microscopic evidence suggests that, in young axanthic axolotls, there may be xanthophore precursor cells. However, the lack of well-defined pigment organelles (or even prepigment organelles) makes it difficult to identify such cells unequivocally as chromatoblasts. Whereas albino and melanoid axolotls are healthy, vigorous animals, axanthics are not. Most axanthics die before reaching sexual maturity, and many are heavily infected with a virus that has been observed in a variety of axolotls, but which predominates in axanthics (Frost et al., 1986c).

In order to understand the interrelatedness of pigment cells and pigment cell differentiation, it is necessary to understand something about pigment biosynthesis. Melanin, a highly complex molecule of as yet undefined structure, is derived initially from the amino acid tyrosine. One of the key reactions in the melanin-biosynthetic pathway is the conversion of tyrosine to DOPA through the action of the enzyme tyrosinase. Purines and pteridines—the pigments of iridophores and xanthophores, respectively—share some obvious metabolic features. For example, pterins all originate from guanosine triphosphate (GTP) (Brown, 1985); guanine crystals are likely candidates for iridophore pigments (Bagnara and Hadley, 1973). Thus, both xanthophores and iridophores use purines and/or purine derivatives as pigments. There is little evidence to suggest what enzymes might be involved in the production of purine/pteridine pigments. However, in recent years emphasis has focused on the enzyme xanthine dehydrogenase (XDH) as a possible key mediator of pigment biosynthesis in xanthophores and iridophores (Bagnara et al., 1978; Frost and Bagnara, 1979). Experiments described in the next section also suggest that XDH activity may modulate (either actively or passively) pigment cell differentiation.

Although the metabolic links between xanthophores and iridophores are obvious, similar connections to melanophores are not at all clear. Possibly tyrosinase, or some other enzyme involved in melanin biosynthesis, requires a pteridine cofactor. Tyrosine hydroxylase, an enzyme similar in activity characteristics to tyrosinase, is known to require a pteridine cofactor (tetrahydrobiopterin) (Nichol et al., 1985) and is an enzyme that functions in many of the neural derivatives of the neural crest.

The best current guess as to what is responsible for the albino phenotype is that there is a tyrosinase inactivator present in albino skin (Presumably localized in melanophores) that shuts down melanin biosynthesis (Harsa-King, 1978, 1980). What this inhibitor might be is not known, but if proteolytic enzymes are added to albino skin extracts, tyrosinase can be activated (Harsa-King, 1978). Tyrosinase is known to exist in other animals in a proenzyme form, and the implications from Harsa-King's biochemical studies are that in albino axolotls the normal cleavage mechanism for activating axolotl protyrosinase is defective.

Although this would explain the lack of melanin pigment in albino axolotls, it still does not address why xanthophore and iridophore pigment biosynthesis appear to be enhanced. At the present time there is no good explanation for why this happens, although we can reiterate the obvious "seesaw" effect that seems to occur with respect to chromatophore differentiation, such that any time melanization decreases, xanthophore and iridophore pigmentation increases, and vice versa.

The notable exception to this observation is axanthic. In this animal there is no increase in melanization even though there is a total lack of xanthophores and iridophores. The absence of these two cell types and the viability problems associated with this mutation suggest that the axanthic gene may be a more fundamental type of metabolic problem—perhaps involving purine and/or pteridine precursor biosynthesis.

Table 12–2 XDH Activity from the Liver and Skin of Wild-type Control, Wild-type Allopurinol-fed, Melanoid, and Axanthic Axolotls

Animal Type	Tissue	XDH Activity[a] (μmol IXP/mg protein/hr)	
Age group I (<150 days postspawning)			
Wild-type control	Liver	59.51 ± 10.24	(*N* = 7)
Allopurinol-fed wild type	Liver	68.82 ± 19.71	(*N* = 5)
Melanoid	Liver	22.64 ± 8.14	(*N* = 6)
Axanthic	Liver	40.66 ± 3.78	(*N* = 4)
Wild-type control	Skin	8.07 ± 2.47	(*N* = 7)
Allopurinol-fed wild type	Skin	11.73 ± 4.94	(*N* = 5)
Melanoid	Skin	3.69 ± 1.11	(*N* = 6)
Axanthic	Skin	5.44 ± 1.13	(*N* = 4)
Age group II (>150 days postspawning)			
Wild-type control	Liver	24.19 ± 5.52	(*N* = 9)
Allopurinol-fed wild type	Liver	34.50 ± 6.45	(*N* = 3)
Melanoid	Liver	7.72 ± 2.92	(*N* = 6)
Axanthic	Liver	20.24 ± 8.95	(*N* = 8)
Wild-type control	Skin	5.19 ± 1.52	(*N* = 9)
Allopurinol-fed wild type	Skin	7.62 ± 2.82	(*N* = 3)
Melanoid	Skin	1.20 ± 1.13	(*N* = 6)
Axanthic	Skin	4.77 ± 1.57	(*N* = 8)

[a]Values are mean ± SD. Assay procedure is based on the fluorometric detection of isoxanthopterin (IXP), a product of XDH activity, from extracts of axolotl tissues. This technique is described in detail in Frost et al. (1986a).

The recent experimental and descriptive studies involving melanoid axolotls have yielded promising insight into the nature of genetic melanism. As discussed in the next section, it now seems likely that melanoid animals are XDH deficient (Table 12–2). Decreased levels of XDH activity in the tissues (including the skin) lead to the complete suppression of iridophore differentiation, the gradual elimination of xanthophores, and the hyperproliferation of melanophores. Although axanthic and melanoid share several phenotypic features (no iridophores and, eventually, no xanthophores), the enzymatic basis for these two defects would appear to be fundamentally different because XDH levels in axanthic tissues are not significantly different from those in the wild type (Table 12–2).

EXPERIMENTAL MANIPULATION OF PIGMENT PHENOTYPE IN VIVO

Even though information is available on these mutants, there is still much to be learned. One fruitful approach to investigating these defects has been to alter pigment phenotypes in the axolotl by feeding wild-type animals certain types of chemicals.

Because of the suspected key role that XDH is thought to play in purine/pteridine pigment biosynthesis, an obvious experiment is to feed axolotls the well-known XDH inhibitor, allopurinol. Allopurinol functions as a suicide substrate for the enzyme by binding tenaciously, but not irreversibly, to XDH. If allopurinol is fed to wild-type axolotls they become very much darker than normal in a relatively short period of time (e.g., phenotypic differences between control and drug-treated animals are visible within 5–7 days after commencement of drug treatment and continue to increase thereafter as long as allopurinol administration is continued). Eventually, allopurinol-fed wild-type axolotls come to look remarkably like melanoid axolotls. Then, for as long as drug treatment continues, the darker-than-normal phenotype persists. Interestingly, if drug treatment is discontinued after several months, the axolotls gradually revert back to the wild-type phenotype, and in time such animals assume a brilliant, glossy-green phenotype indicative of enhanced xanthophore and iridophore differentiation. The reasons for this "more colorful" phenotype can be traced to what happens to XDH activity during allopurinol treatment.

The data in Table 12–2 illustrate that XDH activity in the tissues of allopurinol-treated axolotls is always higher than that found in control tissues. A possible explanation for these findings is that allopurinol-induced inhibition of XDH activity actually stimulates synthesis of more enzyme. Thus, when drug administration is stopped, XDH gradually returns to full activity, but because more enzyme is now present in such animals, increased levels of XDH activity eventually lead to enhancement of xanthophore and iridophore pigmentation. These predictions are consistent with the observed phenotypic changes following subsidence of allopurinol treatment.

Clearly, allopurinol treatment of wild-type axolotls produces phenocopies of the melanoid mutant. Drug-treated animals not only look like melanoid animals, but they share other, more subtle, features as well. For example, the patterns of pteridine biosynthesis observed in allopurinol-treated animals are more similar (statistically) to the patterns observed in melanoid animals than either is to the wild type (Thorsteinsdottir and Frost, 1986). Physically, pterinosomes in xanthophores from allopurinol-treated axolotl skin are smaller than normal (by 6–9%; Table 12–3), as are the pterinosomes in xanthophores from melanoid skin (10–15% smaller than normal) (Frost et al., 1987a). There is nearly twice as much melanin and nearly double the number of melanophores in either melanoid or allopurinol-treated skin than there are in normal wild-type skin.

All of these observations lead to the obvious conclusion that melanoid axolotls are the result of deficient (i.e., defective) XDH activity. Precisely how this enzyme is defective—whether the enzyme itself is defective or whether cofactors or substrates necessary for XDH activity are lacking—is not known.

The possibility that melanoid animals might be deficient in purine/pteridine substrates has, in fact, been examined. A likely candidate for a common substrate link between xanthophores and iridophores is guanosine. If exogenously supplied guanosine is able to get into cells (much like allopurinol has been observed to affect pigment cells), then it might be expected eventually to function in the formation of guanine crystals in iridophore-reflecting platelets or as the common substrate (GTP) for pteridine pigment biosynthesis in xanthophores. Guanosine has been fed to both melanoid and wild-type axolotls, and the results are striking (Fig. 12–4) (Frost et al., 1987).

Guanosine-fed melanoid axolotls are significantly more yellow in color than normal. Xanthophores are clearly visible on the head, flanks, and expecially along the dorsal midline of such animals. Moreover, melanin pigmentation appears to be distinctly reduced in these animals. This is particularly noticeable in the more vascularized areas of the skin such as the gills, head, tailfin, and limbs, wherein areas of "pink" skin, devoid of black pigmentation, appear. This is in distinct contrast to the usually uniform black body pigmentation of melanoid axolotls.

Table 12–3 Summary of the Effects that Allopurinol and Guanosine Have on Pigment Cell Differentiation in the Axolotl.[a]

	Drug Treatment	
Pigment Cell Type	Allopurinol	Guanosine
Melanophores	Increase significantly in number; cells appear morphologically normal and healthy; indistinguishable from individual control melanophores	Decrease slightly in number; cells punctate and necrotic; melanosomes appear to be degenerating and are commonly observed in macrophages
Xanthophores	Pterinosomes shrink 6–9% in size compared to controls; pteridine pigment biosynthesis reduced overall	Pterinosomes are 25–30% larger than normal; yellow pteridine pigment biosynthesis is enhanced
Iridophores	Never appear	Appear at an abnormally early age; contain more reflecting platelets than normal

[a]Data are from Thorsteinsdottir and Frost (1986) and Frost et al. (1987, 1988).

Fig. 12–4. Wild-type control (top) and guanosine-fed wild-type (bottom) axolotls. The guanosine-fed animal is actually bright yellow in color, and there are iridophores organized in clusters along the flank (arrows). The control animal, on the other hand, exhibits no evidence of iridophores in the skin. Both animals are siblings, approximately 10 weeks posthatch.

Clearly, feeding guanosine to melanoid axolotls alters their pigment phenotype, although it by no means corrects the melanoid defect. Consequently, guanosine has also been fed to wild-type axolotls in an attempt to assess what effects this substance might have on the normal phenotype and thus pigment cell differentiation.

Guanosine-fed wild-type axolotls gradually acquire a bright-yellow phenotype (Fig. 12–4). In addition to enhanced yellow pigmentation (which has been shown to result directly from enhanced synthesis of sepiapterin and deoxysepiapterin, two yellow pterin pigments), melanization appears to be suppressed and iridophores develop abnormally early in these animals (Frost et al., 1987). Iridophores in guanosine-treated axolotl skin also contain significantly more reflecting platelets than are normally observed to occur in similar cells from wild-type control skin (cf. Fig. 12–1C) (Frost et al., 1987).

The observed effects that both guanosine and allopurinol have on pigment cells are presented in Table 12–3. In both cases all three pigment cell types are affected in different (sometimes opposite) ways by perturbing the biosynthesis of the purine/pteridine pigments. Clearly, slight perturbations in pigment metabolic pathways may lead to profound changes in pigment phenotype and disruption of the normal pigment pattern. Thus, dietary and other environmental factors may influence patterns and phenotypes, especially as such factors change over time (i.e., with aging).

Pigmentation and pigment cells, unlike other phenomena and cell types, are not static entities fixed forever within the developmental genetic program of an animal. There is instead a degree of plasticity associated with pigment cells, such that patterns may change during development and such changes may be initiated either by external or internal cues.

COLOR VARIANTS

Given the plasticity inherent in the pigmentary system, it is no wonder that patterns can vary widely even between animals of similar genotype. Numerous examples of unusually pigmented animals have been observed and reported, but few of these turn out to have a genetic basis or to be anything more than an isolated incidence of aberrant pigmentation. Exceptions to this are several color variants that are observed with some frequency in axolotls, and at least two of these variants—white and eyeless—are associated with well-characterized genetic mutations.

The white mutant has been around for as long as axolotls have been used in laboratories (see Chapter 1). Unlike the pigment mutants discussed previously, the white mutant is purportedly not a problem intrinsic to the pigment cells, but is a defect believed to be associated with the environment into which pigment cells must migrate (i.e., the dermis of the skin). Thus, white animals start out with some pigment cells developing in the skin, but these gradually disappear so that within a few weeks after hatching young white axolotls exhibit no evidence of skin pigmentation. Later in development, after white animals have reached sexual maturity, they may start to repigment, and the degree to which a white animal may ultimately repigment is highly variable (see Frost and Malacinski, 1980).

Early studies attempting to determine the nature of the white defect were largely inconclusive and various suggestions as to the defect involved have been made, including a pituitary deficiency, an unfavorable dermal/epidermal environment, and dermally/epidermally localized factors that prohibit pigment cell migration into and subsequent survival in the skin (reviewed in Frost and Malacinski, 1980). Recently, renewed interest in the white mutant has led to two novel approaches that may ultimately provide insight into the nature of the white defect.

Keller and co-workers (1982) attempted to define the role that extracellular matrix (ECM) plays in the failure of pigment cells to migrate into white skin. Initially, they repeated the classical experiments of DuShane (1935, 1939) and Dalton (1949, 1950a), which involved reciprocal grafts of crest cells and patches of epidermis and/or mesoderm between white and dark (wild-type) axolotl embroyos. They concluded that epidermis alone could control the migratory patterns of pigment cells; white epidermis did not support migration of either white or wild-type crest cells, whereas wild-type epidermis would support migration of crest cells from either white or wild-type embryos. The use of scanning electron microscopy and advanced techniques in light microscopy including time-lapse cinemicrography for tracing the migration and differentiation of pigment cells added substantially to the characterization of the white defect, even though the conclusions reached by Keller and colleagues (Keller et al., 1982; Keller and Spieth, 1984) were not significantly different than those of earlier workers (DuShane, 1935, 1939; Dalton, 1949, 1950a). Experimental analysis has since focused on attempts to determine whether there are differences in the ECM between white and dark axolotls. These are described in Chapter 9.

This writer has proposed a different approach to the white mutant. The white defect shares a number of similarities with a human skin condition known as vitiligo. In humans, vitiligo is associated with hypothyroidism and is characterized by the failure of the skin to support pigment cells (Lamoreux et al., 1987). Consequently, people with vitiligo develop (among other symptoms) unsightly patches of white skin that can become extensive. Evidence seems to indicate that some forms of human vitiligo may be due to an autoimmune response whereby pigment cells are recognized by the immune system as foreign cells and are selectively killed. There are few animal model systems for studying vitiligo; the axolotl, because of its intrinsic hypothyroid (neotenous) state and because its embryonic immune system can be easily manipulated, may provide a valuable model for studying the pigmentary defect associated with vitiligo.

Unusual pigmentation patterns are associated with other genes in the axolotl as well. For example, the eyeless mutant, which until recently occurred exclusively in association with the white gene, exhibits a much enhanced pigment pattern (Epp, 1972). This enhancement is obvious because of the eyeless–white combination and the fact that the white gene generally does not support the development of pigment cells. White eyeless axolotls, on the other hand, are much darker than white axolotls with eyes. Melanophores, which are especially obvious, are larger than normal, stellate, and may persist in white skin of eyeless animals indefinitely (Fig. 12–5). Epp (1972) speculates that the enhanced pigmentation associated with the eyeless condition may be due to abnormally high levels of circulating melanocyte-stimulating hormone in these animals.

Finally, there is a recurring pigment pattern that has been observed in pre-

Fig. 12–5. Comparison of pigment patterns in a white eyeless larva (right) and a sibling white larva with eyes (left). Note how much larger and more stellate the melanophores are on the dorsum of the eyeless animal as compared to the animal with eyes.

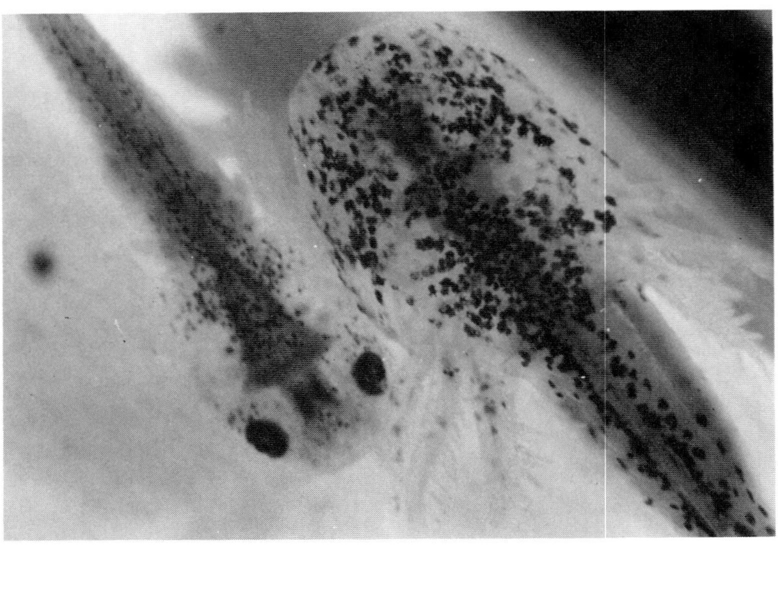

sumed wild-type spawnings from time to time. The phenotype in question is characterized by abnormally large numbers of iridophores in the skin, resulting in a "starburst" pattern of pale yellow-green spots on a dark olive-green background. Unfortunately, this phenotype is not obvious until rather late in development (several months posthatch), when iridophores normally begin in appear. Consequently, whether or not this is a genetic mutation that affects iridophore/xanthophore differentiation is not known, and until someone assumes the unenviable task of raising large numbers of axolotl larvae to near sexual maturity, the genetic nature of this defect will remain unknown.

CONCLUSIONS AND PROSPECTIVE

Pigment cells provide unique opportunities for investigating basic properties of cells such as migration, differentiation, and general mechanisms involved in embryonic development such as pattern formation. The axolotl, in particular, is well suited for such studies, providing investigators with three pigment cell types and several well-characterized pigment mutants. Future studies involving the axolotl pigmentary system will no doubt focus on fundamental problems in developmental biology such as factors that influence the migration and differentiation of crest cell derivatives both in vivo and in vitro, and how pattern formation is controlled both genetically and extrinsically.

ACKNOWLEDGMENTS

Much of the work described in this chapter was supported by grants from the National Science Foundation (PCM80-22599) and National Institutes of Health (AM34478).

13

The *Eyeless* (*e*) Gene: Effects on Embryonic Development

ROBERT CUNY AND GEORGE M. MALACINSKI

Eye tissues are derived from the neuroectoderm (retina), the head epidermis (lens, corneal epithelium), neural crest mesenchyme (sclera, corneal and iris stroma), and the neural crest pigment cells (iris stroma, sclera). If a tissue-specific mutation prevents one of the tissues from growing, the other tissues will also be affected, because of their dependency on tissue interactions (Cuny and Malacinski, 1986). Instances of genetically based eye anomalies are well known. The eyes of the cave salamander *(Typhlotriton spelaeus),* for example, are programmed to degenerate during metamorphosis (Eigenmann and Denny, 1900; Stone, 1964). Also, mutations that lead to eyelessness are known in the rat (Kinney et al., 1982), mouse (Salaün, 1982), and axolotl (Humphrey, 1975). The last mutation is the subject of this review.

GENETICS

The gene *e (eyeless)* (for previous reviews see Epp, 1978; Harris, 1983) was discovered at Indiana University by Rufus R. Humphrey (1969a) in the inbred Wistar stock (Malacinski and Brothers, 1974) of pure gene *d* (white) animals (Malacinski, 1978b). It is a single-locus autosomal nonlethal recessive allele, and its homozygous phenotypic features are (1) the lack of eyes in the adult, (2) gonads that never mature, and (3) a dark-freckled pigmentation (Fig. 13–1a). The genes *e, r* (renal insufficiency, lethal, recessive), and *x* (fragile gills, lethal, recessive) appear to be linked, albeit through a considerable map distance: *e*–24–*x*–22–*r* (R. R. Humphrey and A. J. Brothers, personal communication, cited in Epp, 1978). Genotypes $+/e$, r/r (Epp, 1978) or $+/e$, $+/r$ (Cuny and Malacinski, 1986) have eyes but exhibit retinal malformations, and often the optic cups are duplicated (Fig. 13–1b). The displacement (and often small size) of the otherwise normal lens is probably a secondary consequence of the primary retinal disorder. The gene *x* seems to enhance the effects of the genes *e* and *r* (Epp, 1978), but has no effect on eye development in the absence of *r*.

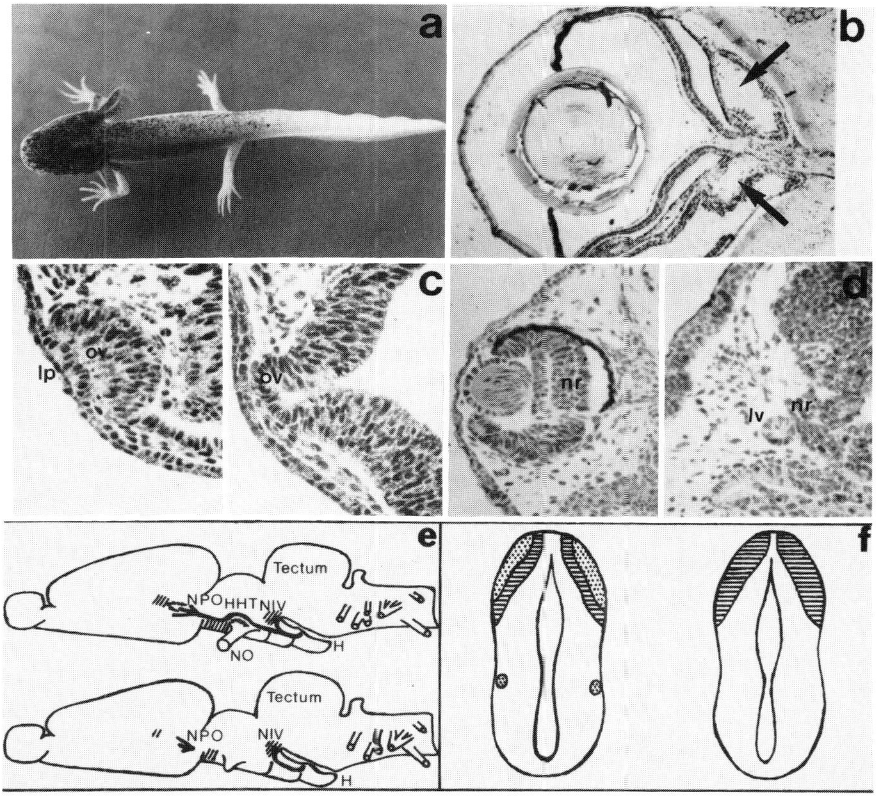

Fig. 13–1. (a) Blind e/e axolotl. (b) Eye of 7-year-old +/e, +/r axolotl. Note the duplication of retinal vesicles (arrows). (c, right) Retarded optic vesicle (ov) (Rabl stage R1a, but smaller), in a 4-day embryo (stage 30). (left) By that time wild-type eyes would have reached the lens placode (lp) stage (stage R2). (d, right) Small lens vesicle (lv) (stage R5) adjacent to disorganized neural retina (nr) cell mass (abnormal stage R1b-4) in a 10-day embryo at hatching (stage 42). (left) Wild-type eyes would be at the rod-and-cone stage (R7) (Cuny and Malacinski, 1986). (e) *Eyeless* (bottom) and wild-type (top) whole brain with PAF$^+$ (black) and GnRH$^+$ (hatched) staining. Note the absence of PAF$^+$ from the diencephalic e/e hypothalamo-hypophyseal tract (HHT). NPO, Nucleus preopticus; NO, nervus opticus; H, hypophysis; NIV, nucleus infundibularis ventralis (Eagleson and Malacinski, 1986). (f) *Eyeless* (right) and wild-type (left) tectal cross section: cholinergic-visual (dotted); serotonergic-somatosensory (hatched) (Gruberg and Harris, 1981).

MORPHOGENESIS

Eye development in axolotls homozygous for *eyeless* arrests at the early optic vesicle stage (Rabl eye stage R1a) (Cuny and Malacinski, 1986) (Fig. 13–1c). Humphrey (1969a), Van Deusen (1973), and Ulshafer and Hibbard (1976, 1979) did not observe any evagination of optic vesicles in *eyeless* axolotl embryos at late tailbud (stage 26), although they all observed small indentations at the site where these vesicles normally form in wild-type animals. Brun (1983), however, found small retinal rudiments in all 2-month-old *eyeless* larvae. The discrepancy between these reports could be due to differences in genetic background (F. Briggs, personal communication, cited in Harris, 1983), but the stages at which the embryos have been examined may also be important. We have found that *eyeless* optic vesicles (stage R1a) establish contact with the head epidermis (Fig. 13–1c) during the initial period of eye development (stages 29–32) (Cuny and Malacinski, 1986). This seems to rule out the possibility often mentioned (Ulshafer and Hibbard, 1976, 1979; Epp, 1978; Harris, 1983; Brun, 1985) that the neural crest mesenchyme spreads early across the orbital space, preventing the interaction of the optic vesicles with the head epidermis. Around hatching (stages 42–43), small disorganized masses of neural retinal cells (Rabl stages R1b–4) (Fig. 13–1d) and, more rarely, some pigmented retinal cells (stage R6) remain at the optic stalk. Differentiation of at least some retinal cell types may indicate that the *eyeless* mutation does not directly interfere with retina cell differentiation. Lenses are rarely formed, and when they are, they remain small (stages R4–6). The head mesenchyme from the neural crest, which migrates in later (R. J. Ulshafer, cited in Epp, 1978), pushes the lenses either close to the retinal rudiments or close to the head epidermis. In 2-month-old larvae both retinal and lens tissues disintegrate (Cuny and Malacinski, 1986), because at that stage the head mesenchyme greatly impairs the epidermal–retinal interaction. Cornea and sclera never develop.

Fewer cells are present in the ventral diencephalon of *eyeless* embryos when compared to wild-type embryos (Ulshafer, 1975; Ulshafer and Hibbard, 1976, 1979). Cell division furrows are more consistently perpendicular to the optic vesicle surface (Ulshafer, 1975; Ulshafer and Hibbard, 1976), possibly because cells are scarce. This prevents three-dimensional cell stacking. At the ultrastructural level, intercellular junctions are seen more rarely in *eyeless* optic vesicles (Ulshafer, 1975), which may be the reason for the disorganized structure of retinal cell masses at hatching (stages 42–43). The basal lamina of the mutant optic vesicle thickens prematurely (Ulshafer and Hibbard, 1979). The basal laminae modify cellular responses to growth factors. Thus, the gene could prevent eye development by a primary defect in the basal lamina of the optic vesicle, although it is equally probable that the lamina thickens prematurely due to a lack of cell proliferation in the optic vesicle cells (Harris, 1983).

Since gene *e* homozygotes are sterile, their reproductive system has also been examined. Both male and female gonads remain immature, although the germ cells home normally to the gonadal ridges (Van Deusen, 1973). Oocytes remain smaller than 350 μm in diameter, and they accumulate neither yolk platelets nor melanin pigment granules. Mutant oocytes reach the pachytene stage of the first meiotic cycle (Van Deusen, 1973). Oviducts are short and narrow (Van Deusen, 1973). In males, the testes are narrow and the vasa deferentia slender and straight. The seminal tubules are lined only with premeiotic spermatogonia (Van Deusen, 1973). It appears as if the gonads are normal but insufficiently stimulated by pituitary hormones.

The anterior lobe of the pituitary gland of *eyeless* axolotls is smaller than that of the wild type. The neurohypophysis is atrophied and paraldehyde/fuchsin (PAF)-stainable secretory material is rarely observed (Eagleson and Malacinski, 1986). The anterior hypothalamus perikarya have only a small quantity of immunoreactive gonadotropin-releasing hormone (GnRH). Both effects on pituitary gland and anterior hypothalamus can be simulated in frogs by the embryonic removal of the optic vesicles (see Eagleson and Malacinski, 1986). Hence, these effects could be secondary consequences of the lack of visual input. There are, however, additional defects in the *eyeless* brain. Although paired nuclei preoptici are formed, their PAF^+ emanating fiber tracts are missing (Eagleson and Malacinski, 1986). The preoptic recess is reduced or completely lacking (Eagleson and Malacinski, 1986) (Fig. 13–1e). These defects could be directly caused by the mutant *e* gene. The visual cortex of the tectal neuropil is almost completely absent in the *eyeless* axolotl, and somatosensory serotonergic nerve fiber endings reach the tectal surface (Gruberg and Harris, 1981) (Fig. 13–1f). The cholinergic visual cortex can be restored, however, by grafting a wild-type eye rudiment into the contralateral side of the mutant head at stage 30 (Harris, 1982).

PRIMARY TISSUE FOCUS

The survey given above of morphological defects caused by the gene *e* indicates that the most severe malformations are associated with the retina and anterior portion of the hypothalamus. The cells of both those tissues are derived from the same (possibly clonal) cell cluster in the median line of the anterior neural plate (Adelmann, 1929; Eagleson and Malacinski, 1986), although later the retinal cells move laterally into the neural folds (Woerdeman, 1929). Therefore, the retina and anterior hypothalamus may both be parts of the primary target of gene *eyeless* expression (Van Deusen, 1973; Cuny and Malacinski, 1986). Alternatively, tissues that normally interact with the developing retina and hypothalamus may not produce sufficient quantities of growth factors. The degeneration in the retina and hypothalamus may thus be a secondary effect, rather than the primary result of the action of the gene *e* (R. Briggs, 1973; Van Deusen, 1972).

The first tissue interaction needed for paired eye development occurs between the prechordal plate mesoderm and the neuroectoderm (Adelmann, 1929). Using heterozygous *eyeless* spawnings with 25% eyeless embryos, Van Deusen (1971, 1973) transplanted the blastula roof (Fig. 13–2a) of the late blastula (stage 8), or dorsal lip (Fig. 13–2b) of the early gastrula (stage 9), between mutant *e* and wild-type embryos. The former region gives rise to the retina–hypothalamus tissues, while the latter region forms the prechordal plate. Eyeless larvae developed when the blastula roof originated from an *eyeless* embryo, while both wild-type and *eyeless* dorsal lips stimulated paired wild-type retina development. These results demonstrate that the genetically *eyeless* prechordal plate mesoderm participates in a normal way in eye development.

The second tissue interaction is that of head epidermis and anterolateral neural plate or optic vesicle. When wild-type head epidermis or anterolateral neural plates are explanted into organ culture, they only form skin or brain tissue, respectively. When they are cultured together, however, they form neural retina and frequently lenses as well. Combinations of wild-type neural plate with *eyeless* epidermis also leads to retina development, while combinations of *eyeless* neural plate with wild-type epidermis yield at the most relatively small retinas (in up to 33% of the cases)

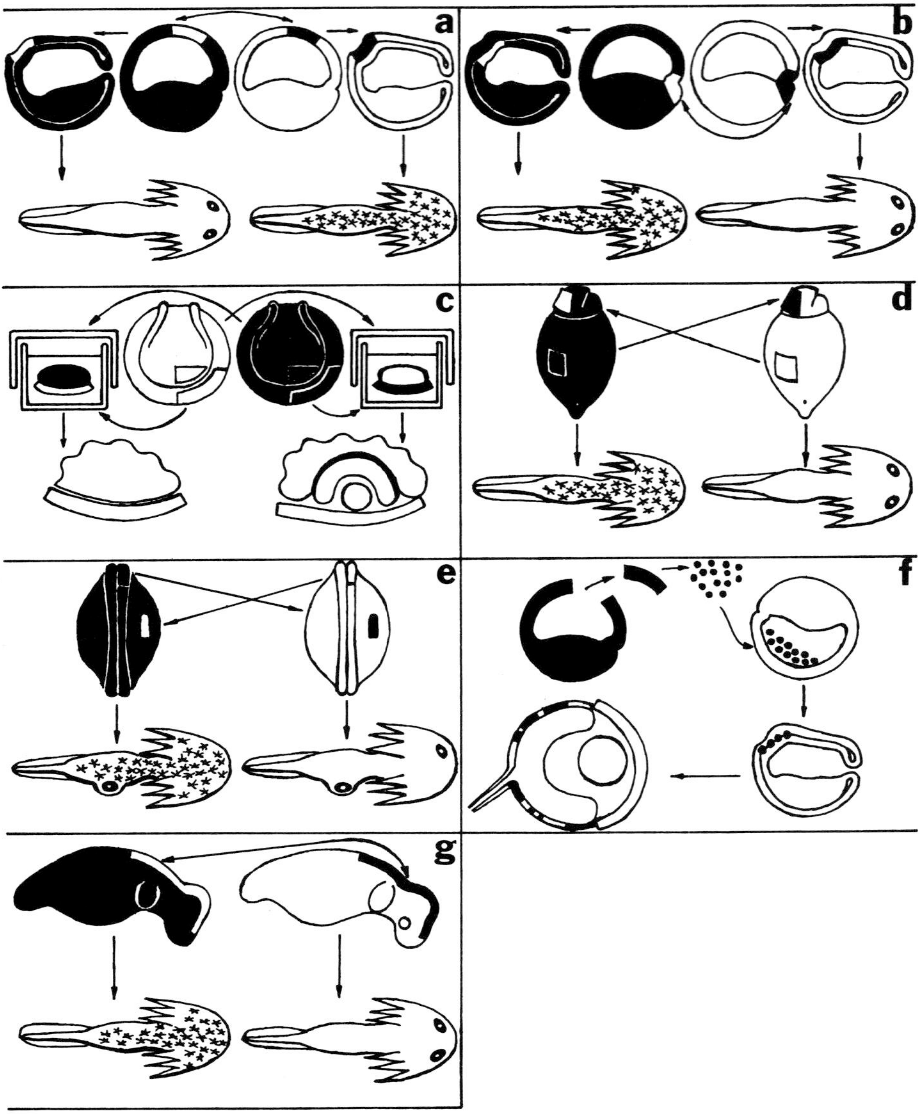

Fig. 13–2. Transplantation of retina- and hypothalamus-forming blastula roof (a) or prechordal plate-forming dorsal lip (b) between *eyeless* (black) and wild-type (white) embryos. In either series, axolotls with *eyeless* blastula roofs were blind (Van Deusen, 1973). (c) Organ culture of sandwiches of lens-forming head epidermis and retina-forming anterolateral neural plate (stage 13). Most tissue sandwiches with *eyeless* anterolateral neural plate lacked eye tissues (Cuny and Malacinski, 1986). (d) Flank epidermis grafted over optic vesicles of tailbud embryos. Mutant embryos with wild-type epidermis grafts remained eyeless, and wild-type embryos with mutant epidermis developed eyes in most cases (Brun, 1978). (e) Anterior neural fold complexes grafted under the flank epidermis. Both e/e and wild-type neural folds developed eyes even under mutant epidermis (Brun, 1978). (f) Injection of *eyeless* pigmented retina-forming blastula roof cells into wild-type eyed inverted albino blastulas. Mosaic eyes formed with patches of black and transparent pigmented retina (Van Deusen, 1973). (g) Reciprocal transplantation of neural crest between wild-type and *eyeless* embryos (stage 30). Eye development was independent of the genotype of the neural crest, and *eyeless* embryos remained eyeless (Epp, 1972).

(Cuny and Malacinski, 1986) (Fig. 13–2c). Brun (1980) attempted to transplant neural plates between *eyeless* and wild-type neurulas. However, as he realized (Brun, 1987), the results remained inconclusive due to problems with tissue cross-contamination. When Brun (1978) grafted wild-type flank epidermis over presumptive optic vesicles of *eyeless* embryos at stages 21–23, only 33% formed externally visible retinas (Fig. 13–2d). When *eyeless* flank epidermis was used over wild-type optic vesicles, pigmented retinas were recognized in 64% of the cases (Brun, 1978). These results suggest that the presumptive neural retina of the mutant embryo is intrinsically defective, while mutant epidermis most likely secretes the normal level of retina growth factor.

Brun (1978) also grafted mutant anterior neural folds, which contain presumptive brain and retina tissue, neural crest, and head epidermis, under the flank epidermis of wild-type embryos at stages 19–21. In 86% of the cases, recognizable pigmented retinas developed (Fig. 13–2e). When wild-type neural fold complexes were grafted under mutant flank epidermis, 74% of the cases exhibited pigmented retinas (Brun, 1978). Lenses were seen more rarely in all experiments. These results, in both cases positive, seem to indicate that wild-type flank epidermis can rescue the gene *e* eye tissues. It should be noted, however, that these experiments were performed after the onset of eye tissue interaction, and tissue types were not purified. *Eyeless* pigmented retinal cells may fully differentiate and function normally when surrounded by wild-type cells. This was demonstrated by injection of dissociated melanin-synthesizing *eyeless* blastula roof cells into an eye-forming albino blastula (Fig. 13–2f). The blastula was then inverted to allow the dissociated cells to migrate into the blastula roof. Mosaic eyes were obtained with patches of pigmented and nonpigmented tapetum nigrum (Van Deusen, 1973). This demonstrates that pigmented retinal cells of mutant origin may behave quite normally when in a wild-type tissue environment.

Epp (1972) performed reciprocal transplantations of neural crest between wild-type and *eyeless* embryos at the dorsal fin stage (stage 29–30) (Fig. 13–2g). Mutant embryos with wild-type neural crest implants remained eyeless, and wild-type embryos developed two eyes when mutant neural crest was implanted. The pigment pattern of the skin was unchanged. This confirms the morphological observation that neural crest mesenchyme and pigment cells do not prevent the early optic vesicle–epidermis tissue interaction in the mutant, and hence play a secondary role in generating the mutant phenotype.

Van Deusen (1971, 1972, 1973) successfully transplanted whole heads of wild-type axolotls to *eyeless* bodies, and mutant heads to wild-type bodies (Fig. 13–3a) at the tailbud stages (stages 21–25). While axolotls with wild-type bodies and mutant heads were sterile, those with mutant bodies and wild-type heads were fertile after 1 year of development. Thus, spawnings with 100% mutant (*e*/*e*) progeny were obtained. When mutant anterocentral neural plates of early neurulas (stage 13–14) were exchanged with wild-type tissue that contained the presumptive hypothalamus cells, a similar result was obtained (Fig. 13–3b). Mutant axolotls with wild-type hypothalamic implants were fertile, while wild-type axolotls with mutant implants were not (Van Deusen, 1971, 1973). Van Deusen (1973) also exchanged the gonad-forming region of tailbud-stage embryos (stages 21–25). Mutant gonads in wild-type axolotls were fertile, while wild-type gonads in mutant axolotls were

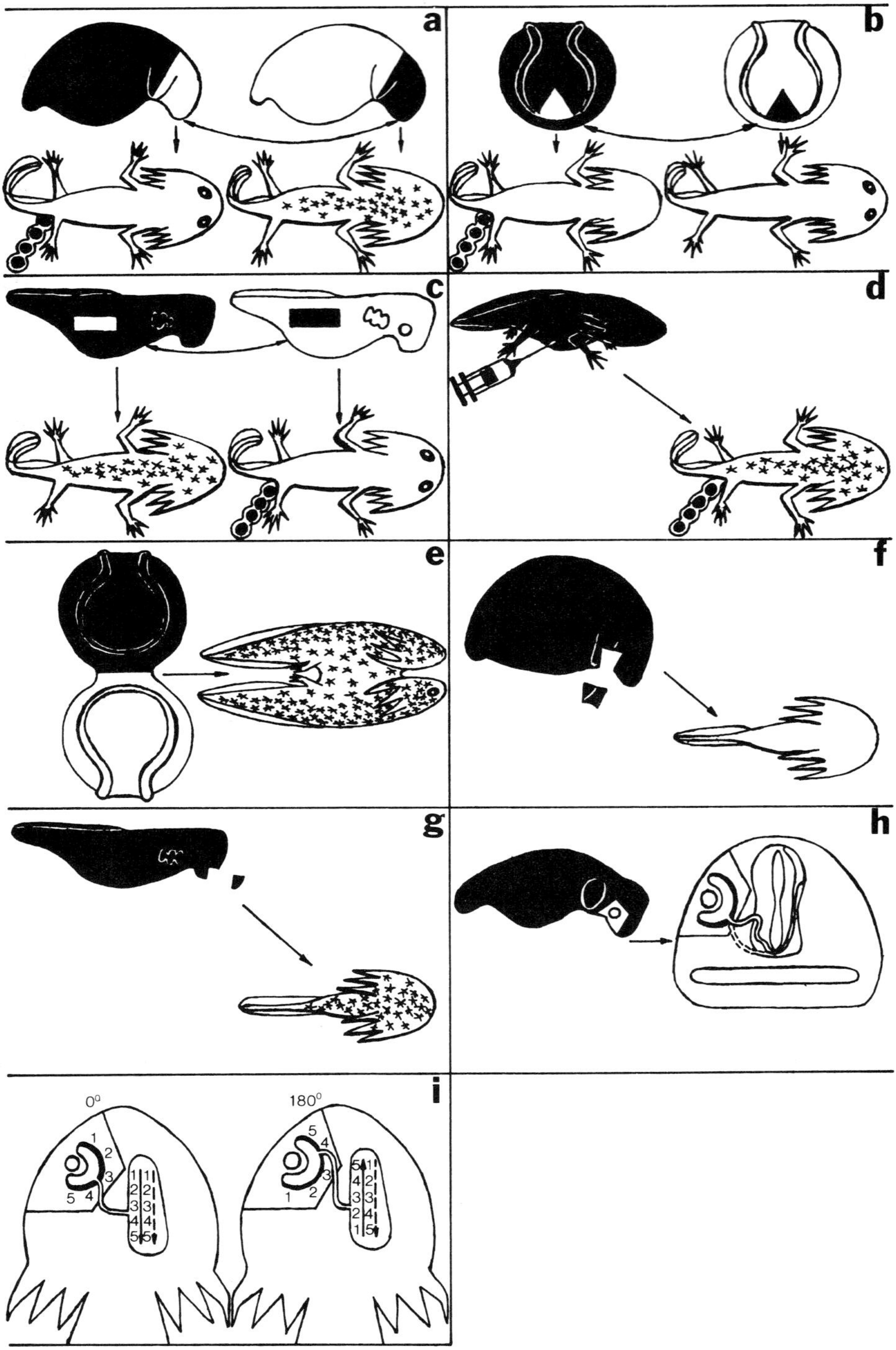

Fig. 13–3. (a) Reciprocal transplantation of whole heads between wild-type and *eyeless* embryos at tailbud stages (stages 23–25). Wild-type recipients of *eyeless* heads were blind, sterile, and dark-freckled (Van Deusen, 1973). (b) Reciprocal transplantation of hypothalamus-forming anterocentral neural plates between wild-type and *eyeless* embryos at the neural plate stage (stage 14–15). Wild-type hosts of *eyeless* anterocentral neural plates had eyes, were sterile, and were

sterile (Fig. 13–3c). Injections of 0.5–1.0 mg of bovine follitropin (bFSH) and 1 mg of human chorionic gonadotropin (hCG) (Fig. 13–3d), once a week, started at the age of 4 months, stimulated oocyte maturation with yolk incorporation, pigmentation of the animal hemisphere, and spawning in female axolotls homozygous for *eyeless* (R. Cuny, unpublished observation). Injections of 0.1–2.0 mg/week of synthetic GnRH increased the average size of oocytes and the size of the anterior lobe of the pituitary gland; however, oocytes remained immature and no spawnings were obtained. The timing of GnRH injections may be critical. If optimal injection schedules were established, it is conceivable that spawnings could also be stimulated by GnRH. These results confirm that the mutant gonads can function normally if they receive sufficient pituitary hormone stimulation.

Parabiotic pairs of mutant and wild-type embryos did not survive long enough to permit gonad maturation. However, the pigmentation of both embryos was dark-freckled (Fig. 13–3e), due to maximal skin melanocyte expansion. That observation provides evidence for the spread of a mutant characteristic to wild-type embryos via the circulatory system (Van Deusen, 1971, 1972; Epp, 1972). Rathke's pouch forms the anterior lobe of the pituitary gland. When it is excised (Fig. 13–3f) at the late-tailbud stage (stage 25–26), mutant embryos are not dark-freckled, but rather are as white as wild-type hypophysectomized embryos (Epp, 1972). Epp (1972, 1978) concluded that the mutant embryos overproduce melanocyte-stimulating hormone (MSH) in the pars intermedia of the pituitary gland. That could be due to insufficient secretion of MSH-inhibiting factor in the mutant hypothalamus (Harris, 1983). Melatonin is secreted by the pineal gland and the neural retina in a lighted environment. Melatonin counterbalances the effect of MSH and stimulates the contraction of skin melanocytes. However, the parabiosis experiments reveal that two wild-type eyes and a wild-type pineal gland cannot secrete enough melatonin to balance the mutant MSH overproduction. Epp (1972) excised the pineal gland area of late-tailbud-stage embryos (stages 26–27) (Fig. 13–3g). Survival was poor. However, mutant embryos were dark-freckled, while eyed embryos remained white. This confirms the view that the epiphysis does not play a key role in the expression of the *eyeless* gene.

Eye rudiments of wild-type embryos implanted into the head of mutant embryos at the straight-body stage (stage 30) restore full vision (Epp, 1972; Hibbard and Ornberg, 1976; Schwenk and Hibbard, 1977), although without nerve guidance

---

lightly pigmented (Van Deusen, 1973). (c) Reciprocal transplantation of gonad-forming belly mesoderm of tailbud embryos (stage 25) in heterozygous spawnings. *Eyeless* gonads implanted into wild-type axolotls were fertile (Van Deusen, 1973). (d) Injection of 1 mg bFSH and 1 mg hCG every week in gene e females, beginning at the age of 4 months, stimulates oocyte maturation and spawning at the age of 1 year. (R. Cuny and G. M. Malacinski, unpublished observation). (e) Parabiosis between gene e and wild-type embryos. The mutant embryo was eyeless; both were dark-freckled (Van Deusen, 1973). (f) Excision of Rathke's pouch corrects the overpigmentation in *eyeless* embryos (Epp, 1972). (g) Excision of the epiphysis had no effect on pigmentation. *Eyeless* axolotls remained freckled (Epp, 1972). (h) Implantation of wild-type eye rudiment at straight-body axis stage (stage 36) into *eyeless* embryos restores full vision unilaterally in the *eyeless* axolotl (Epp, 1972). Novel wild-type optic nerve entry into brain of an *eyeless* axolotl after implantation of a wild-type eye rudiment (Harris, 1982). (i) Wild-type eye anlage implantation in 180°-inverted position (5–1) produces inverted register of optic input (5–1) in the tectum when compared to the somatosensory input (1–5) (Harris, 1982).

the optic nerves enter the diencephalon or mesencephalon walls at novel ectopic sites, depending on the positioning of the implant (Harris, 1983) (Fig. 13–3h). The vision of these animals was confirmed in a rotating black-and-white-striped optokinetic drum (Epp, 1972; Hibbard and Ornberg, 1976). The optic nerve endings were detected in the superficial layer of the tectum by autoradiography after radioactive proline had been injected into the implanted eye (Schwenk and Hibbard, 1977; Harris and Cole, 1984), by horseradish peroxidase (HRP) tracing (Harris, 1984), and by extracellular electrical recording after flash stimulation of selected regions of the implanted wild-type retina (Hibbard and Ornberg, 1976; Harris, 1982). When the eye anlage was implanted in a normal position, the electric retinal signals in the optic neuropil were in correct register with the somatosensory signals in the deeper layer of the tectum (Harris, 1982). When the wild-type eye was implanted in a 180°-inverted position, the electrical signals from the retina were also inverted in the optic tectum, while the somatosensory input remained in normal somatotopic register (Fig. 13–3i) (Harris, 1982, 1983). Wild-type eye implantation corrected the skin pigmentation of the mutant (Epp, 1972, 1978), but only when the optic nerve endings had established contact with neurons in the optic tectal layer (Harris 1982, 1983). This indicates that pigmentation is corrected via neuronal pathways to the hypothalamus coming from the tectum. However, mutants with implanted wild-type eyes remained sterile (Epp, 1972, 1978), due to residual primary defects in the anterior hypothalamus. When both eye rudiments are removed from wild-type dorsal fin-stage embryos (stage 30), the adults will have a dark-freckled skin pigmentation, but they are fertile (Epp, 1972). This confirms the notion that pigmentation is controlled via innervation of the tectum, and also indirectly by the anterior hypothalamus. Fertility is dependent on the secretion of pituitary hormones, which is controlled by the hypothalamus.

MODELS AND FUTURE TASKS

One line of experimentation revealed that, in the eye, it is the retinal component that is directly affected by the gene *eyeless* (Cuny and Malacinski, 1986). In the hypothalamus, it seems to be the anterior portion, especially the preoptic recess or hypothalamo-hypophyseal tract (Eagleson and Malacinski, 1986). Pigmented retinal cells can differentiate from mutant cells (Van Deusen, 1973; Cuny and Malacinski, 1986). However, full differentiation of neural retinal cells has not yet been demonstrated. It would be interesting to monitor neural retina cell differentiation in a Van Deusen–type experiment (1973), in which mosaic *eyeless* wild-type eyes are produced by injection of blastula roof cells of mutant embryos into the blastocoel of wild-type embryos. It is probable that only one of the many cell types in the neural retina is directly targeted by the *eyeless* gene. To date, individual cell types have not been monitored. For example, the photoreceptor cells of *eyeless* retinas could be examined using rhodopsin and associated proteins and glycoproteins as cell markers. The accumulation of various synaptic transmitters, such as acetylcholine, 5-hydroxytryptamine, glutamine, glycine, substance P, epinephrine, γ-aminobutyric acid, and also taurine, melatonin, and thyrotropin-releasing hormone (TRH) could be traced using histochemical and immunochemical methods.

Glial fibrillary acidic protein, myelin basic lipoprotein, neurofilament proteins, vimentin intermediate filaments, and galactocerebroside could be used to identify glial cells in the mutant retina rudiment. It would also be important to develop very early neural retinal cell makers. Most of the early markers are, however, shared with the brain, such as the neuronal cell adhesion molecules N-CAM and Ng-CAM. It might be possible to immunize mice with axolotl anterolateral neural plate material and to select monoclonal antibodies that react only with presumptive retinal cells.

The premature thickening of the basal lamina of the optic vesicle of mutant embryos (Ulshafer and Hibbard, 1979) should be studied further. Fibronectin and laminin are glycoproteins of basal laminae of growing tissues, while collagen IV is present in mature basal laminae of postmitotic epithelia. The presence of these glycoproteins in the basal laminae of mutant optic vesicles could be followed during development using immunofluorescence. Sandwiches of wild-type epidermis and mutant neural plates could be cultured. Would the addition to the culture medium of high concentrations of fibronectin or laminin or low concentrations of collagenase restore eye development in organ culture? The basal lamina may act as an adsorbant or filter for growth factors. Unfortunately, the retina-stimulating growth factor of the head epidermis is not yet purified (Cuny and Malacinski, 1986). Hence, it is not yet possible to answer the question whether *eyeless* optic vesicles have enough cell surface receptors for this growth factor, and whether these receptors are masked by the early thickening of the basal lamina.

Fewer gap junctions are found in *eyeless* optic vesicles (Ulshafer, 1975; Ulshafer and Hibbard, 1976). Gap junctions are no doubt needed for vertebrate eye development, because injection of antibodies against gap junction protein into retina-forming early blastula roof cells suppresses eye formation (Warner et al., 1984). Radioactively labeled gap-junctional proteins could be injected into the lumen of the neural tube, or if they are not automatically incorporated into the plasma membrane of optic vesicle cells, they might have to be injected into neural plate cells. These are undoubtedly difficult experiments, but the degenerating mutant retina might be rescued by gap-junctional protein supplies. Immunofluorescence using antibodies against gap-junctional proteins could be used to determine whether optic vesicles not only display a decreased amount of gap junctions, but also exhibit a depleted intracellular stock of gap-junctional proteins.

Several general enzymes or cytoskeletal proteins needed for the basic retinal cell metabolism may shift from an embryonic form to a retina-specific form, whereby the embryonic gene is turned off and the retina-specific gene is turned on during retina development. If the retina-specific form is malfunctioning or unstable, retinal cells would die. This kind of defect would, unfortunately, be very difficult to unveil by current methodology, since most physiochemical and immunological properties would remain very similar.

The ventral portion of the retina is normally derived from the contralateral body side (M. Jacobson and G. Hirose, 1978). There is a possibility that the crossing over to the contralateral body side of ventral retinal cells does not occur normally in the mutant, and the presumptive hypothalamic cells, which also lie in this general area, might be affected as well. HRP could be injected into retina-forming anterior neural plate cells to determine to which body side they move. HRP is then revealed by its enzymatic activity on color dyes. In this connection, the recently proposed bilateral "anteroposterior morphogenetic system hypothesis" of Brun (1985, 1987) should be mentioned. Brun claims that implantation of wild-type right-body side eye-forming anterolateral neural fold complexes into the right-body side statocyst-forming posterolateral neural fold region of mutant neurulas stimulates normal eye formation on the left body side (in 79% of mutant embryos). Those transplants consisted of tissue mixtures. Double-eye formations on the right body side were not considered, and eyes were scored in vivo only. Clearly, more work is needed to elucidate the migration of presumptive retinohypothalamic cells in the neural plate.

In order to study the molecular genetics of the *eyeless* gene, direct isolation of the gene might be attempted as follows. An eye tissue cDNA library (wild type) could be screened with cDNA prepared from normal and mutant target tissue mRNA, on replica filters ("differential screening"). Clones that are (+) with normal cDNA, but (−) with mutant cDNA would be identified. They could in turn be used to screen a genomic DNA library. The difficulty is the huge amount of DNA per nucleus in axolotls. A large proportion of this DNA appears, however, to consist of repetitive intervening sequences, which probably could be removed with endonucleases. If at least the chromosome on which the *eyeless* gene is located were known, the studies could be restricted to the DNA of one chromosome. Cuny and Malacinski distinguished 11 of 14 chromosomes of interspecific hybrids between tiger salamanders *(Ambystoma tigrinum)* and axolotls by their banding patterns (Cuny and Malacinski, 1985). The chromosome segregation in the F_2 hybrid generation could be correlated with the phenotypic eyelessness. The identified gene *e*-bearing chromosome could be isolated by density gradient centrifugation or microexcision, and its DNA could be cloned and studied as proposed above. Since the *eyeless* gene has served as a model for gene action in eye pattern formation, its localization at the cellular and DNA levels would be most important for future progress in studies of genetic regulation of retina development.

14

Growth Control in Limb Regeneration

KEN MUNEOKA, SUSAN V. BRYANT,
AND DAVID M. GARDINER

The regeneration of limbs has for many years intrigued developmental biologists, and a great deal has been learned about the role of different tissues involved in this process. The axolotl has proven to be a particularly important organism for studies of this problem because of its ability to regenerate limbs faithfully (as well as a wide variety of other body parts) throughout life. Although other urodeles regenerate, the axolotl is unique because it breeds readily in captivity, a number of mutants are available, and heritable markers have been developed for experimental use (see Chapter 22). From studies of axolotl limbs, a number of different hypotheses or models have been developed to account for regeneration. In recent years, these hypotheses have led to a more cellular level of analysis due primarily to the development of cell marking techniques. This chapter highlights what is known about the cellular aspects of regeneration in the axolotl, emphasizing the mechanisms by which growth is controlled during this process.

OVERVIEW OF REGENERATION

The regenerated limb arises from a blastema composed of epidermal tissue surrounding a mass of rapidly proliferating mesenchymal cells. The morphological stages through which the blastema progresses are shown in Fig. 14–l. Following amputation of the limb, the wound surface is quickly covered by epidermis that later thickens to form an apical epidermal cap (AEC). The exact role of the epidermis during regeneration is still unkown; however, it clearly allows for continued growth of the regenerate (Stocum and Dearlove, 1972), possibly by providing a division signal for blastema cells (Globus, 1978). Epidermal cells themselves do not participate in the formation of the blastemal mesenchyme (Riddiford, 1960; Hay and Fischman, 1961; O'Steen and Walker, 1961), and there is no evidence that the AEC conveys patterning information to the underlying blastemal cells (Carlson, 1975b). Thus, the epidermis appears to influence the conditions necessary for maintaining cells in the cell cycle, but not the eventual fate of those cells.

The mesodermal cells of the blastema are thought to arise from the nonepi-

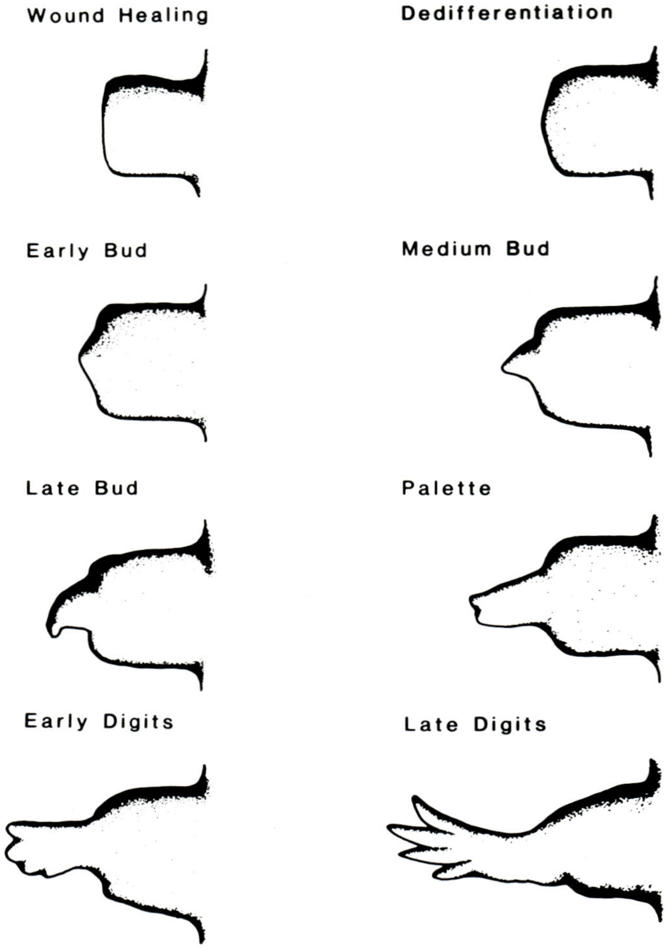

Fig. 14–1. Stages of limb regeneration in axolotls. Drawings of a dorsal view of a left forelimb amputated through the distal third of the upper arm. The stages are those of Iten and Bryant (1973) for regeneration in newts. (Redrawn from Tank et al., 1976.)

dermal tissues of the stump through dedifferentiation. Although the origin of all the cells in the blastema has not been precisely accounted for, we know that cells derived from the dermis contribute substantially. In contrast, cells derived from skeletal tissue and the nerve sheath make a relatively minor contribution (Muneoka et al., 1986b; K. Muneoka, unpublished data). The heterogeneity of blastema cells has been demonstrated with monoclonal antibodies (Kintner and Brockes, 1984). Thus, the blastema—while appearing to be a homogeneous mass of cells—is composed of at least two subpopulations of cells with distinct molecular characteristics. This reflects either the propensity of blastema subpopulations to differentiate along particular pathways (Cameron et al., 1986) and/or their cellular origins within the stump (Brockes, 1984). Once the blastema is established, subsequent outgrowth leading to the formation of a limb proceeds by mechanisms analogous to those utilized during limb development (Muneoka and Bryant, 1982, 1984b). Blastema cells proliferate (see Tassava et al., 1974; Maden, 1976; A. R. Smith and A. M. Crawley, 1977; Connelly and Bookstein, 1983), the blastema elongates along the proximodistal axis, and differentiation occurs primarily in a proximal to distal sequence.

BLASTEMA GROWTH

A number of different factors that influence the normal course of growth during regeneration have been described. The neurotropic influence on regeneration is probably the best studied of these factors, and it is now well established that denervation of the regenerating urodele limb prior to the medium-bud stage (Fig. 14–1) results in the inhibition of further growth of the blastema and the arrest of cells in G_1 of the cell cycle (see Tassava and Olsen, 1985). It has also been established that such growth-arrested blastema cells resume cycling when reinnervation occurs (Olsen et al., 1984). Recently it has been shown that denervation of the limb does not affect all blastema cells in a similar manner. Kintner and Brockes (1985) found that the subpopulation of cells that label with the antibody 22/18 was 7-fold more sensitive to denervation than the 22/18-negative subpopulation. This variability in the response of blastema cells to innervation reinforces the concept that the blastema is composed of distinct subpopulations of cells. The evidence from numerous studies is consistent with the hypothesis that nerves provide some factors that are important for blastema cell cycling (see Singer, 1978; Brockes, 1984; Tassava and Olsen, 1985; Carlone and Mescher, 1985). Efforts to identify neurotropic factors (NTF) for regeneration have resulted in a number of possible candidates, including glial growth factor (Brockes and Kintner, 1986) and fibroblast growth factor (Mescher and Gospodarowicz, 1979).

In recent years position-specific cellular interactions have been implicated in the control of growth during blastema formation and subsequent limb outgrowth. The concept that regenerative growth and patterning may be causally linked was first proposed in the polar-coordinate model (French et al., 1976; Bryant et al., 1981). The underlying theme of the polar-coordinate model is the proposal that limb cells possess positional information (Wolpert, 1969) and that these cells have the ability to recognize and respond to positional disparities by division. This position-specific stimulation of growth, termed intercalation, is thought to occur in an organized manner, with newly formed daughter cells acquiring intermediate positional values (Fig. 14–2). Hence, growth is stimulated and maintained by positional disparities and is terminated when all such disparities are resolved.

Other factors known to influence blastemal growth include a number of hormones such as thyroxine, growth hormone, insulin, prolactin, and hydrocortisone (see Liversage et al., 1985), bioelectric currents (see Borgens, 1985), the AEC (see Stocum, 1985), and the immune system (see Sicard, 1985). Various models emphasizing the role of one or more of these factors have been proposed. For example, bioelectric currents are thought to stimulate neuronal sprouting, thereby ensuring an adequate supply of NTF (Borgens et al., 1977). Alternatively, it is possible that

a

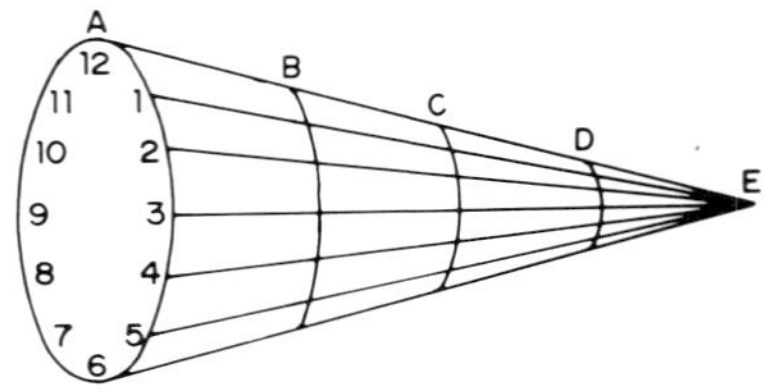

b

9 * 4

9,8,7,6,5,4

9 * 2

9,10,11,12,1,2

AB * E

ABCDE

Fig. 14–2. Intercalation as proposed in the polar-coordinate model. (a) Polar coordinates of positional information in a limb as viewed from the side. Each cell is assumed to possess information specifying its position around the limb circumference (1–12) and along the proximodistal axis (A–E). Positions 12 and 0 are identical; thus, the sequence of circumferential values is continuous. (From Bryant et al., 1981.) (b) Intercalation of missing circumferential and proximodistal positional values. When normally nonadjacent values come into contact (*), the missing values are restored by growth and insertion of the intermediate values (underlined). Circumferential intercalation (left) generates confronted values by the shorter, rather than the longer, of the two possible routes (the "shortest intercalation rule"). (From Bryant et al., 1987.)

bioelectric currents are involved in directing the aggregation of limb-forming cells (Borgens, 1984), thereby creating the proper environment for intercalary growth to proceed. The role of many of these factors in limb regeneration has not been clearly characterized, and it is still uncertain whether their effects on blastemal cell proliferation are direct or indirect.

GROWTH CONTROL MECHANISMS

It has been recognized for many years that the mechanisms by which regenerative growth is controlled are fundamental to the regeneration issue (see Singer, 1952). To understand regenerative growth, it is essential to determine which of the many factors that influence regeneration are controlling growth by dictating its onset, maintenance, and/or termination, and which represent requirements necessary to sustain growth-competent cells. Achieving this goal is a difficult task, and progress has been impeded by the exclusive use of inhibition assays. In such assays, if removal of the source of a factor (e.g., nerves or AEC) is found to inhibit regeneration, then that factor is determined to be growth promoting. It is impossible to determine from the results of this test whether the factor actually controls growth directly, or is merely required for growth. This distinction can be best illustrated with a hypothetical example. Suppose that the putative NTF is a mitogenic growth factor (such as glial growth factor or fibroblast growth factor) and that it affects blastema cells via membrane-bound receptors. If NTF is indeed a controlling factor, then changes in growth would occur in response to fluctuations in the availability of NTF independently of the number of NTF receptors. However, if NTF is only a growth *requirement,* then its effect on blastema cell division would be dependent on intrinsic changes in the availability of NTF receptors. In the latter case, the factor actually controlling growth would be the one that regulates the number of NTF receptors on cells. It should be possible to distinguish experimentally between growth-*controlling* factors and growth requirements by applying the criteria used to identify growth-controlling factors in vitro: (1) inhibition of growth when absent and (2) excessive growth when present in high amounts. In the above example, if NTF is growth controlling, then excessive amounts of NTF might result in overgrowth leading to either larger regenerates or regenerates with additional limb structures.

Many of the growth-promoting factors discussed earlier satisfy the first criterion; hence the key to evaluating their role with respect to regeneration depends on their ability to induce overgrowth. With respect to NTF, hyperinnervation of the urodele limb has not been tested; however, partial denervation does not alter the normal course of regeneration (see Tassava et al., 1985). From these data we can conclude that NTF is normally present in excessive amounts, and—according to our second criterion, nerves are not providing growth-controlling factor(s) but are providing factors necessary for growth-competent cells to make the transition from G_1 to S phase of the cell cycle.

Reducing the normal levels of hormones inhibits limb regeneration, but animal viability is also compromised. These studies have been interpreted to suggest that regenerative growth is controlled by the hormonal milieu (Liversage et al., 1985) within the animal. This interpretation seems unlikely, since Tassava (1983) has shown that well-fed animals survive hypophysectomy and also regenerate their limbs. The implication of this result is that the inhibition of regeneration observed previously is related to general metabolic deterioration, rather than to any direct influence of the reduced hormonal levels on blastema cell growth.

Although the AEC has been shown to be required for regeneration, sufficient data are not available to determine whether it has a role in controlling regenerative growth. As mentioned previously, removal of the AEC results in the inhibition of regeneration in vivo and, in addition, causes precocious cartilage differentiation during in vitro organ culture of blastemas (Globus et al., 1980). It has been proposed that the AEC provides a diffusible signal that stimulates cell division and inhibits differentiation during normal regeneration (Globus, 1978). According to this model, as the blastema enlarges, growing cells become farther removed from the signal and therefore differentiate. Hence, the AEC would be controlling blastema growth via the division signal, and enhanced growth should be stimulated if additional AECs are added. Thornton and Thornton (1965) showed that supernumerary outgrowths could be stimulated by supernumerary AECs; however, it is unlikely that the grafted AECs were free of contaminating blastema cells (see Jasch, 1979).

The factors underlying the intrinsic growth control mechanism (intercalation), which is an integral part of the polar-coordinate model (French et al., 1976; Bryant et al., 1981), satisfy both of the criteria established above for factors controlling regenerative growth. According to this view, a growth response results from cell–cell interactions at positional disparities, and it is presumed that the factors involved are locally confined signaling molecules probably located on the cell surface. Numerous experiments have been performed to test the hypothesis that growth is a function of positional disparities. The results of such experiments demonstrate that when tissues are rearranged so as to increase the probability of interactions between cells with disparate positional values, overgrowth occurs, resulting in the formation of supernumerary limbs. Conversely, when rearrangements are made so as to decrease the probability of interactions between cells with disparate positional values, regenerative growth is severely curtailed resulting in incomplete truncated or tapered regenerates. The remainder of this chapter will highlight experiments that support the idea that position-specific cellular interactions resulting in intercalary growth are controlling the onset, maintenance, and termination of growth during limb regeneration.

OVERGROWTH: SUPERNUMERARY LIMBS

A wide variety of tissue-grafting experiments demonstrate that supernumerary limb structures (overgrowth) result from increasing positional disparities between cells during regeneration (see Tank and Holder, 1981). The best studied of these is the interaction between anterior and posterior cells following blastema grafting between contralateral limbs. In these experiments, the grafted blastema is positioned on the contralateral host stump so that dorsal and ventral regions of the graft–host interface are normally aligned, whereas anterior and posterior regions confront one another (Fig. 14–3). At each confrontation, the interactions between anterior and posterior cells result in the localized stimulation of growth (intercalation), the formation of a supernumerary blastema, and the eventual regeneration

Fig. 14–3. Supernumeary limbs. (a) Diagram of the grafting procedure, shown here as viewed from the dorsal aspect. Forelimb blastemas at the stage of palette were exchanged contralaterally between diploid (left) and triploid (right) sibling axolotls and oriented on the host to appose anterior and posterior limb tissues. Dorsal and ventral tissues were positioned in correct axial alignment with the stump. The inset shows a closer dorsal view of a left triploid blastema grafted onto a right diploid stump. a, Anterior; p, posterior; d, dorsal. (b) Skeletal preparation of an experimental limb as viewed from the dorsal aspect. This limb resulted from a graft of a left triploid blastema onto a right diploid stump. Two supernumerary limbs formed, one anterior (A) and the other posterior (P) to the grafted limb. The grafted limb maintained its original handedness, and both supernumerary limbs are of stump handedness. The overall digital sequence from anterior to posterior is 1, 2, 3, 4 (anterior supernumerary limb); 3, 2, 1 (grafted limb); 1, 2, 3, 4 (posterior supernumerary limb). (c) The cellular contribution boundaries of supernumerary limbs resulting from anteroposterior contralateral grafts of blastemas are superimposed onto a summary diagram of an idealized transverse section at the level of the digits. Contribution from anterior and posterior tissues was found to be approximately equal in all 20 supernumerary limbs analyzed. (From Muneoka and Bryant, 1984a.)

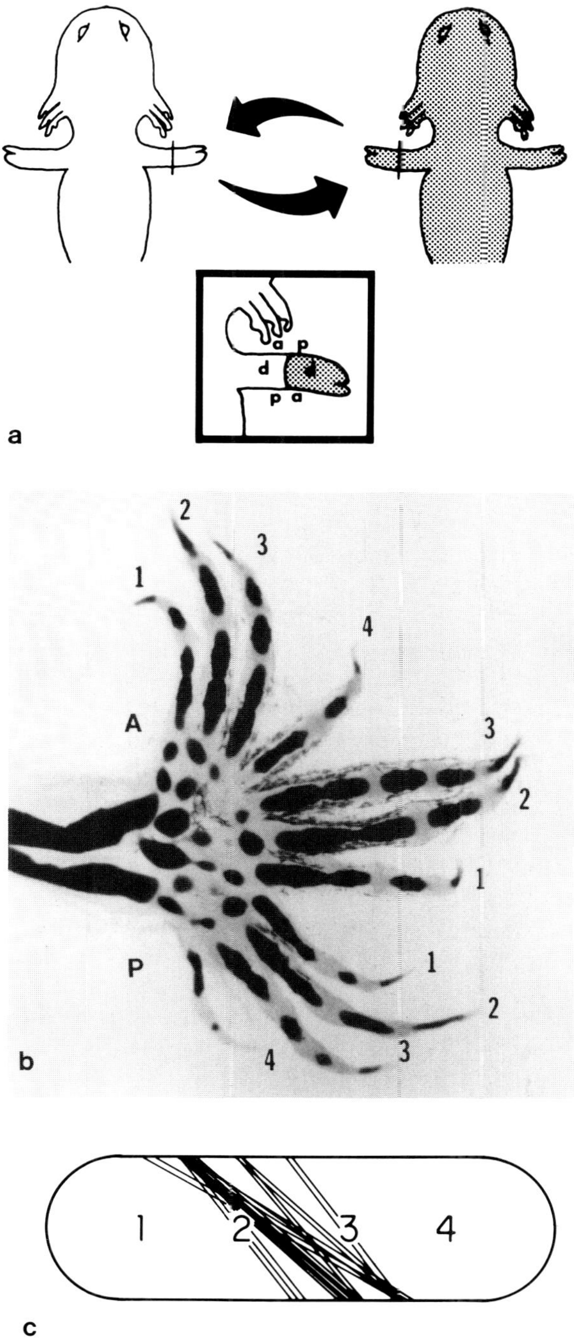
a p
d
p a
a
2 3
1
4
A
3
2
1
P
1
2
4 3
b
1 2 3 4
c

of a supernumerary limb. Although the final limb pattern is grossly abnormal, the neighbor relationships between cells is normal. Thus, although digit 2 is present three times in the final limb, the cells in digit 2 are adjacent to their correct neighbors (i.e., cells in digit 1 and digit 3). Control experiments in which blastemas are grafted without altering the spatial relationships of cells do not lead to the formation of supernumerary limbs. Thus the stimulation of additional growth cannot be attributed in any way to nonspecific effects of the grafting operation, such as wounding. These results were instrumental in the development of the polar-coordinate model for the control of growth and patterning, and subsequent data on the cellular origins of supernumerary limbs support this view. Experiments utilizing the triploid/diploid cell marker in the axolotl (see Chapter 22) demonstrate that supernumerary limbs are composed of approximately equal numbers of anterior- and posterior-derived cells that meet at a distinct boundary (Muneoka and Bryant, 1984a). Thus, the interaction between positionally disparate cells is mutual and reciprocal as would be expected from intercalation.

The formation of supernumerary limb structures that result from the spatial rearrangement of limb tissues has been used as an assay to investigate which tissue types are involved in position-specific growth and patterning. These studies demonstrate that some limb tissue rearrangements have no effect on the pattern (nerve sheath—K. Muneoka, unpublished; cartilage—Goss, 1956; epidermis—Carlson, 1975b), whereas others have a profound effect (dermis—Carlson, 1974; 1975b; Tank, 1981; Rollman-Dinsmore and Bryant, 1982; muscle—Carlson, 1975a,b), causing regenerates with supernumerary limb structures. Hence, position-specific growth stimulation (intercalation) is not a general property of all cells, but is a property of cells present in dermal and muscle tissues. On the basis of evidence from a variety of studies, we deduce that intercalation is a property of limb fibroblasts (see Bryant et al., 1987). First, fibroblasts are common to both dermis and muscle tissues: in dermis fibroblasts predominate (Holder and Glade, 1984), and in muscle both fibroblasts and myogenic cells are present. Second, experiments on the development of myogenic cells indicate that they form a distinct lineage that arises from somitic tissue at pre-limb bud stages. These cells are not involved in the formation of the limb pattern; rather, their pattern is dictated by other cells in the limb (see Gumpel-Pinot, 1984). Third, the behavior of fibroblasts during the early stages of limb regeneration supports the idea that these cells play a major role in limb patterning (see next section). Hence, on the basis of the exclusion of other cell types and on the behavior of dermal fibroblasts during regeneration, we deduce that fibroblasts possess positional information and interact with one another in a position-specific manner to stimulate intercalation and thus control growth during regeneration.

The position-specific interactions between fibroblasts that result in overgowth and supernumerary limb formation occur under the same physiological conditions that support regeneration of a single limb. This observation implies that many of the general growth-promoting factors in the blastema must be present in at least 3-fold greater amounts than that required for normal limb regeneration. Therefore, according to the criteria just proposed, many of these factors cannot be controlling growth. In the case of the nerve requirement for regeneration, this interpretation is in agreement with the results of the effect of partial denervation on limb regeneration (see earlier).

The only known exogenous factors that enhance the amount of growth during regeneration are the retinoids (see Maden, 1985). Retinoids stimulate the regeneration of limbs with structural duplications in the anteroposterior (Stocum and Thoms, 1984), dorsoventral (D. L. Stocum, personal communication), and proximodistal axes (Maden, 1982). The effect of retinoids on limb regeneration is dependent on both dose and treatment time. Interestingly, the direct effect of retinoids on the regenerating limb is not to enhance growth but to inhibit growth during the period of exposure, with overgrowth occurring after the removal of retinoids (Maden, 1983). One hypothesis for the effect of retinoids on regenerating cells is that it alters the positional information of blastema cells (Maden, 1982). For example, cells present at a distal amputation plane are thought to be altered to become more proximal by exposure to retinoids. Once the retinoid is removed, these altered cells regenerate as if they were at a more proximal level. According to this interpretation, the alteration of positional information from a distal level to a more proximal level alters the amount of intercalary growth required to regenerate the final limb pattern.

NORMAL REGENERATION

The study of intercalation and the formation of supernumerary limbs has led to the idea that normal regenerative growth is controlled by position-specific interactions between fibroblasts. These interactions result in intercalation and the generation of a template of the limb pattern that subsequently dictates the pattern of other cell types present in the limb (see Bryant et al., 1987). A number of predictions arising from this view have been tested experimentally using the triploid/diploid cellular marker in the axolotl. First, if position-specific interactions between fibroblasts are stimulating regenerative growth, then the population of cells in the resulting blastema would be composed of a large proportion of cells of fibroblast origin. This prediction has been tested by analyzing the cellular contribution from the dermis to the medium-bud blastema (Muneoka et al., 1986). Our results demonstrate that dermally derived cells (fibroblasts) make up about half (43%) of the blastema and that they are overrepresented by more than 2-fold when compared to their availability in the stump. Hence, a likely possibility is that the growth of fibroblasts during blastema formation is enhanced in relation to that of other stump cells. The fact that dermal fibroblasts represent half of all stump fibroblasts (Tank and Holder, 1979; Muneoka et al., 1986b) raises the possibility that virtually all of the early blastema (86%) is composed of fibroblast-derived cells.

A second prediction of this view is that fibroblasts must alter their normal spatial relationships during the regeneration process in order to stimulate intercalary growth. In experimental situations where intercalation has been best studied, the normal spatial relationship of cells is altered by grafting as in the stimulation of supernumerary limbs described previously. During normal regeneration such alterations of spatial relationships have been proposed to come about by the migration of dermally derived cells during the initial stages of regeneration (Fig. 14–4; Bryant et al., 1981). Indirect evidence for cellular displacement during regeneration can be found in a variety of studies (Rollman-Dinsmore and Bryant, 1984; Tank et al., 1985; Muneoka et al., 1985). Direct evidence for the migration of dermal

a

b

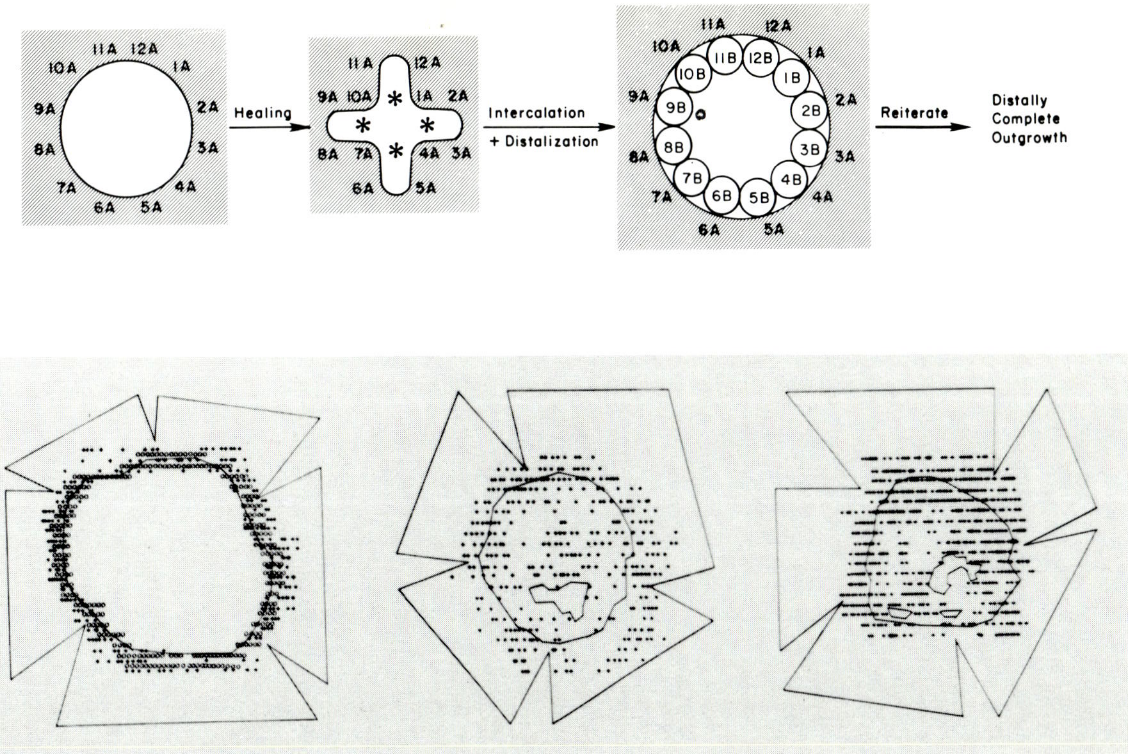

Fig. 14–4. Early stages of regeneration. (a) The polar-coordinate model for distal outgrowth from a wound surface after amputation of levels B–E (left). It is proposed that during the early stages of regeneration, migration brings cells from different circumferential positions into contact (middle). Interactions between cells with disparate positional values result in circumferential intercalation (*) and the generation of the next most distal level B (right). Reiteration of this process generates the complete limb. (From Bryant et al., 1981.) (b) Computer-assisted plots of the distribution of subepidermal cells of dermal origin in whole-mount preparations of early regenerates. Each symbol (+) represents a minimum of one trinucleolate cell in a region measuring 35 × 125 μm. The edge of the wound area and cell-free regions within the wound area are outlined by solid lines. The migration of dermal cells across the wound surface is just beginning at 5 days postamputation (left). At 10 days postamputation, dermal cells have migrated across the entire wound surface and have lost their normal neighbors (middle). By 15 days, the number of marked cells present on the wound surface has increased dramatically (right). (From Gardiner et al., 1986.)

fibroblasts has recently been acquired by Gardiner et al. (1986). In their experiment the migratory behavior of dermal cells was analyzed during the early stages of blastema formation in chimeric limbs created by grafting triploid dermis onto diploid stumps. The results from that analysis demonstrate that dermal cells migrate over the wound surface, between the wound epidermis and the underlying limb stump, during the initial stages of regeneration (between 5 and 10 days postamputation; Fig. 14–4). In addition, the migration of dermally derived cells occurs at a time coincident with the first measurable increase in mitotic activity during regeneration (Kelly and Tassava, 1973; Tassava et al., 1974; Maden, 1978). Taken together, the early events following amputation support the idea that positional rearrangements resulting from fibroblast migration provide the conditions necessary for intercalation.

GROWTH INHIBITION: SYMMETRICAL LIMBS

An important prediction arising from the idea that regenerative growth is controlled by intercalation is that decreasing the positional disparities present within the blastema should decrease the amount of growth. A number of studies have been performed that validate this prediction (see Bryant et al., 1981, 1982; Tank and Holder, 1981). The experimental design of these studies involves changing the normally asymmetric stumps to symmetrical stumps by grafting half-limbs to create double-anterior, double-posterior, double-dorsal, and double-ventral limbs. Such grafts are intended to reduce positional disparities by allowing for interactions between cells with similar positional values thereby inhibiting intercalation. The results of regeneration from all types of symmetrical limbs are similar and form a continuum from symmetrical regenerates to no outgrowth. The degree of regeneration (as measured by the number of skeletal elements in the regenerate) is found to be inversely proportional to the length of time that symmetrical limb grafts are allowed to heal. Thus, limbs amputated immediately following grafting regenerate symmetrical limb structures, whereas following long healing times limbs regenerate poorly or not at all. In all cases, regeneration is initiated and blastemas form. In addition, these blastemas are well innervated and possess AECs. In limbs amputated shortly after grafting, each half appears to behave autonomously and a proximal-distally complete but symmetrical regenerate results. Histological examination of the early stages of regeneration from such limbs indicate that each half is spatially isolated from the other (D. M. Gardiner and S. V. Bryant, unpublished observations), and it is therefore unlikely that cellular interactions occur across the plane of symmetry. This could account for the apparently autonomous behavior of each limb half. Hence, experiments that involve short healing times do not provide an appropriate test for the consequences of reduced positional disparities on regenerative growth.

More relevant to the issue of growth control is the observation that symmetrical limbs amputated after long healing times are characterized by reduced outgrowth. This growth inhibition cannot be attributed to an inability of blastema cells to participate in distal outgrowth, since symmetrical blastemas grafted onto asymmetric limb stumps participate in the formation of distally complete supernumerary limbs (Holder and Tank, 1979). In addition, limbs subjected to sham surgery and control double-grafted reversed stumps (Bryant and Baca, 1978) regenerate in a normal fashion, demonstrating that the inhibition of regeneration is not in any way an artifact of the grafting procedure. It is our thesis that the observed growth inhibition is a direct consequence of reduced positional disparities within the blastema. After long healing periods the tissue integrity of the stump is stablilized, and during the early stages of regeneration, cells from different regions of the stump migrate across the amputation surface and are free to interact across the middle of the stump (Fig. 14–5). However, since no positional disparities exist between cells on either side of the plane of symmetry, intercalary growth is not stimulated and regeneration is inhibited.

SYNOPSIS

The long-term goal in studying limb regeneration in axolotls is to understand the developmental mechanisms controlling limb formation. Central to these studies is the ability to define the parameters associated with the stimulation of blastema cell proliferation. This involves recognizing factors important for the control of growth as well as those that are merely required for growth. We define controlling factors as those associated with the cellular decision to proliferate and requirements as those factors needed for cells to continue through the cell cycle. We propose that controlling factors may be identified on the basis of ability to cause overgrowth and reduced growth as well as normal growth. To date, only those factors associated with the position-specific growth response termed intercalation can be definitively categorized as controlling growth. This conclusion is based on the results of pattern formation experiments where overgrowth or reduced growth occurs following grafting to alter the normal spatial arrangement of cells. An important conclusion drawn from several lines of evidence is that a single cell type, the fibroblast, is responsible for the intercalation growth response. It is likely that the characterization of factors that underlie intercalation will represent a major step in elucidating the cellular mechanisms controlling limb development, regeneration, and normal and abnormal growth in vertebrates.

We anticipate that future research will identify and characterize the cell surface-associated molecules that are necessary for the positional signaling between cells that leads to pattern formation and growth. This task will be facilitated by the identification of the cell type responsible for patterning and by information about the behavior and characteristics of this cell type. As a consequence, it should be possible for the first time to develop relevant in vitro assays for positional responses. The molecules involved in such responses may already be known from apparently unconnected studies such as those on the effects of growth factors on cell proliferation in vitro. In fact, surprising evidence is accumulating that molecules with homology to known growth factors are not only involved in crucial developmental decisions in *Drosophila* (Wharton et al., 1985) and nematodes (Greenwald, 1985), but are also likely to be cell surface molecules that could be involved in direct cell–cell interactions. Although we are a long way from understanding the

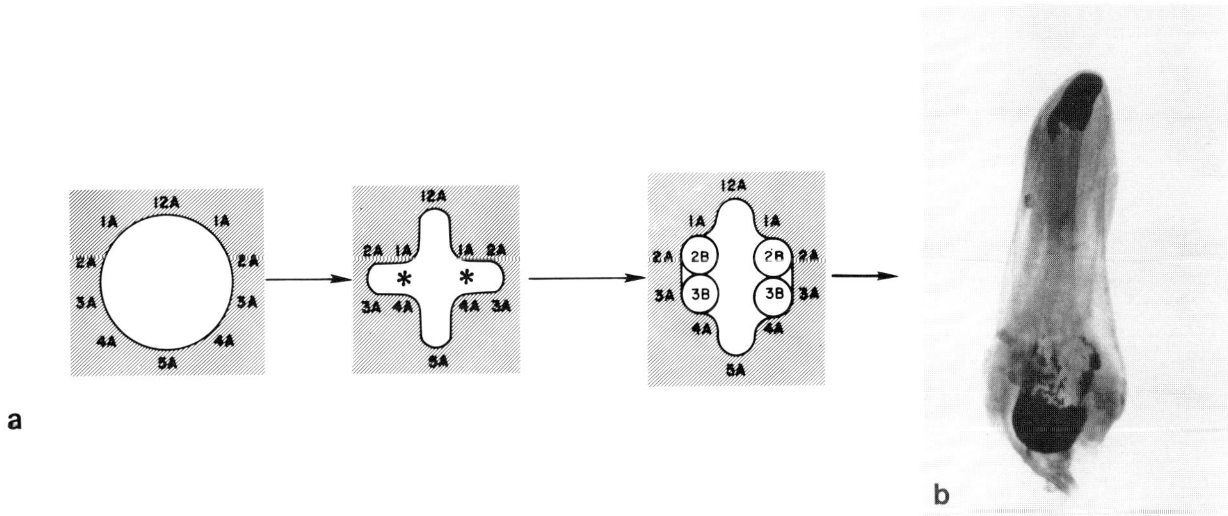

Fig. 14–5. Inhibition of regenerative growth. (a) The polar-coordinate model for growth inhibition in symmetrical limbs after amputation of levels B–E (left). During the early stages of regeneration, migration brings some cells with similar positional values into contact, and growth is not stimulated (middle), leading to early cessation of growth (right). (From Bryant et al., 1981.) (b) Skeletal preparation of a double-posterior limb stump that failed to regenerate, made 2 months after amputation. (From Bryant, 1976.)

molecular basis of growth control and patterning in limb regeneration, it is likely that a convergence of information from a variety of fields could herald an era of rapid progress.

ACKNOWLEDGMENTS

We thank Rebecca Shaddock for help in preparing the manuscript. Research was supported by PHS grant HD06082 and a gift from the Monsanto Company.

15
Neurobiology

WILLIAM A. HARRIS

It is debatable whether the most exciting feature of the axolotl for the study of neurobiology is the primitive and simple structure of its brain or the accessibility of its embryo to experimental manipulation. Axolotls offer the neurobiologist at least three other distinctive advantages: First is the remarkably large size of individual nerve cells at all stages of development; second is the ability of both the peripheral and central nervous systems (CNS) to regenerate; and third is the limited but increasing availability of mutant stocks that show abnormal neural development. Because of these significant attractions, over the last century many researchers have turned their attention to the salamander's brain. In a short chapter one cannot hope to cover all of this work; a full review of axolotl neurobiology still needs to be written. Instead, my intention here is to illustrate the usefulness of the axolotl in this area by providing a few examples of the development, function, regeneration, and genetics of its nervous system.

NEUROGENESIS AND THE NEURAL PLATE

During gastrulation, the involuting chordomesoderm contacts the overlying ectoderm and induces the neural plate. It is well known that urodelean embryos of differing pigmentation were used by Spemann and Mangold (1924) to demonstrate the phenomenon of neural induction, and the axolotl still serves as a useful and classic model system in which to study this event (see Chapter 6). After induction, perhaps in part through self-organization, the neural plate becomes organized into subdomains each of which is fated to give rise to a different part of the adult nervous system. Using vital staining techniques, C. O. Jacobson (1959) carefully mapped out the presumptive regions of the brain in the axolotl neural plate. This map (Fig. 15–1) serves as an extremely useful guide for those who want to study the development of the CNS with microsurgical techniques There are some inadequacies in Jacobson's map, partially because it is based on a single stage of development whereas the neural plate is constantly changing. It has thus proved useful for certain investigations to map particular neural primordia at a higher degree of resolution in time and space. For example, Brun (1981) has mapped the presump-

A

B

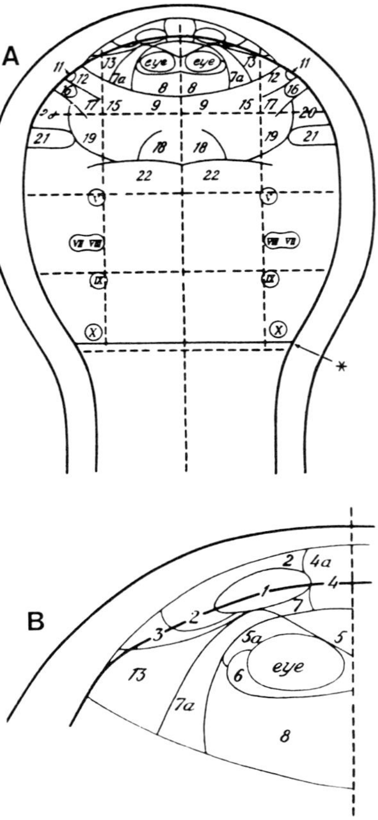

Fig. 15–1. (A) The presumptive brain areas of the axolotl plotted onto the neural plate. (B) Detail of plan in (A) representing the left side of the anterior neural plate. 1, Area olfactoria primitive; 2, primordia piriforme; 3, primordium hippocampi; 4, septum (primordium); 4a, septum ependymale; 5, nucleus praeopticus, pars anterior; 5a, nucleus praeopticus, pars posterior; 6, chiasma ridge; 7, area strioamygdaloidea; 7a, cranial part of pars ventralis thalami; 8, pars ventralis hypothalami; 9, pars dorsalis hypothalami; 10, pars buccalis of the hypophysis; 11, primordium epiphyse; 12, pars intercalaris diencephali; 13, primordium habenulae; 14, pars dorsalis thalami; 15, pars ventralis thalami; 16, commissura posterior (emintia); 17, commissura posterior (nucleus); 18, tuberculum posterius (pedunculus); 19, tegmentum dorsale; 20, tectum mesencephali; 21, regio cerebellaris; 22, tegmentum isthmi. Roman numerals indicate nuclei of cranial nerves. Asterisk marks limit between presumptive brain and presumptive spinal cord. (Reprinted by permission from C. O. Jacobson, J. Embryol. Exp. Morphol 7:1–21. Copyright © 1959, The Company of Biologists Limited.)

tive neural retinas as they migrate from the neural plate into the neural fold using time-lapse photography and transplantation techniques. Cleine and Slack (1985) used orthotopic grafts from fluorescently labeled whole embryos to map the fates of cells in younger gastrula embryos. The sophisticated fate map of the neural plate certainly makes one wonder about the poorly understood mechanics of its organization.

Clearly, even as the neural plate forms it becomes progressively determined.

When small pieces are transplanted to ectopic locations, they retain enough information about their origin to give rise to particular neuronal structures characteristic of their original fates and orientations (C. O. Jacobson, 1964). This is not to say that regulative mechanisms are not still at play, for extirpated pieces of the neural plate can be replaced by surrounding plate cells, isolated fragments display fates broader than expected, rotations of pieces can lead to replications of cell types, and translocated pieces can be influenced by surrounding tissue to produce some of the local cell fates. For example, Model (1982) has shown that prospective forebrain neural plate tissue, when transplanted posteriorly, is sometimes capable of giving rise to the giant Mauthner neurons of the hindbrain. Thus it would seem as if the spatial determination of the plate were only gradually coming into coincidence with cellular fate, or as C. O. Jacobson (1959) might put it, at this stage "prospective potency" is wider than "prospective signficance."

How cellular commitments are made within the induced neural plate is entirely unclear, but it is likely that cell–cell interactions within either the neural plate or neural tube are critical. In the axolotl it has been possible to study the electrophysiological properties of neural plate cells. Warner (1973) has shown that neural plate cells are electrically coupled, presumably through gap junctions, throughout neural plate stages. Thus gap junctions could mediate important specification events. These intracellular recordings also reveal that as the neural tube closes, the membrane potential of the presumptive neural cells and neuroblasts becomes progressively more depolarized, primarily as a result of activation of the sodium–potassium pump (Messenger and Warner, 1979). Experimental blockade of the pump using strophanthodine interferes with subsequent neuronal differentiation (Messenger and Warner, 1979). In *Xenopus* it has been shown that uncoupling single blastomeres using intracellular injections of antibodies to gap junctions at early cleavage stages can result in deformations of pattern and organogenesis (Warner, et al., 1984). It could be revealing to follow the fates of single neural plate cells injected with these antibodies.

As neurons begin to emerge in the neural tube, they gain the ability to generate action potentials. The maturation of the action potential in axolotls is being studied by patch-clamping single neural plate cells differentiating in culture (Barish, 1986). The development of new outward potassium conductances appears to be a critical feature in shaping the action potential for a function in communication. Although some cells of the CNS are already born in neural plate stages, others continue to divide as the neural tube forms in the tailbud embryo. In certain proliferative regions of the growing larva and adult axolotl, neurons are still being added to the brain (Richter and Kranz, 1981).

NEURAL DIFFERENTIATION, PATHFINDING, AND THE TAILBUD AXOLOTL

The tailbud ambystomatid embryo, because of its large size and its surgical resilience, has been one of the great model systems in which to study neural differentiation, particularly with regard to the formation of neural connections. By this stage of development, extirpation of pieces of the nervous system can no longer be compensated for by the regulatory mechanisms that replace missing pieces of the neural plate. It is also during these stages that neurons send out growth cone-tipped axons that are responsible for wiring up the neural circuits subserving behavior.

Coghill (1929) realized the potential of the tailbud embryo for yielding insights into the neural basis of particular behaviors. He reasoned that correlative neuroanatomical and behavioral observations might reveal the developing circuitry underlying newly acquired behaviors. For example, the first movement a tailbud axolotl makes is the bending of its head away from a light touch to the skin. As the animal continues to develop, this flexure spreads down the body, so that when touched, the tailbud throws itself into a tight coil. A little later the embryo is capable of generating the S shapes that are the elements of undulatory locomotion.

The early-tailbud embryo, though nonmotile, has functional muscle cells contacted by motor nerves, and the skin is already innervated by sensory neurons. Correlated in stage with the first flexures of the neck is the differentiation of a set of anteriorly located neurons that sit on the floor of the spinal cord (and are therefore called "floor plate neurons"), which complete the circuit by joining the sensory elements on the one side of the cord to the motor neurons on the other (Fig. 15–2). Muscular contractions are carried down the body along the developing unilat-

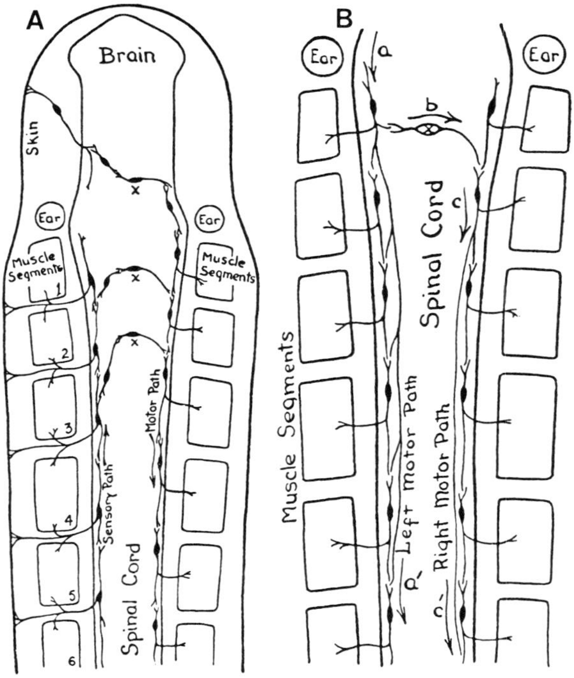

Fig. 15–2. (A) A diagram of the mechanism that accounts for the coil reaction. A stimulus in front of the ear would excite an afferent neuron, which, in turn, would excite a commissural cell in the floor plate (X, floor plate cell). From there the signal is transmitted to the motor tract of the other side. (B) Diagram of the neuromotor mechanism of swimming in *Ambystoma* embryos. The signal a–a′ travels down the left motor column and causes the first flexure. Behind the ear, a motor neuron transmits an impulse through the floor plate pathway b, which then excites the second flexure through the right motor column c–c′. (From *Anatomy and the Problem of Behaviour* by G. E. Coghill. Copyright © 1929, Cambridge University Press. Reprinted with permission.)

eral motor paths in the cord. As the animal begins to make "S" movements, new connections appear to be formed from anterior motor neurons to the floor plate cells. In this way a motor neuron firing on one side of the body can—with a small delay—trigger firing of motor neurons on the other side; thus the beginning of the first flexure is immediately followed by a second, opposite flexure. This succession generates the posteriorly propagating and self-perpetuating series of S shapes that are the basis of swimming (see Fig. 15–2).

As behaviors become more sophisticated the neural circuits underlying them become increasingly complex, but it may still be possible using such a correlative approach to find the neural connections responsible for the behavior. For instance, it is not until early larval stages that the axolotl first begins to orient to a pattern of rotating stripes. In our lab, K. Fahrner (unpublished observations) has shown, using injections of tritiated proline into the developing eye, that fibers from the retina first grow into the basal optic root at approximately the same stage that this optomotor response develops, implicating the basal optic root in this behavior. Of course, such correlations do not constitute a proof, and verification of functions suggested by these neuroanatomical-behavioral coincidences is needed.

The most important aspect of the tailbud axolotl with respect to developmental neurobiology is the ease with which microsurgical procedures can be used to ask particular experimental questions. Detwiler, Twitty, Harrison, and many others have turned to the ambystomatid tailbud embryo for their experiments. Several of their landmark experiments examined neuron and target interactions through transplantations. In these studies, either pieces of neural tissue or their targets were transplanted either to unusual locations on the embryo (heterotopic transplants) or to an embryo of a different species (heteroplastic transplants). Detwiler (1936), for example, used heterotopic grafts of the tailbud nervous sytem to show that higher centers influence the number of neurons in the spinal cord. Adding a supernumerary medulla leads to hyperplasia in the cord, while isolating the spinal cord from descending inputs leads to a reduction in cell proliferation. Similarly, Twitty (1932) showed that increased retinal input to the tectum, caused by the heteroplastic transplantation of optic cups from the large-eyed *Ambystoma tigrinum* to the smaller eyed *Ambystoma punctatum,* leads to cellular hyperplasia in the tectum. Similar studies showed that the peripheral sensory field could also influence the number of cells in the innervating neural tissue. For instance, spinal ganglia adjust their size and cell number to the amount of their peripheral targets (Detwiler, 1936). Through studies like these on *Ambystoma,* it became obvious that trophic influences were responsible to a large extent for controlling the morphogenesis and cell number in most regions of the nervous system. We now know that such trophic influences may exert their effects either by changing the rate of cell proliferation or, more often, by preventing naturally occurring cell death (Berg, 1982).

In 1937, Twitty observed that when an optic cup was transplanted from the California newt *(Taricha torosa)* to an axolotl embryo, the host embryo became temporarily paralyzed. In parabiotic joinings of the two species, the newt twin grew normally, swam, and ate, while the axolotl twin remained immotile and grew at the expense of the newt's digestion. By transplanting different pieces of tissue back and forth between the two embryos, Twitty (1937) demonstrated that the newts concentrated a toxin in their yolk that acted on the axolotl nervous system. Twenty-five years later this toxin was chemically analyzed and found to be the same as that found in the Japanese puffer fish known as tetrodotoxin. This toxin acts as a specific inhibitor of the voltage-gated sodium channels that are essential for action potentials. We used this information to test the idea that impulse activity was involved in early phases of neurogenesis and differentiation. In a first experiment (Harris, 1980), optic cups of tailbud axolotls were transplanted as extra eyes to newt embryos of the same stage. Such eyes, though bathed in blocking concentrations of tetrodotoxin throughout the period of retinal cell proliferation, differentiation, and axogenesis, formed normal, properly layered retinas that sent a well-ordered axonal projection to the tectum. Thus neural impulse activity did not seem to be involved in any of these features of neural development. In a second experiment (Harris, 1984), we tested the possibility that impulse activity originating from the host brain could be responsible for guiding incoming neurons to their appropriate targets. Eye cups were transplanted heterotopically from one axolotl to another, and then these tailbud embryos were parabiotically joined to newt embryos. In these cases, although the transplanted eye as well as the entire host nervous system were under the influence of the toxin throughout these early stages of brain differentiation, there was no observed derangement of the retina or the ability of retinal fibers to find their targets.

THE FUNDAMENTAL VERTEBRATE PLAN: THE AXOLOTL BRAIN

One of the key features of the axolotl as an experimental animal for neurobiological research is that it has a simple nervous system. It has relatively few cell groups and a fundamental circuitry. It seems to lack the highly specialized neural machinery that would take it off the main line of vertebrate evolution. This may be partly because the urodeles, of which the axolotl is a member, are relatively early and slowly evolving vertebrates. Since the first urodeles had to attain neural adaptations for terrestrial locomotion, predation, and a variety of other challenges that life on land provided, the nervous systems of their least-altered modern descendents cannot represent the most primitive vertebrate form. Yet among the terrestrial species that most neurobiologists study, the urodele is in a uniquely primitive position. The second key to the primitive nature of the axolotl brain may be a derived feature, having to do with paedomorphosis. The axolotl, because it is neotenic, becomes sexually mature at a relatively immature stage of somatic development. Its ancestors were capable of a metamorphosis that enabled them to live as highly terrestrial individuals. The axolotl line of evolution has lost this ability and, compared to ancestral forms, retains the larval features of its fully aquatic existence throughout its life. This means that its brain may also belong to a more embryonic, and therefore generalized, vertebrate pattern and physiology. Axolotls can be forced to metamorphose by introducing thyroxine into the aquarium water (Chapter 18). As a result they develop more mature and terrestrial somatic traits including particular pathways in the brain. For instance, Serling and Braendle (1982) found that the indirect ipsilateral projection from the retina to the tectum that may be involved in the binocular vision of a terrestrial predator, expands under induced metamorphosis.

The embryonic simplicity of the adult axolotl brain is reflected in the fact that neurons in almost all areas of the CNS, with the exception of the cerebellum, are paraventricular. The adult neurons, instead of migrating away from the zone of their birth as happens in higher vertebrates, retain their position at the germinal zone. Thus the embryonic mammalian brain looks very much like the adult axolotl brain, both in whole-mount views and in cross section.

Detailed neuroanatomical studies on the brain of the tiger salamander continue to be a major source of information about the axolotl brain. Except for the fact that the adult tiger salamander progresses through metamorphosis and thereby becomes more specialized, the two brains are nearly identical. Figure 15–3 shows the *Ambystoma tigrinum* brain. Herrick's book, *The Brain of the Tiger Salamander* (1948), is a classic source of information on the basic circuitry of the ambystomatid brain. Since Herrick's work there have been a series of anatomical, physiological, and neurochemical papers on the organization of the ambystomatid brain to which

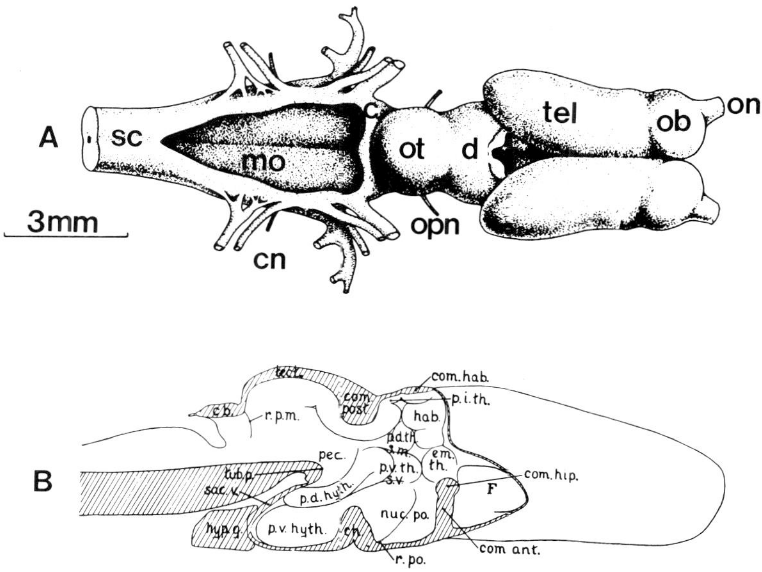

Fig. 15–3. (A) Dorsal view of the adult brain of *Ambystoma tigrinum.* on, Olfactory nerve; ob, olfactory bulb; tel, telencephalon; d, dorsal thalamus; opn, optic nerve; ot, optic tectum; c, cerebellum; mo, medulla oblongata; cn, cranial nerves; sc, spinal cord. (From R. G. Northcutt and E. Kicliter, Organization of the amphibian telencephalon. In *Comparative Neurology of the Telencephalon.* Copyright © 1980, Plenum Press. Reprinted with permission.) (B) Median view of the brain of an adult *A. trigrinum.* cb., Cerebellum; tect., tectum; com. post., posterior commissure; com. hab., habenular commissure; r. p. m., recessus posterior mesencephali; p. d. th., pars dorsalis thalami; p. i. th., pars intercalaris thalami; hab., habenula; ped., peduncle; s. m., sulcus medius thalami; em. th., emenentia thalami; p. v. th., pars ventralis thalami; s. v., sulcus ventralis thalami; com. hip., hippocampal commissure; F, forebrain; com. ant., anterior commissure; nuc. po., nucleus preopticus; r. po., recessus preopticus; ch., optic chiasm; p. d. hyth., pars dorsalis hypothalamus; p. v. hyth., pars ventralis hypothalamus; tub. p., tuberculum posterius; sac. v., hypophysis pars nervosa; hyp. g., hypophysis pars glandularis. (From *The Brain of the Tiger Salamander, Ambystoma tigrinum,* by C. J. Herrick. Copyright © 1948, The University of Chicago Press. Reprinted with permission.)

the interested reader is also referred (Opdam and Nieuwenhuys, 1976; Kokoros and Northcutt, 1977; Northcutt and Kicliter, 1980; Inham and Güldner, 1980; Sims, 1977; Gruberg and Harris, 1981; Fritzsch, 1981a).

Besides serving as a model for the generalized vertebrate brain, some parts of the salamander CNS, because of their large cells or accessibility, have served as important material for functional studies of the nervous system. This is particularly clear in the case of the retina, where urodele species including the axolotl (Waloga and Pak, 1978; Grabowski and Pak, 1975; Dvorak, 1984; Attwell et al., 1985b, 1987) have afforded opportunities for intracellular recordings from all the cell types in the retina. This has led, on the one hand, to a basic understanding of the functional circuitry of the retina, and on the other to a better understanding of the cellular physiology of individual cell types. Similarly, because the cells are large and accessible, axolotls and their relatives have been used in pioneering studies on sensory transduction in hair cells of the lateral line (see Chapter 17) and olfactory receptors (Trotier, 1986).

Munz et al. (1984) have shown, using single-unit recordings, that the axolotl possesses sensory neurons in the ampullary organs around the head that detect weak electrical signals in the water. This electric sense may thus be an ancestral vertebrate adaptation that most "higher" vertebrates seem to have lost. Further investigations into the transduction apparatuses in these cells should prove very interesting.

REGENERATION IN THE MATURE NERVOUS SYSTEM

The very fact that the salamander is primitve and paedomorphic may be part of the reason why the animal has tremendous powers of regeneration. Regeneration of CNS pathways and tissues would be a medical miracle for an adult human, but for an adult axolotl it is routine. Studying what it is about this lower vertebrate that allows regeneration of nerve fibers might one day lead to practical benefits. One dramatic example of the regenerative capacity of the adult axolotl is its ability to regenerate its telencephalon. After the ablation of nearly the entire cerebral hemisphere, including severance of the olfactory nerve, the axolotl's forebrain regrows from the caudal cut end to the anterior olfactory bulb, which then reestablishes functional connections with the olfactory nerve stumps (Winkelman and Winkelman, 1970; Kirsche and Kirsche, 1964, 1968; Richter, 1968; Mellinger, 1975). New olfactory glomeruli form. If a piece of telencephalon from another animal is substituted by transplantation for an ablated piece, the transplant can heal into place and establish a relatively normal-looking cerebral hemisphere. Other parts of the CNS can also regenerate. The retina re-forms through the proliferative activity of any remaining pigmented epithelial cells (Svistunov and Mitashov, 1983, 1984; Yew, 1985).

In addition to being able to regenerate certain missing pieces of the nervous system by cell proliferation, the adult axolotl is capable of regrowing severed peripheral nerves and central tracts. For instance, the regeneration of an optic nerve or a limb nerve is a capacity that all lower vertebrates seem to possess. Only the salamanders, however, can regenerate entire lost limbs in the adult (Chapter

14). As a result, there has been a significant amount of attention paid to the innervation and reinnervation of the axolotl limb. One amazing feature of regenerating limb nerves in the urodele is that particular motor neurons are able to reconnect with great precision to the same muscle targets that they originally supplied (Freeman and Davey, 1986). This happens not only when the nerve alone is cut but also when the entire limb is amputated (reviewed in Landmesser, 1980). There appear to be several mechanisms that help motor neurons reestablish their appropriate connections. Local cues along the pathway from the spinal cord to the limb muscles may guide regrowing nerves directly to their targets. Evidence for this is that the nerve pathways of a regenerating limb nerve are very similar to the original stereotyped pathways (Piatt, 1957), and even in the absence of a limb or in cases where a limb has been rotated or supernumerary limbs formed, the proximal pathways at least appear to be essentially normal (M. R. Bennett and P. A. McGrath, 1980). If a nerve bundle is forced to traverse limb territory that contains none of its normal pathway cues, then these nerves are likely to innervate inappropriate muscles (Stephens and Holder, 1987). Longer distance cues may also operate, for when nerves or particular muscles are misplaced in the axolotl limb, motor neurons are able to find and innervate their original muscles (Fig. 15–4) (Grimm, 1971; Wigston, 1986; Wigston and Kennedy, 1987). This suggests that the muscles themselves must somehow elicit appropriate reinnervation. Finally, it has been shown that when an inappropriate nerve is forced to innervate a particular muscle, it does so only poorly (Holder et al., 1982), and when the displaced original nerve finds its way back to this muscle receiving foreign innervation, it can functionally suppress the synaptic efficacy of the foreign nerve (Wigston, 1980), and may eventually drive off the foreign synapses altogether (reviewed in Mark, 1980). All of these mechanisms appear to be involved in the remarkable specificity of regenerating motor axons. Still, we do not understand the cellular and molecular basis of any of them. Further study of the regenerating axolotl limb nerves should continue to shed light on this intriguing process.

GENETICS AND NEURAL DEVELOPMENT

One of the main reasons the axolotl stands out among other urodeles as a model system for neural development is a growing collection of mutant strains. Some of these mutations, like those that alter pigmentation or nucleoli, do not affect the nervous system directly but can nevertheless be used as cellular markers (see Chapter 22) in studying neurogenesis. Others, like the *eyeless* mutation discovered by Humphrey (1969a), affect the nervous system and either can be studied with the goal of understanding the primary action of the gene in question, or can be used as tools for other experiments. The etiology of the *eyeless* phenotype has been addressed in Chapter 13. Here, I would like to discuss how the *eyeless* mutant has been used to show that retinal axons can travel to the tectum along unusual pathways. Eye cups from normal animals were transplanted to ectopic locations on *eyeless* host embryos (Schwenk and Hibbard, 1977; Harris, 1982). Thus, retinal axons that grew out of the transplanted eyes entered the host brain at regions off the optic tract and were unable to contact normal optic axons that could guide them to their

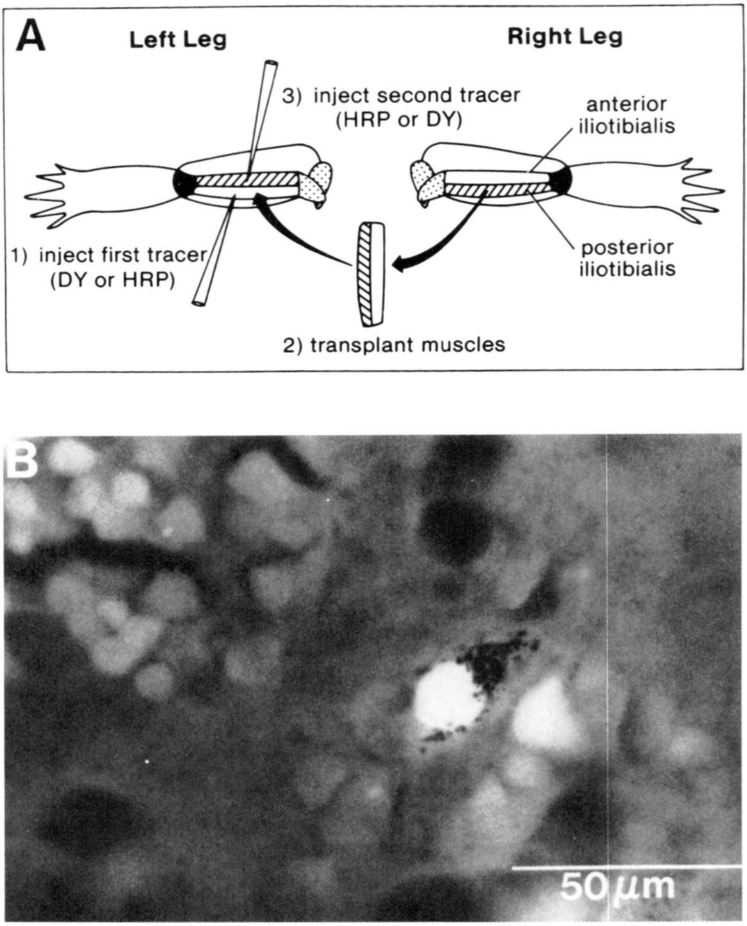

Fig. 15-4. (A) Diagram of an experiment to determine at a cellular level the specificity of motor neuron reinnervation of limb muscles. (1) The first tracer (diamino yellow or horseradish peroxidase) was injected into the left posterior iliotibialis (ILT). (2) After 5–10 days, the left anterior and posterior ILT muscles were replaced by the same pair of muscles from the right side, but in reversed orientation. (3) After reinnervation (50–90 days), the former posterior ILT muscle, now in the normal position of the anterior ILT, was injected with the second tracer. (B) Double-labeled motor neuron in the spinal cord resulting from the experiment. The dark grains are the HRP label; the fluorescent circle is the DY label. (From D. J. Wigston and P. R. Kennedy, J. Neurosci. *7*:1857–1865. Copyright © 1987, The Society of Neuroscience. Reprinted with permission.)

targets. Nevertheless, the axons from the transplanted eyes were able to home in on the host tectum and other visual centers. Sight could thus be restored to these animals, which showed various visual reflexes (Hibbard and Ornberg, 1976). Therefore, in the CNS the cues that axons follow may be dispersed throughout the brain.

In a similar experiment (Harris, 1982), *eyeless* axolotls were used to demonstrate that the coordinated mapping of sensory systems in the tectum is largely developmentally rather than interactively achieved. Eyes transplanted to *eyeless*

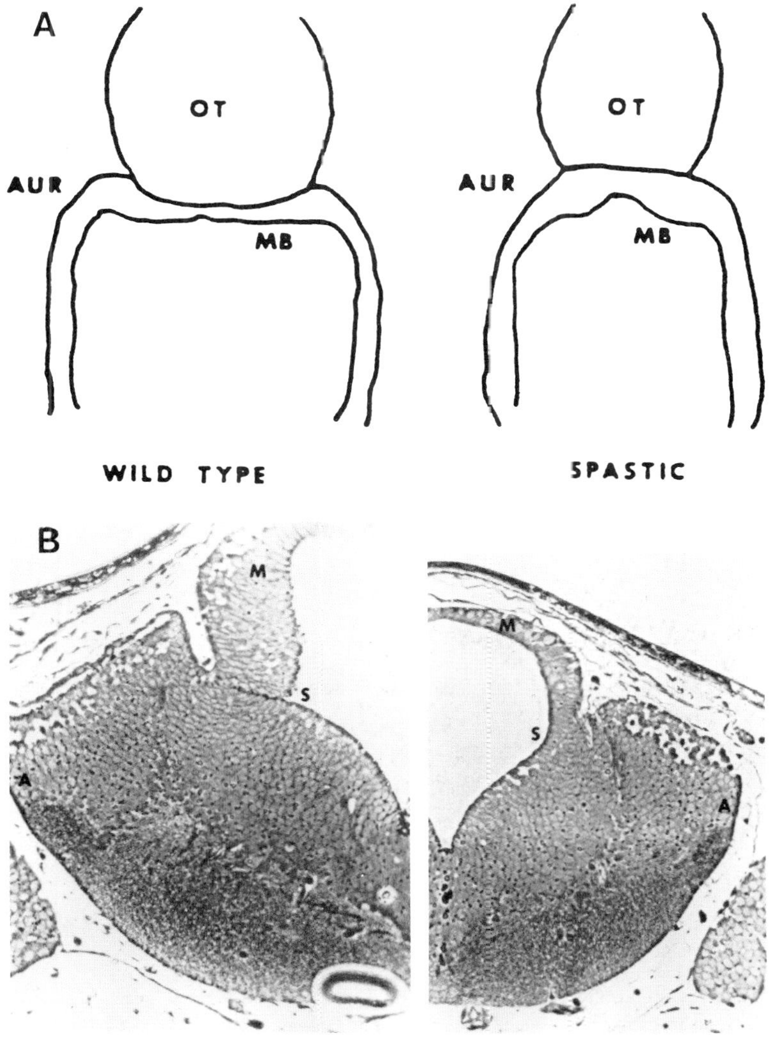

Fig. 15–5. Neuroanatomy of the *spastic* mutant. (A) Dorsal view of whole-brain sketches. In the wild type the cerebellar auricle (AUR) flares rostrally, whereas that of the mutant slopes caudally. OT, optic tectum; MB, median body of the cerebellum. (B) Histological section through the cerebellar region of normal (left) and *spastic* larvae. The posterior midbrain roof (M) is smaller, the isthmic sulcus (S) is nearly absent, and the auricle of the cerebellum (A) is less extended laterally. (From C. F. Ide et al., Analysis of *spastic:* a neurological mutant of the Mexican axolotl. In *Cellular Neurobiology,* Copyright © 1977, Alan R. Liss, Inc. Reprinted with permission.)

hosts in rotated orientations mapped onto the tectum in a rotated sense yet did not disturb the basic orientation of the underlying somatosensory map with which the visual map was now out of register. These two normally aligned sensory maps must therefore form by means that are functionally independent of each other.

Spastic is another interesting axolotl neural mutant, discovered by Humphrey and studied mostly by Ide and his colleagues. Instead of swimming in straight lines when disturbed, *spastic* embryos coil and spin onto their backs (Ide and Tompkins, 1975), apparently as a result of an equilibrium problem. Interestingly, this abnormal behavior starts to appear just at the time Coghill (1929) reasoned that cerebellar functions are beginning to coordinate muscle action. In confirmation of this, anatomical evidence shows a gross morphological deficiency in the lateral extension of the cerebellum (Fig. 15–5); (Ide et al., 1977). A histological analysis of this region shows a disorganization of ependymal cells and neurons. Electrophysiological records demonstrate a shift in the placement of vestibular units that respond to tilting movements (Ide, 1977). Bilateral eighth-nerve lesions or undercutting of the cerebellar auricle in normal axolotls leads to a phenocopy of the behavioral defect in *spastic* (Elbert et al., 1983). The underlying cellular defect that leads to these neurological problems is not clear. Ide et al. (1977) suggest faulty morphogenetic movements or cell–cell interactions during development as possible primary sites for the mutant action. Further work is needed to evaluate these possibilities.

CONCLUSIONS

If one thing emerges from this review, it is that the axolotl has a long-standing relationship with the frontiers of neurobiological research, from the early years of the twentieth century to the present. Because of its special features, the axolotl will undoubtedly remain a central and leading contributor to our understanding of the formation, the function, and the regeneration of vertebrate nervous systems. The future of axolotl neurobiology will mesh with progress in molecular and cellular technology and the current issues of evolutionary and developmental biology.

16
The Amazing Mauthner Cell

PAT G. MODEL

Working with the extraordinary Mauthner cells of the axolotl has been fun, exciting, and enlightening. I am happy to document their exceptional story for you here.

Mauthner cells (M-cells) constitute a unique class of central neurons: they occur as a single pair in the medulla of premetamorphic amphibians and fish, and are precisely located at the level of entry of the vestibular cranial nerve (nVIII) (Fig. 16–1). They are interneurons whose job is to integrate the escape reflex. The time of origin of the axolotl M-cell is late gastrulation (Model, 1978a); its determination occurs soon thereafter, in the very early neurula (Schlenoff and Model, 1987). Because the M-cell is striking in size and distinctive in form, it is easily distinguished morphologically. The axolotl M-cell is polarized: it has a very large lateral dendrite and a large medial one, and its axon always leaves the cell at its mediodorsal margin (Fig. 16–2). The long axis of the M-cell is oriented almost perpendicular to the long axis of the CNS, and thus the origin of the lateral dendrite defines its lateral pole and that of the medial dendrite and axon, the medial pole. With respect to afferent connections, there is a regional organization of synapses on the surface of the M-cell and thus there is topographic order at the level of a single cell. For example, a localized group of dendritic branches projects from the proximoventral surface of the lateral dendrite (Fig. 16–2D); these branches receive synapses from the ipsilateral nVIII. Afferent connections appear during the period of M-cell differentiation; the earliest are those of the vestibular axons (Leghissa, 1941). Later, by early feeding stages, these synapses are identifiable in the light microscope (Model, 1978a) and in the electron microscope (Kimmel and Schabtach, 1974; Model and Wurzelmann, 1982) as large club endings (Fig. 16–3). M-cells differentiate early (Kimmel and Model, 1978) and in the axolotl embryo may be approached experimentally during many stages of development (e.g., Model, 1978a, 1982; L. A. Goodman and P. G. Model, 1988).

THE NEURAL PLATE AS A MORPHOGENETIC FIELD FOR THE CNS

A problem of considerable interest to the developmental neurobiologist is that of the generation of spatial pattern in the early CNS. The establishment of the neural

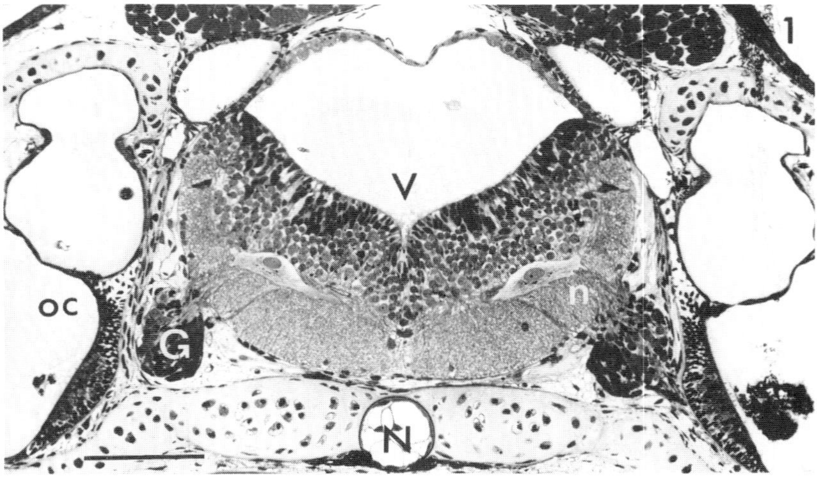

Fig. 16–1. Light micrograph of an Epon-embedded, 4 μm thick, toluidine blue-stained cross section through the medulla of a control 21-mm larva at the level of the M-cells, nVIII ganglia (G), and otic capsules (oc). n, nVIII; V, fourth ventricle; N, notochord. Bar = 200 μm. (From L. A. Goodman and P. G. Model, 1988.)

axis is a consequence of the morphogenetic movements of gastrulation. Thus, in the amphibian embryo, as in other vertebrates, the mesoderm that comes to underlie the prospective CNS ectoderm is responsible not only for its neuralization but for its spatial patterning as well (Spemann, 1931). By the end of gastrulation, the prospective neuroepithelium is committed to form neural tissue, but the time at which the spatial pattern of the various parts of the CNS is specified within the neuroepithelium itself remains unclear (Cooke, 1980). In fact, there are a number of contradictory observations in the literature regarding the degree of determination of the neural plate in amphibian embryos. For example, in some hands (Roach, 1945; C. O. Jacobson, 1976), unilateral rotations of pieces of neuroepithelium through 180° results in the inversion of the parts of the nervous system formed from the grafts. In the hands of others (Sladecek, 1955; Detwiler, 1940; Model, 1978b), however, the result is opposite: a rotated piece of neuroepithelium reorients to develop normally with respect to the nervous system, indicating that the ability to regulate remains. Positive findings of regulatory capacity at any stage must outweigh negative findings, since they are the more meaningful result (Cooke, 1980). One aspect of the determination of the CNS *is* its axial polarization, an event that has not yet occurred by midneurula stages in the axolotl embryo (Model, 1978b). The polarization of an embryonic tissue can be interpreted to mean its compartmentalization, or loss of equipotentiality, and, in fact, regionalization of the developing CNS has not occurred in the early midneurula either (Model, 1982). While there is considerable controversy with respect to the time at which compartmentalization actually occurs (e.g., Cooke, 1985 versus M. Jacobson, 1983), the above observations indicate that it must happen at some point after the neural folds begin to close.

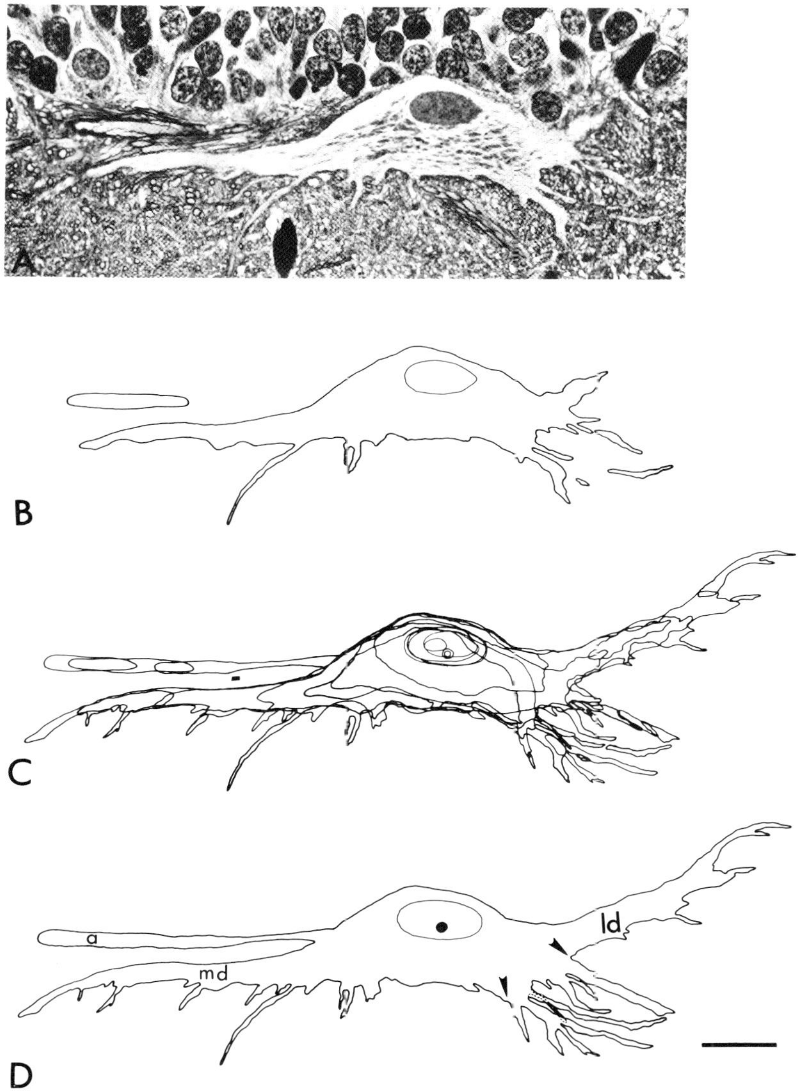

Fig. 16–2. M-cell reconstruction. (A) One of a series of light micrographs of cross sections through an M-cell. (B) Perimeter outline of the M-cell shown in (A). (C) Superimposed contours of the M-cell in the series of light micrographs. (D) Perimeter of the superimposed contours in (C) showing the proximoventral surface and branches of the lateral dendrite (ld) that receive nVIII terminals (between arrowheads). md, Medial dendrite; a, axon. Bar = 40 μm. (From L. A. Goodman and P. G. Model., 1988.)

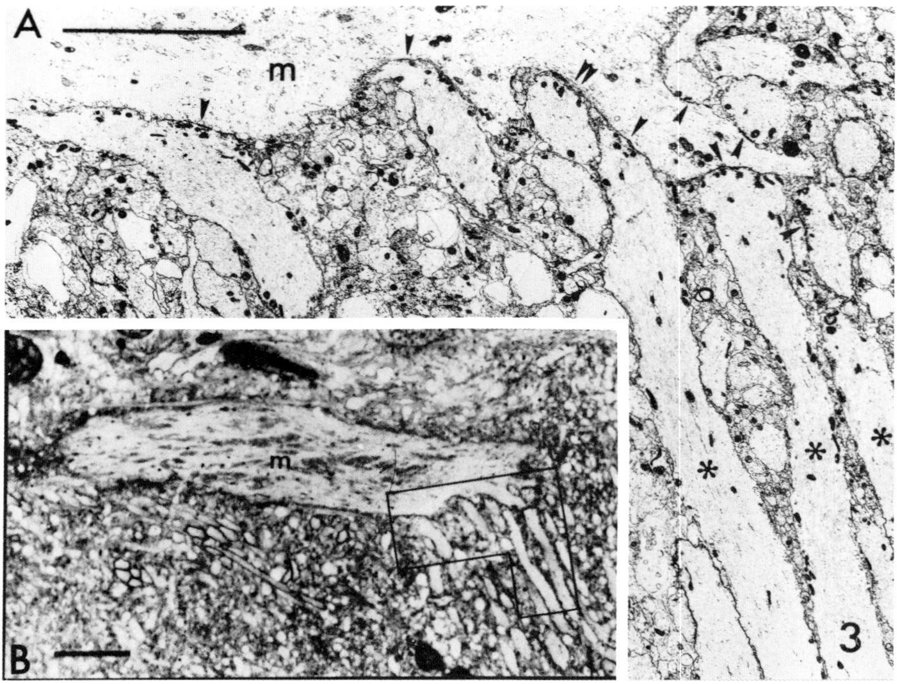

Fig. 16–3. (A) Electron migrograph of the region of vestibular contacts on the M-cell shown in the light micrograph (B). m, M-cell; *, nVIII axons. Vestibular club endings on the M-cell (arrowheads) are easily identified because they are very large (3–12 μm in diameter; Model and Wurzelmann, 1982), they make morphologically mixed (chemical and electrical) synaptic contacts (Model and Wurzelmann, 1982), they have mitochondria grouped close to the contact region, and they contain densely packed, highly organized neurofilaments. Bars = 10 μm. (From L. A. Goodman and P. G. Model, 1988.)

Embryonic regulation occurs within populations of cells that are organized as sheets; it is defined as the ability to compensate for rearrangements or perturbations within the sheet by a change in the constituent cells' fates to give rise to a normal pattern. This ability is lost as the progression to the determined mosaic or compartmentalized state proceeds. M-cells ordinarily originate in the embryonic hindbrain, but can arise from prospective forebrain–midbrain that has been reprogrammed to differentiate as medulla by unilaterally substituting it for prospective hindbrain in the neural plate of the axolotl neurula (Model, 1982). Spatial pattern *is* present in the CNS at neural plate stages—e.g., pieces of neuroepithelium when transplanted to a non-neural region of an embryo differentiate more or less as they would have had they been left in place, according to their original developmental programs (Von Woellwarth, 1952)—but that extensive regulation can occur in the neuroepithelium of the early midneurula (Model, 1982) indicates that the spatial pattern has not yet been irrevocably fixed. The ability to regulate is a property of morphogenetic fields, and thus the neural plate can be thought of as a field for the

CNS. It is widely held that the fate of uncommitted cells depends on their positions within a field, and Wolpert's theory of positional information (1971) suggests a means by which position could affect determination. Accordingly, it is reasonable to suppose that the differentiation of M-cell-containing medulla from prospective forebrain–midbrain that has been translocated to the prospective hindbrain region of the neural plate (Model, 1982) is due to the different positional signals that it receives from the surrounding tissues by virtue of its new location in the CNS field. Thus, the generation of M-cells in such medullae is likely to be a consequence of the position-dependent determination of uncommitted cells. Despite the fact that the M-cell is specified early, prior to the time that the neural plate has become a definitive structure (Schlenoff and Model, 1988), those signals responsible for its specification must remain within the neuroepithelium beyond the time at which they appear to be required (Model, 1982).

THE FORMATION OF SPECIFIC NEURONAL CONNECTIONS

The spatial patterning of synapses in the vertebrate CNS is highly ordered: the projection of retinal ganglion cells to the optic tectum in fish (Attardi and Sperry, 1963) and amphibians (M. Jacobson, 1968) is an example of this specification. How specific connections are formed during development remains unknown, but there are several possibilities as to the nature of the mechanisms by which neurons form appropriate interconnections. Some investigators believe that the surfaces of neurons are uniquely labeled (after Sperry, 1963), whereas others believe that a well-ordered temporal and spatial sequence can account for the apparent accuracy with which neurons connect (as argued by Kimmel and Schabtach, 1974). While it is unlikely that any one factor is responsible for the specificity of neuronal connections, it seems clear that an affinity between neurons is essential to their interaction (e.g., C. S. Goodman, et al., 1984).

Sperry's chemoaffinity hypothesis of neuronal specificity (1963) suggests a mechanism by which the patterning of connections might occur: the acquisition of a unique cytochemical label by a differentiating neuron depends on its prior position within a morphogenetic field, and the later formation of neuronal connections is attributed to specific cytochemical affinities between neurons. Sperry's hypothesis can be extended to include the possibility that regional localization of synaptic input on individual neurons is due to regional specification of the cell surface. In any case, the accumulating evidence (e.g., C. S. Goodman, et al., 1984) supports an interpretation of neuronal specificity based on a mechanism of intercellular recognition, with specificity-conferring molecules being located on the surfaces of the interacting cells.

As already noted, M-cells receive input from several groups of neurons whose axons terminate at specific loci on its cell body, initial segment, or dendrites. For example, axolotl nVIII synapses are restricted to the proximoventral surface and branches of the cell's lateral dendrite, and lateral line terminals are confined to the dendrite's dorsolateral branches. In addition, different afferent populations are restricted to growth along particular tracts of the brain: normally, the nVIII tract intersects the ventral branches of the lateral dendrite, while the ventral-most lateral line tract intersects its dorsolateral branches; neither of these afferents projects to the contralateral M-cell (David and Model, 1984). Mechanisms other than one based soley on position–time appear to be responsible for synaptic site specificity: disruption of spatial (David and Model, 1982a,b) and temporal (Leber and Model, 1981, 1982) relations between the developing M-cell and its afferents, prior to the time of nerve outgrowth, has no effect on the subsequent patterning of synaptic connections. Further experiments (David and Model, 1984) have shown that following the displacement of the developing vestibular primordium and associated nVIII and lateral line ganglia along the anteroposterior axis, axons from both ganglia enter the brain at ectopic locations, course rostrocaudally in tracts indistinguishable from those of controls, and terminate on the appropriate target regions of the ipsilateral M-cell. The data indicate that axon growth paths are restricted and may organize connectivity patterns by limiting the access of afferents to particular targets (mechanism of axonal segregation). Chemoselective processes operative at the target site could then contribute to selective synaptogenesis (mechanism of chemoaffinity or target recognition). While this hypothesis does not require the M-cell to have a chemically mosaic membrane, it is clear that some target recognition must occur; thus, whether target recognition refers to the M-cell as a whole or to individual regions of its surface remains unknown.

Fig. 16–4. Contours of a reconstructed pair of M-cells. The upper cell was superinnervated by the implantation of an extra vestibular primordium; the lower cell is its contralateral control. The branching pattern of the proximoventral surface of the lateral dendrite of the experimental M-cell (between arrowheads) is greatly increased over that of the control. md, Medial dendrite; a, axon; ld, lateral dendrite. Bar = 50 μm. (From L. A. Goodman and P. G. Model, 1988.)

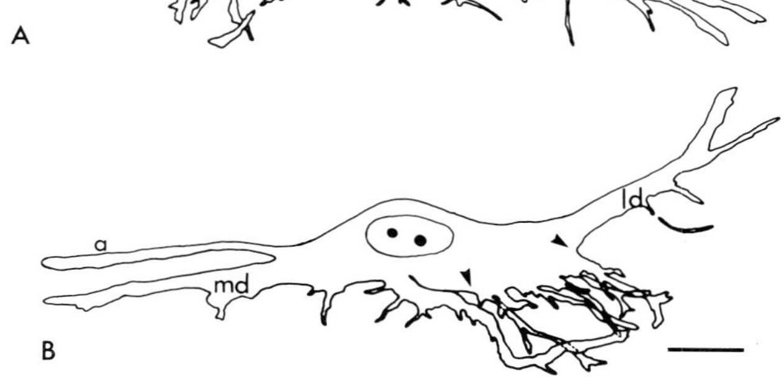

TRANSSYNAPTIC STIMULATION AND M-CELL MORPHOGENESIS

Another unresolved question concerns the degree to which the morphogenesis of a neuron depends on extrinsic factors. For example, there is a correspondence in time between the growth of the M-cell lateral dendrite and the formation of the first synapses (which are of vestibular origin) upon its surface (Leghissa, 1941). Other observations suggest that synaptic contacts induce local growth of dendrites: during periods in which the lateral dendrite elongates rapidly and many ventrally directed branches are formed, numerous synapses of a variety of types appear (Jacoby and Kimmel, 1982). As it turns out, ingrowing axons *do* have an essential role in regulating the elaboration of a normal dendritic branching pattern by target neurons during development: superinnervation of the M-cell, brought about by the implantation of an extra vestibular primordium, results in a large increase in the number of appropriately localized club endings and a significant enhancement of dendritic branching in the region of the lateral dendrite that normally receives nVIII input (Fig. 16–4) (L. A. Goodman and P. G. Model, 1988). nVIII deprivation, on the other hand, results in the reduction or loss of dendritic surface in the same region (Kimmel et al., 1977; L. A. Goodman and P. G. Model, 1988). Afferent innervation is also necessary for the maintenance of the M-cell's dendritic branches: Following transection of nVIII in young axolotl larvae, degeneration of vestibular afferents brings about a loss of postsynaptic dendritic surface in the region deprived of club endings (Covell and Model, 1984). Restoration of the proximoventral branches occurs only when regenerating nVIII axons reestablish synaptic contact with the M cell (L. A. Goodman et al., 1988); thus, regenerating axons stimulate postsynaptic dendritic growth.

EPILOGUE

I hope the reader will have become convinced that the axolotl and its amazing Mauthner cells provide a unique and remarkably advantageous system for the study of problems having to do with developmental neurobiology as well as with developmental biology in general. Obviously there is much to be done; let's get on with it.

ACKNOWLEDGMENT

Were I a poet, I would write a lyric poem about the axolotl smile.

17

The Lateral Line System

MICHAEL J. LANNOO AND STEVEN C. SMITH

The lateral line system of fishes and aquatic amphibians forms part of the VIIIth cranial nerve complex. It contains both mechanoreceptors (neuromasts), designed to detect water perturbations, and electroreceptors (ampullary organs; see Fritzsch 1981b) that detect low-frequency direct-current fields. Workers in fields as diverse as neuroanatomy, sensory physiology, and evolutionary biology have found the lateral line system ideal for their studies (e.g., Dijkgraaf, 1962; M.V.L. Bennett, 1967; Flock, 1971; Russell, 1976; Northcutt, 1986a, Lannoo, 1987a,b, 1988; see also Chapter 15).

The general development of the lateral line system was first elucidated in a series of classical experiments by Harrison (1904) on *Rana* tadpoles and by Stone on *Ambystoma maculatum (Amblystoma punctatum)* in the 1920s and 1930s. Some subsequent work has been done on the comparative embryology (Holmgren and Pehrson, 1949), migration of axonal growth cones (Metcalfe, 1985; Metcalfe et al., 1985), evolutionary and developmental origins of mechano- and electroreceptors (Bullock et al., 1982; Northcutt and Gans, 1983; Northcutt, 1986b), and response of neuromasts during regeneration (M. R. Wright, 1947, 1951) and metamorphosis (Shelton, 1970, 1971; Fritzsch et al., 1984; Wahnschaffe et al., 1985). Additionally, Winklbauer and Hausen 1983a,b, 1985a,b) have described a developmental model of the lateral line system of *Xenopus laevis* based on rates of cell division and the location of stem cells. This model is both simple and elegant, and if it can be easily modified to explain phylogenetic diversity, it may provide evolutionary biologists with a model system for studying a variety of heterochronic events, which are believed to be important in lateral line evolution (Webb, 1985).

Aside from these studies, little work has been done on the development of the lateral line system. Several general questions that remain unanswered relating to the migratory nature of lateral line nerve and receptor primordia, and the establishment of polarity differences in neuromasts within and between lines, appear to be best studied in urodeles. These questions should prove to be important in understanding some of the fundamental processes of vertebrate development.

ANATOMY AND PHYLOGENETIC DIVERSITY

As mentioned previously, the lateral line system of fishes and aquatic amphibians is composed of mechanoreceptive neuromast organs and electroreceptive ampullary organs (see reviews in Bullock and Szabo, 1986; Northcutt, 1986a). Neuromasts are better understood in terms of their sensory physiology and phylogenetic distribution (Northcutt, 1986a); ampullary organs have only recently been discovered in amphibians (Hetherington and Wake, 1979; Fritzsch, 1981b; Fritzsch and Wahnschaffe, 1983; Fritzsch et al., 1984). Neuromasts are directionally sensitive; they tend to be elliptical when viewed on the body surface and are usually most sensitive in a direction parallel to their long axis (Flock, 1965, 1967; see Fig. 17–1). In early fishes and amphibians, neuromasts appear to have been organized into single rows and embedded in covered bony canals (Romer, 1966; Schmalhausen, 1968; Lannoo, 1988). In these early fishes, as in modern fishes with neuromast canals, neuromasts must have been oriented with their long axis *parallel* to the long axis of the canal in order to function (Lannoo, 1988).

The lateral line system of the axolotl is composed of several descrete lines (Fig. 17–2). Basically, there are three on the head—supraorbital, infraorbital, and man-

Fig. 17–1. Schematic diagrams of mechanoreceptive lateral line receptors. (A) A longitudinal section of an individual hair cell, through the single kinocilium (k) and the center of the stereocilia bundles, and its support cells (sc). This hair cell is maximally sensitive to water displacements in the plane of the paper. (After Flock, 1967, which also contains details of the histology.) (B) An oblique view of a neuromast, illustrating four hair cells with their cilia embedded in a gelatinous cupula. Note that kinocilia are located at opposite poles of hair cells, and positioned along the long axis of the neuromast. The direction of maximum neuromast sensitivity is *parallel* to its long axis, indicated by the arrow. (C) Three neuromasts combined to form a stitch. Stitches are formed from the secondary budding of primary neuromasts, which are formed embryonically. In *Ambystoma,* stitches are aligned with their long axis perpendicular to the long axis of their component neuromasts, a morphology that we call transverse stitches.

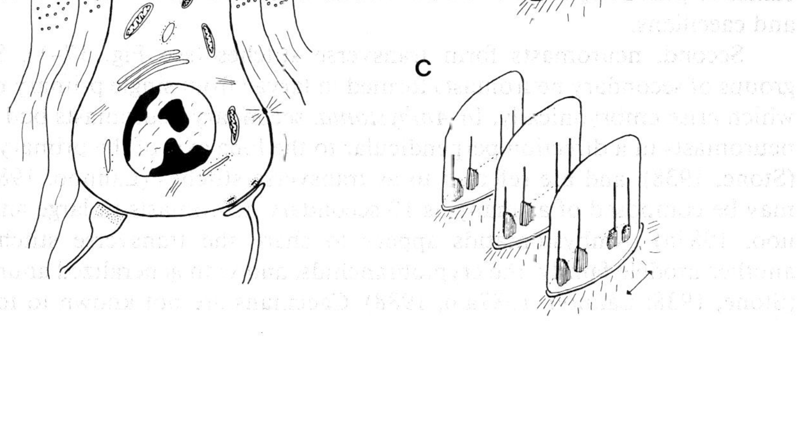

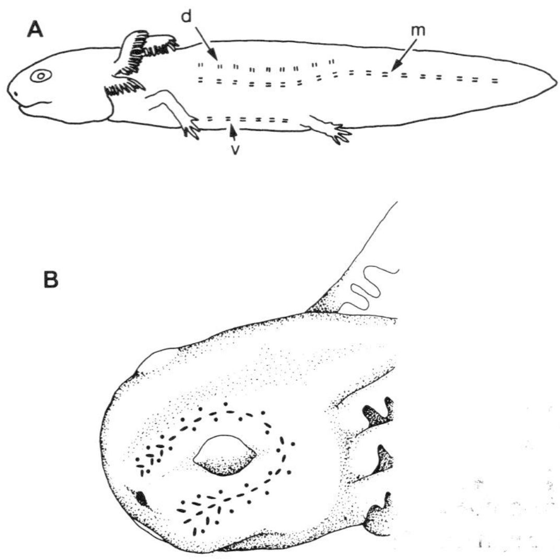

Fig. 17–2. Schematic diagrams showing the location of lateral line neuromasts on the body (A) and head (B) of *Ambystoma*. Body lines shown in (A) are d, dorsal; m, middle; and v, ventral. Individual neuromasts are indicated as lines, with the long axis of each line corresponding to the long axis of the neuromast, and therefore to its axis of maximum sensitivity. Neuromasts form stitches; here each stitch contains two neuromasts. Note that neuromasts, and therefore stitches, in the dorsal and middle body lines together form orthogonally oriented couplets on the trunk, and that the middle body line alone extends onto the tail. In (B), neuromasts are again illustrated as lines, but for clarity we have not illustrated stitches; ampullary organs are illustrated as circles (for more detail of head neuromast arrangement in *Ambystoma* see Lannoo, 1985). Note that circumorbitally, neuromasts form a single row, but that anteriorly, neuromasts form double rows and orthogonal couplets along the nasal snout (supraorbital line), and triple rows and orthogonal couplets along the maxillary snout (infraorbital line). Among amphibians, multiple neuromast rows forming orthogonal couplets are characteristic of urodeles.

dibular (not shown in Fig. 17–2; see Lannoo, 1985, for more detail)—and three on the trunk—dorsal, middle, and ventral. Ampullary organs are associated with these neuromast lines (Fig. 17–2). As with other species of *Ambystoma,* axolotls appear to exhibit three derived lateral line features compared with early vertebrates. First, they have receptors that are epidermally located, rather than restricted to bony canals or pits. They share this trait with all modern amphibians: urodeles, anurans, and caecilians.

Second, neuromasts form transverse stitches (see Fig. 17–1). Stitches are groups of secondary neuromasts formed in larvae from single primary neuromasts, which arise embryonically. In *Ambystoma,* secondary neuromasts bud off primary neuromasts in a direction perpendicular to the long axis of the primary neuromast (Stone, 1938), and are referred to as transverse stitches (Lannoo, 1985). Stitches may be composed of as many as 10 secondary neuromasts in large animals (Lannoo, 1987a). Ambystomatids appear to share the transverse stitch trait with another urodele family, the cryptobranchids, and with generalized anuran tadpoles (Stone, 1938; Lannoo, 1987a,b, 1988). Caecilians are not known to form stitches

(Hetherington and Wake, 1979; Wahnschaffe et al., 1985, Fritzsch and Wake, 1986). Another urodele family, proteids (mudpuppies), form longitudinal stitches (Malbranc, 1876; Lannoo, 1987a,b). Unlike transverse stitches, which are frequently papillate, longitudinal stitches occur in epidermal grooves.

The third derived lateral line feature is that ambystomatids have multiple rows of neuromasts on their nasal and maxillary snout, and along the oral margin of their mandible (Lannoo, 1985, 1987a; see Fig. 17–2). Adjacent neuromasts among rows tend to be orthogonally oriented. Because neuromasts are directionally sensitive, orthogonal couplets constitute the minimum number of neuromasts necessary to detect water movements through 360°. This morphology may therefore enhance the directional discrimination of the neuromast system (Lannoo, 1985). *Ambystoma* appears to share this trait with all urodeles (Lannoo, 1987a).

It appears that all urodeles and caecilians have retained ampullary organs; anurans are not known to have ampullary organs and have apparently lost them concomitant with developing specializations for herbivory (Fritzsch et al., 1984; Lannoo, 1987b).

DEVELOPMENT

Some basic aspects of the lateral line system, such as receptor morphology and head and body line placement, do not vary substantially among fishes and amphibians (Northcutt, 1986a). Likewise, early developmental events appear to be consistent across taxa (Holmgren and Pehrson, 1949; Metcalfe, 1985), allowing results obtained in distantly related species to be generalized.

Lateral line receptors form embryonically from ectodermally derived pre- and postotic placodes (Landacre, 1921; Stone, 1922). While most peripheral nerves are derived from neural crest cells (see Hörstadius, 1950; Weston, 1970; Le Douarin, 1982 for reviews), the ganglia and peripheral nerves that supply the lateral line are derived from the same ectodermal placodes as the receptors (Landacre, 1921; Stone, 1922, 1931). This is a trait that the lateral line nerves share with other cranial nerves, which are also derived from ectodermal placodes (Hörstadius, 1950; Le Douarin, 1982). It is unknown how these placodes arise from the cranial ectoderm. Hörstadius (1950) suggested that neural crest cells induce the overlying ectoderm during their own migration, but the experimental evidence appears inconclusive (Raven and Kloos, 1945).

Once formed, placodes migrate subectodermally, distally from the otic region across the body surface: preotic placodes migrate across the head, while postotic placodes migrate across the trunk and tail (Fig. 17–3; Stone, 1931, 1933, 1938). The number of placodes involved has not been determined with confidence, but Northcutt (1986a) estimates that there are as many as 13 placodes unilaterally in primitive vertebrates.

Experiments on amphibian embryos, especially *Ambystoma maculatum,* have pioneered our understanding of the development of the lateral line system. Harrison (1904) provided the first experimental study of amphibian lateral line development. He produced chimeras by exchanging heads and tails of *Rana palustris* and *Rana sylvatica,* which have different pigment densities and therefore have tis-

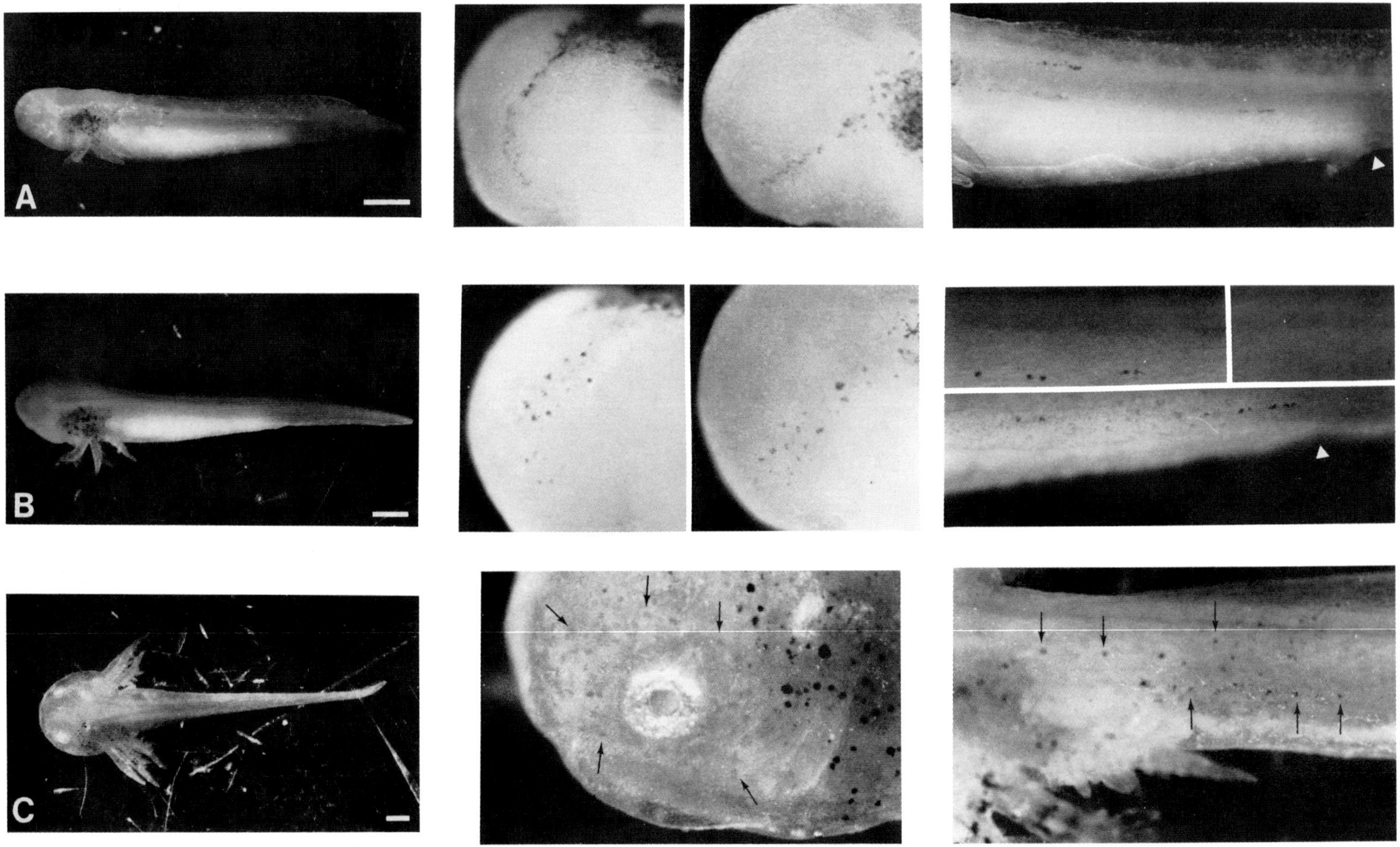
A
B
C

sues that can be distinguished during subsequent development, to show that tissue originating in the otic region migrates across the trunk and tail to form neuromasts. Later, Stone (1922) confirmed Harrison's findings in *Ambystoma.* Stone (1922) defined the positions of several pre- and postotic lateral line placodes by transplanting tissue from embryos vitally stained with Nile blue sulfate into normally pigmented hosts and observing whether the blue primordia migrated. Stone (1922, 1931) determined that the proximal portion of each placode (i.e., nearest the otic region) eventually formed the ganglion cells and peripheral nerves, while the distal portion migrated to form the receptors.

Stone (1924, 1928b,c, 1929) also determined that pre- and postotic lateral line placodes could be substituted for each other without appearing to affect their migratory pathways. Substituted placodes behaved slightly differently, however. While a postotic, middle body line placode appeared to migrate normally in place of a preotic supraorbital placode, a supraorbital placode in place of a middle body line placode would not migrate caudally all the way to the tailtip. Transplanted supraorbital placodes produced, at most, 18 organs along the anterior part of the trunk and, interestingly, Stone (1928c) observed that "the posterior half of the graft shows a somewhat irregular arrangement of the organs." It appears to us from Stone's drawing (1928c, Fig. 12) that the most distal half of this line could contain orthogonal neuromast couplets, normal for the supraorbital line but abnormal for the trunk. Additionally, Lannoo (1985) found that 17 *Ambystoma maculatum* larvae had a mean of about 56 supraorbital neuromasts, about three times Stone's maximum of 18 neuromasts, suggesting either that Stone did not transplant all of the placode, that placodal cell division was inhibited by the surgery, or that the intact supraorbital line is formed by multiple placodes, of which Stone managed to excise only one.

During the course of this series of experiments (Stone 1924, 1928b,c, 1929), Stone (1928c) also determined that lateral line placodes substituted for general

Fig. 17–3. Micrographs showing the results of our transplantation experiments to determine the time course of placodal migration and neuromast development. The otic region of albino and albino axanthic embryos was replaced with the normally oriented otic region of pigmented embryos at Bordzilovskaya and Dettlaff (1979) stage 28. (A) Stage 37, lateral view. Column 1, whole embryo showing pigmented transplant. Column 2, supraorbital (left) and infraorbital (right) neuromasts migrating distally along the head. Note the curved supraorbital line. Column 3, dorsal (top) and middle (lower) body line neuromasts as they migrate caudally. Note that the dorsal body line lags about a quarter body length behind the middle body line. Cloaca is indicated by arrowhead. (B) Stage 38, dorsolateral view. Column 1, whole embryo. Column 2, further development of the supraorbital and infraorbital neuromasts. Note that, compared to (A), neuromasts have become more superficial [i.e., unlike (A), ectoderm does not distort focus], larger, and have spread distally and laterally. Along the body, the dorsal body line has migrated almost to the cloaca (arrowhead); the middle body line has migrated past the cloaca and has moved dorsally. (C) Stage 44 (feeding); the lateral line system is fully developed. Column 1, dorsal view. Column 2, left side of the head showing irridescent retina surrounded by light-colored neuromasts (a few are indicated by arrows), which are apparently losing their embryonic pigment. Note developing melanophores, the result of transplanting some neural crest cells along with the placodes. Column 3, dorsal and middle body line neuromasts (arrows) that have retained their embryonic pigment. This pigment was also lost with time. Bars in column 1 = 1 mm.

cutaneous placodes retain their lateral line function. As well, he found that lateral line placodes, transplanted into the ventral surface of the embryo, where lateral lines do not normally appear, formed ganglion cells but failed to migrate. Eventually the ganglion cells atrophied (Stone, 1928c). Lateral line placodes transplanted into the path of other lateral line primordia migrated only when the migrating host primordia reached the graft (Stone, 1928a, 1929, 1938). These primordia appeared to fuse. The host primordia then ceased or greatly decreased their migration, piling up organs at the proximal boundary of the graft while the transplanted primordia migrated distally to complete the line.

When a supraorbital placode was transplanted in place of a host's dorsal and middle body line placode, remnants of the host placodes would sometimes begin migrating down the body along the path of either the dorsal or the middle body line. The course the transplanted supraorbital line took was always the path *not* taken by the host primordium (Stone, 1928c). Additionally, Harrison (1904) and Stone (1928c) both emphasized that no structures exist within the substrate the placodes migrate through that determine the location of organ deposition. Stone (1928c) stated: "There are no preexisting structures in the pathway over which the migratory lateral-line primordium passes that determine the final positions of the organs of the line."

Stone (1933, 1938) investigated the process of stitch formation and budding from terminal tail neuromasts during regeneration and concluded that they were both the results of the same process. He found that in both cases, new neuromasts were formed from the support cells of existing neuromasts, and that the direction in which the new neuromasts bud is determined by the location of the support cells that supply the new organ. Therefore, when a secondary neuromast is formed dorsal to a primary neuromast (see Fig. 17–1), it is the dorsal support cells that supply the tissue.

Several other conclusions can be drawn from Stone's work. First, transplants into the location of cutaneous placodes show that by the time lateral line placodes appear as ectodermal bulges, their fate as lateral line primordia has already been determined. In this respect, it is informative to compare Stone's results in *Ambystoma* with Knouff's results (1935) in *Rana.* Whereas Stone (1922, 1933), concluded that the primordia for each line arise separately, Knouff (1935) found that the numerous pre- and postotic placodes arise from a single otic placode that subsequently breaks up to form the primordia for the various lines. Separate placodes may arise the same way in *Ambystoma,* but the larger combined placodes may not form an observable bulge.

A second conclusion that emerges (Stone, 1928c), is that reciprocal lateral line placode transplants migrate to form the organs and peripheral nerve connections correct for their new location. However, Northcutt (1986b) has shown that preotic lateral line placodes contain both neuromast and ampullary organ anlagen. Stone was unaware that ampullary organs existed in urodeles, and his conclusions may have to be modified.

A third discovery was that transplants to the ventral surface, an area of the body that does not normally have neuromasts, failed to migrate, and ganglion cells were resorbed (Stone, 1929, 1931). This strongly suggests that there are pathways either subectodermally or on the surface of the subadjacent mesoderm, that guide the migrating primordia. In addition, lateral line placodes are able to plug these pathways. Placodes transplanted in the path of host primordia cause the host tissue to stop migrating, and remnants of host placodes exclude transplanted placodes from pathways that they occupy. Although Harrison (1904) and Stone (1928c) both argued that there is nothing in the substrate that influences organ deposition, these results suggest that the substrate may form pathways for the placodes.

Finally, whereas Stone (1928c) showed that supraorbital placodes transplanted to middle body line locations migrate only partway down the body and only form 18 organs, the key observation may be that *the organs in the distal half of the line are irregularly placed.* This arrangement, illustrated in Stone (1928c, Fig. 12), suggests the normal arrangement of supraorbital, but *not* body neuromasts. Supraorbital neuromasts form a single line above the eye but form orthogonal couplets more distally on the nasal snout (see Fig. 17–2). These couplets can appear "irregularly placed" when observed with a light microscope.

To summarize, these results suggest that the ectodermal placodes contain both lateral line receptor (neuromast and, preotically, ampullary organ) anlagen and peripheral nerve anlagen. In addition, the placodes are believed to contain directional cues, a programmed number of receptor primordia, and perhaps information about receptor polarity. The substrate may contain pathways that the primordia follow while laying down receptors. Stone's conclusions (1928c) on the similarity of pre- and postotic placodes will have to be modified to accommodate the origin and fate of ampullary organs in transplants of preotic placodes to postotic regions.

CONTEMPORARY PROBLEMS INVOLVING THE LATERAL LINE SYSTEM

One of the morphological features of the lateral line system that may be attractive to developmental biologists is that it is polarized; this polarity is easily determined and accessible experimentally. Placodes are polarized rostrocaudally, and they produce directionally sensitive hair cells that cluster to form directionally sensitive neuromasts (Fig. 17–1). Neuromast polarity changes between and within lines: middle and ventral body line neuromasts are oriented rostrocaudally; dorsal body line neuromasts are arranged dorsoventrally (Fig. 17–2). Supraorbital neuromasts around the eye are arranged in one row and are oriented tangential to the eye margin. More rostrally, supraorbital neuromasts form at least two rows of orthogonally oriented neuromasts (Fig. 17–2). Infraorbital neuromasts are similar to supraorbital neuromasts around the eye, but form three rows of neuromasts on the maxillary snout (Fig. 17–2). In this line neuromasts in the two outside rows are oriented orthogonally to each other; the center row is rostrocaudally polarized and is simply an anterior extension of the single neuromast row circumorbitally.

Some questions arise. For example, how is this neuromast polarity achieved? Are the placodes that form lines containing orthogonally oriented neuromasts different from placodes that form straight lines, as Stone's results (1928c) suggest, or are placodes the same, with polarity determined by the substrate? Are the mechanisms, either placodal or substrate, that determine orthogonal couplets derived from separate primordia (i.e., dorsal and middle body lines) the same as orthogonal couplets generated by a single primordium (i.e., rostrally within the supraorbital and infraorbital lines)? Also, do preotic placodes containing presumptive ampullary organs differ from postotic placodes that will not form ampullary organs, or does the substrate act on migrating primordia to determine organ type?

Dorsal and middle body lines tend to occur in the same dorsoventral plane and together produce orthogonal couplets on the trunk (Fig. 17–2). During placodal migration in the axolotl, the dorsal body line lags about one-quarter body length behind the middle line (Fig. 17–3A). At the level of the cloaca the middle body line turns dorsally, then continues down the tail in the dorsal line position, while the dorsal line ceases migrating, apparently when it encounters the middle line. How does the dorsal primordium manage to leave one neuromast almost exactly above each middle line neuromast if, as Harrison (1904) and Stone (1928c) argued, the substrate does not play a role in the positioning of the organs? Some sort of timing mechanism might produce this correspondence if the placodes started migrating at the same rostrocaudal position, or if the dorsal line began migrating at the position of a middle line neuromast. However, Stone (1933) has argued that migration of the primordia is not a smooth process, but rather proceeds in fits and starts, in a discontinuous, jerky fashion. Therefore, in contrast to what Harrison (1904) and Stone (1928c) thought, the positioning of neuromasts may be entirely dependent on the substrate through which the primordia migrate. Indeed, it is known that the segmental arrangement of spinal ganglia is dependent on the influence of somites (Detwiler, 1934).

The fact that placodes placed in the pathway of other primordia do not begin to migrate until the host primordium reaches them suggests, additionally, that substrate changes in some manner allow migration to continue. It is interesting to speculate that this change may be propagated in the form of a morphogenetic wave, as discussed in Chapter 8. Indeed, a series of waves may be necessary to allow for the differences in the timing of the migration of different primordia. It is possible that these changes may be restricted to the immediate pathways, however. Stone (1928a) mentioned, but did not elaborate on, a series of experiments where placodes were transplanted along the body near the future pathway of the middle body line placode. Again, these placodes did not begin migrating until the host's placode reached them, but the direction they migrated in was dependent on their orientation. Transplanted placodes migrated in a direction normal for them, *not the host.* This suggests that placodes are perhaps triggered by a morphogenetic wave, but do not "ride the wave." However, Stone (1928a), atypically, did not fully develop these results, suggesting either than only a few animals were observed, or that he otherwise doubted these findings. Stone (1931) repeated these experiments and concluded that if the placodes are transplanted early enough, then the host substrate can determine placode orientation. This work should probably be redone in the context of the morphogenetic wave hypothesis.

Other questions arise as well. As mentioned previously, the peripheral nerves and ganglia that supply the lateral line are of ectodermal (placodal) origin (Landacre, 1921; Stone, 1922, 1931; Hörstadius, 1950), and are not derived from the neural crest. Whether placodes are induced by neural crest cells (as suggested by Hörstadius, 1950) or whether neural crest cells and placodes are independently derived from a common precursor tissue (Northcutt and Gans, 1983), it is interesting to note that the migration is different for the two tissue types. Placodes migrate rostrocaudally, parallel to the neural tube and *perpendicular* to the dorsoventral route followed by neural crest cells, which form the majority of peripheral nerves. Only neural crest cells that form the enteric ganglia move rostrocaudally, inside the mesenchyme of the gut wall (see Le Douarin, 1982). In addition, white mutant axolotls appear to form normal lateral lines. The white mutants, however, are believed to be incapable of supporting the subectordermal migration of pigment cell precursors—that is, neural crest cells (Keller et al., 1982; Spieth and Keller, 1984; Chapter 9). Taken together, these results suggest that entirely separate mechanisms may be responsible for neural crest and lateral line migrations. However, very recent studies have shown that neural cell adhesion molecules (NCAM) appear in otic placodes only after neural crest has migrated through the area (Balak et al., 1987). These results and the subsequent questions that they generate suggest that the lateral line system has the potential to serve as an excellent model system for contemporary developmental biology.

SUMMARY

The lateral line is a system of electro- and mechanoreceptors found on fish and aquatic amphibians. Both the receptors and the nerves that supply them are derived embryonically from ectodermal placodes. These placodes form the lateral lines by migrating distally from their origin near the otic vesicle.

Although it has been generally neglected in recent years, studying of the development of the urodele lateral line system—and here axolotls are most convenient to study—may prove valuable for several reasons. First, neuromasts have a definite polarity once formed, but it remains unclear how, when, and where this polarity is established. The lateral line system is easily accessible experimentally, and these studies may provide valuable clues relating to how polarization is achieved during development.

Second, lateral lines form through migration. This migration occurs in a direction perpendicular to the migration of neural crest cells and in areas (the subectodermal space of white mutant axolotls) where neural crest cell migration is restricted. This suggests that the nerves and receptors of the lateral line migrate via signals unrelated to signals followed by neural crest cells. Lateral line migration may therefore be controlled by a novel set of developmental signals.

Third, lateral line placodes grafted in the pathway of another line fail to migrate until the host primordium reaches it. Once placodes are triggered to begin migrating, the direction that they migrate in appears to be determined by cues intrinsic to the placode. Morphogenetic waves may trigger the start of placode migration, but placodes may have an internal polarity that determines their direction of migration. Placodes can migrate in a direction opposite to a morphogenetic wave, and therefore cannot be passively riding that wave. A new series of experimental placode transplants, as pioneered by Stone, should answer many of these questions.

ACKNOWLEDGMENTS

We thank J. B. Armstrong and W. S. Fletcher at the University of Ottawa axolotl colony for providing us with albino, albino-axanthic, and pigmented embryos. Ann King illustrated Fig. 17–2. This research was supported by Natural Science and Engineering Research Council of Canada grants to R. Wassersug and J. B. Armstrong, and a Medical Research Council of Canada grant to L. Maler.

18
Metamorphosis

E. R. KÜHN AND G. F. M. JACOBS

In 1828 Cuvier described *Ambystoma mexicanum* as the larvae of an unknown salamander. Subsequent studies during the nineteenth century revealed that many urodeles may become sexually mature while retaining many of the characteristics of larvae. This was long before the importance of the thyroid gland in anuran metamorphosis was discovered by Gudernatsch (1912), who observed that ingestion of horse thyroid tissue by tadpoles caused premature metamorphosis.

OBLIGATE AND FACULTATIVE NEOTENES

Some salamanders, described as "obligate neotenes," rarely if ever undergo metamorphosis either in nature or in the laboratory, whereas facultative neotenes commonly do. The obligate neotenes *Proteus anguinus* and *Necturus* spp. are insensitive even to massive doses of thyroid hormones. However, they possess thyroid glands, the products of which can induce metamorphosis in other species (Lynn, 1961). Dent (1968) noted that *P. anguinus* skin did not respond to thyroxine (T_4) *in situ*, but could undergo a form of transformation if transplanted into metamorphosing urodeles of other species.

Some species appear never to metamorphose in nature, but undergo some metamorphic change under laboratory conditions following administration of thyroid hormones. Norris (1985) has described these species as "inducible obligate neotenes." They include *A. mexicanum* and the plethodontids *Eurycea tynerensis* and *E. neotenes* (Kezer, 1952), *Gyrinophilus palleucus* (Dent et al., 1955), and *Typhlomoge rathbuni* (Dundee, 1957).

In facultative species such as the tiger salamander *(Ambystoma tigrinum)*, transformed and untransformed animals may be found side by side. Tiger salamanders metamorphose naturally in warm lakes, but when derived from cold lakes, metamorphose only when maintained under laboratory conditions (Jenkin, 1970). This emphasizes the fact that some species may undergo metamorphosis autonomously, depending on environmental and endocrinological conditions. Neoteny may be favored not only by cold, but also by harsh terrestrial conditions and the absence of predators in the water (Sprules, 1974). The presence of permanent bodies of water favors neoteny in *A. gracile,* whereas *A. laterale,* which lives in temporary ponds, metamorphoses early (Wilbur, 1971).

Recent studies have clearly implicated activation of the hypothalamic-pituitary-thyroid axis in anuran metamorphosis and, whatever the ecological reason for the maintenance of neoteny, some modification of the thyroidal axis is obviously involved.

THYROID HORMONES AND SPONTANEOUS METAMORPHOSIS

How do thyroid hormones affect free-living salamanders that depend on seasonal variations in their environment for their reproduction and metamorphosis? One such species is *Ambystoma tigrinum;* both metamorphosed and aquatic neotenic animals may be found breeding in the same pond. Neotenic tiger salamanders in Colorado spawn between February and May, and spawning is followed by gonadal recrudescence (Norris et al., 1977). Spontaneous metamorphosis has been observed in this population after the spawning period (in May–June), and also later, in November (Platt, 1976).

In 1977, Norris et al. measured T_4 levels in wild *A. tigrinum.* To measure circulating T_4 they used a commercial kit, which, however, was not sensitive enough to measure T_4 directly. In order to ensure that the concentration of T_4 fell on the standard curve, a known amount of T_4 was added to each sample, resulting in relatively high plasma T_4 values compared to other studies. They reported that prespawning levels of T_4 were 5.5 ± 0.4 ng/ml plasma in March, and that levels fell to a minimum level of 2.2 ± 0.31 ng/ml in September, correlating with a maximum gonadal development.

A correlation between gonadal development and decreased plasma T_4 concentration has also been observed in *Rana ridibunda* (Kühn et al., 1985a) and in female *Dicroglossus occipitalis,* a tropical anuran (Kühn et al., 1987). However, T_4 increases in the thyroid, indicating that these variations may be the result of thyroid activation by the peripheral conversion of T_4 into triiodothyronine (T_3). This hypothesis is supported by the finding that 5′-monodeiodinating (5′-MDI) activity in the kidney increases sharply in male *R. ridibunda* during reproduction (Vandorpe et al., 1987).

The highest incidence of spontaneous metamorphosis in *A. tigrinum,* which is observed following spawning (Platt, 1976), may be correlated with high endogenous levels of T_4 and may be due in part to an increased sensitivity to thyroid hormones, since testosterone may enhance metamorphosis induced by thyroid-stimulating hormone (TSH) (Norris et al., 1973). Thyroid glands of larvae examined in September showed reduced sensitivity to ovine TSH (Norris et al., 1977). Sexually mature neotenic tiger salamanders are more sensitive to hypothalamically administered thyroid hormones than are immature larvae (Norris and Gern, 1976).

Regard et al. (1978) have developed a radioimmunoassay (RIA) for T_3 and T_4 with minimum detectable levels of 500 pg/ml for T_4 and 50 pg/ml for T_3. In spontaneously metamorphosing *A. tigrinum* they found significant levels of circulating thyroid hormone only during metamorphosis (Table 18–1; Larras-Regard et al., 1981). T_3 concentrations were high in the middle of metamorphosis (up to 2 ng/

Table 18–1 Plasma Levels of T_3 and T_4 During Spontaneous and Induced Metamorphosis

	T_3 (ng/ml)			T_4 (ng/ml)		
	Before	During	After	Before	During	After
Spontaneous						
Ambystoma tigrinum	ND[a]	1–2	ND	ND	5–15	ND
Ambystoma gracile	—	—	—	3.0	25	3
Eurycea bislineata	ND	1–2	ND	ND	5–8	ND
Induced						
Ambystoma mexicanum						
Exp. 1 (10 $\mu g T_4$)	—	—	—	0.053	1.1	0.171
Exp. 2 (2 × 20 μg T_4)	—	—	—	0.369	—	2.12
Exp. 3 (2 × 20 μg T_4)	0.013	—	0.063	0.152	—	0.402

[a]ND, Below limits of assay.

ml), and lower at the end of metamorphosis (maximal 1 ng/ml), whereas T_4 was lower at midmetamorphosis (5 ng/ml) and higher at the end of metamorphosis (15 ng/ml).

In other species, similar results have been obtained. In general, extremely low levels of T_3 and T_4 are observed prior to the onset of metamorphosis, and in metamorphosed animals, whereas metamorphosis itself is characterized by increased plasma concentrations. An earlier elevation in plasma T_4 has also been reported during hindlimb development in *A. mexicanum* (Rosenkilde et al., 1982).

In two populations of *A. gracile* raised in the laboratory, Eagleson and McKeown (1978) demonstrated that thyroid activity as assessed by radioiodide uptake in the thyroid, and serum T_4 as measured by RIA (Table 18–1), are greatest during metamorphosis. In *Eurycea bislineata,* Alberch et al. (1986) observed that metamorphosis was accompanied by a sharp increase in the circulating plasma levels of T_3 (1–2 ng/ml) and T_4 (5–8 ng/ml). However, levels in both larvae and adults were below the limits of detection of their RIA (1 ng/ml for T_4 and 200 pg/ml for T_3).

Several factors may be responsible for the discrepancy in the data of various authors. These include diurnal variations in plasma T_4 concentration, which have been described in *A. tigrinum* (Norris et al., 1981), and the methodology itself (see Larras-Regard, 1985).

THYROID HORMONES AND INDUCED METAMORPHOSIS IN THE AXOLOTL

In facultative neotenes, metamorphosis can be stimulated with relatively small doses of T_3 (Swanberg and Norris, 1972; Norris and Platt, 1974), T_4 (Norris et al., 1973; Norris and Platt, 1974; Darras and Kühn, 1983, 1984), or mammalian TSH (Norris et al. 1973; Taurog, 1974; Regard and Taurog, 1977). Since plasma concentrations of T_4 in both neotenic and metamorphosed axolotls appear to be too low for direct RIA, an extraction procedure has been used to effect a 20-fold concentration of the samples (Darras and Kühn, 1982).

T_4 plasma concentrations in premetamorphic *Ambystoma mexicanum* are very low (53 ± 3 pg/ml; Darras and Kühn, 1984). A single injection of 10 μg of T_4 is sufficient to start metamorphosis, as judged by a significant reduction in body weight from day 2 and a significant reduction in external gill length and tail height from day 6 (Darras and Kühn, 1984). In animals treated in this way, basal levels of T_4 are still high 1 week after the injection (1094 ± 138 pg/ml) (Fig. 18–1). One could argue that the injected T_4 was metabolized slowly and therefore was responsible for the 13-fold increase observed. This has also been suggested for a 5-μg injection of T_3 (Sawin et al., 1978). However, in the same experiment an injection of 5 mU of TSH raised T_4 levels 4-fold in neotenic animals, and in the metamorphosed animals a 50-fold or greater increase was seen after 4 hours. Basal levels were obtained 24 hours later in both group of animals (Fig. 18–2). Therefore, it seems unlikely that the high concentrations of T_4 found in the plasma of these metamorphosing animals was caused entirely by the injection of T_4. Moreover, if T_4 has a

Fig. 18–1. Variations in external gill (percentage of initial length) in *Ambystoma mexicanum* following intraperitoneal injection of frog Ringer's (●) or 10 μg T_4 (○). Vertical bars represent plasma concentrations (±SEM; numbers represent numbers of animals) of T_4 before (day 0) and 1 and 5 weeks after this injection (Darras and Kühn, 1984).

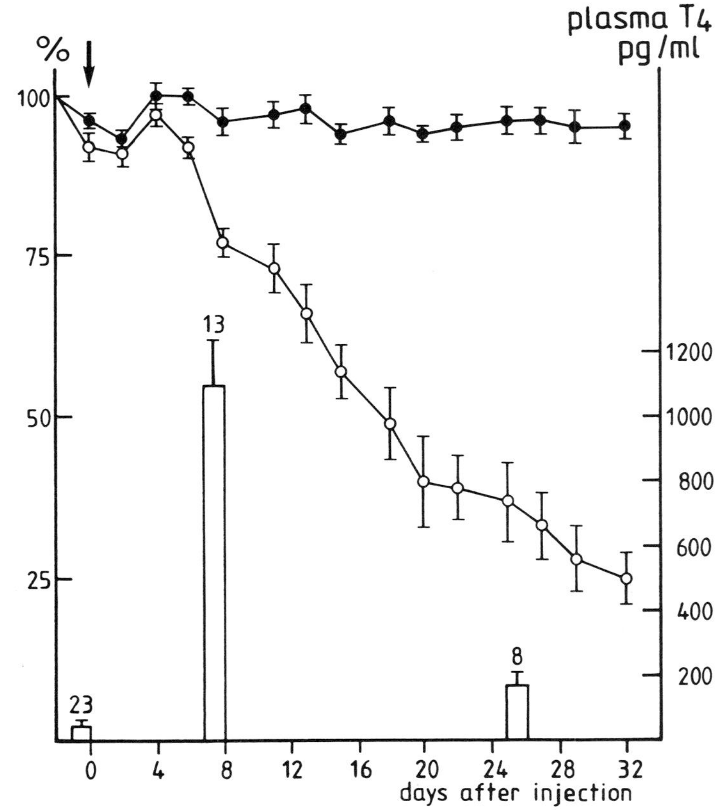

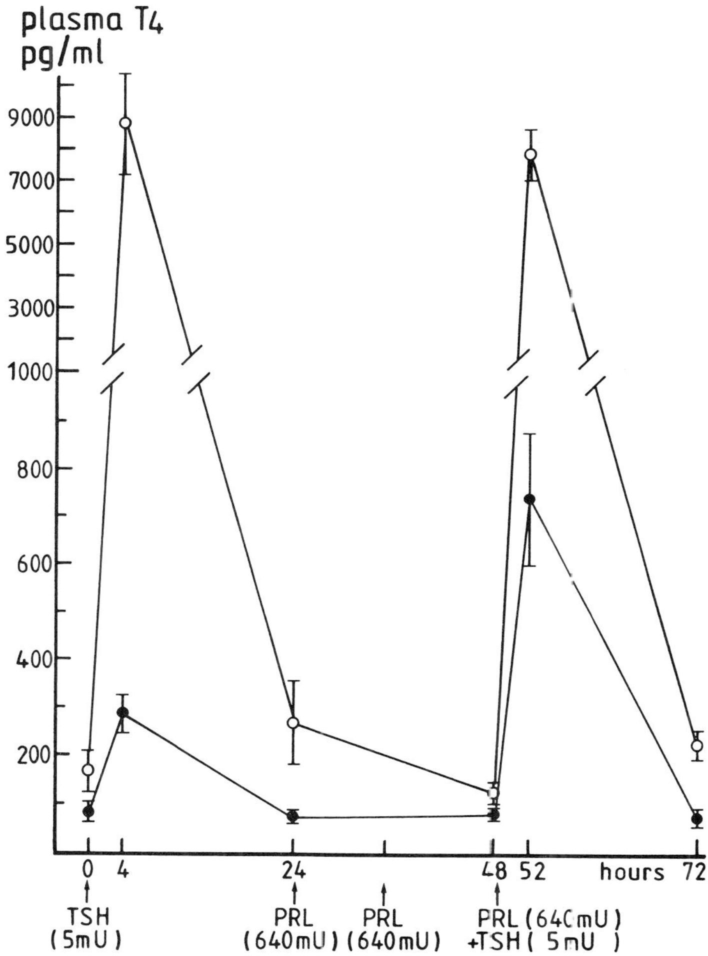

Fig. 18–2. Plasma concentrations (vertical bars, ±SEM) of T_4 in neotenic (●) and metamorphosed (○) axolotls following injection of TSH and prolactin (PRL) as indicated (Darras and Kühn, 1984).

half-life of 14 hours, as in the neotenic axolotl, virtually all of the hormone would have been cleared from the animal after 8 days (Rosenkilde, 1976).

Five weeks after the T_4 injection, body weight was reduced by 30%, gill length by 75%, and tail height by 55%. At that point, plasma T_4 concentrations were down again (171 ± 39 pg/ml), but were still somewhat higher than the original levels (Fig. 18–1). In two later experiments, higher plasma concentrations were found in both neotenic (369 ± 30 and 152 ± 10 pg/ml, respectively) and metamorphosed animals (2120 ± 400 and 402 ± 32 pg/ml). In the latter experiment, plasma concentrations of T_3 were measured using an extraction procedure described by Darras and Kühn (1982), in which the plasma sample is concentrated 4-fold. In neotenic animals only 12.5 ± 0.8 pg/ml T_3 was found. This is at the detection limit of our

RIA. Following metamorphosis, the T_3 level rose to 63.2 ± 7.4 pg/ml (Jacobs et al., 1988).

PERIPHERAL CONVERSION OF T_4 INTO T_3

In anurans, the 5′-MDI of T_4 into T_3 appears to be a fundamental step in the thyroid hormone induction of metamorphosis (see Buscaglia et al., 1985). Premetamorphic tadpoles that had been chemically thyroidectomized by perchlorate were treated *in vivo* with a competitive inhibitor of 5′-MDI activity, and a dose of T_4 adequate for metamorphosis. There was a significant reduction in peripheral T_3 production and delayed metamorphosis.

In the axolotl, evidence for a role of peripheral 5′-MDI is equivocal. Schultheiss (1980) observed that, in neotenic axolotls, the T_4 content of the thyroid glands was always higher than the T_3 concentration, in contrast to the serum. In our experiments, the ratio of T_3 to T_4 in the plasma was much higher than that in the thyroid for both neotenic and metamorphosed animals (Table 18–2). The plasma ratio was highest for metamorphosed animals, perhaps indicating that the peripheral 5′-MDI activity is more pronounced after metamorphosis (Jacobs et al., 1988). We have been able to measure the 5′-MDI activity in the kidney and skin of *Rana ridibunda* (Vandorpe et al., 1987). However, 5′-MDI activity could not be detected in liver or kidney homogenates from either neotenic or metamorphosed axolotls.

During metamorphosis in *Ambystoma tigrinum,* T_3 reaches maximum plasma concentrations earlier than T_4. This observation has been used to support the hypothesis that T_3 is derived from the thyroid gland and not from peripheral deiodination during metamorphosis (Larras-Regard et al., 1981). A similar pattern seems to occur during metamorphosis of *Eurycea bislineata* (Alberch et al., 1986).

TSH-INDUCED RELEASE OF THYROID HORMONES

The thyroid gland of the neotenic axolotl is not deficient in thyroid hormones and may even contain large amounts of T_3 and T_4 (Table 18–2), comparable to those of *Rana ridibunda* in summer (Kühn et al., 1985a). Metamorphosed animals have even more T_4 than neotenic ones.

In neotenic *Ambystoma mexicanum* and *A. tigrinum* a lack of stimulation of the thyroid gland by TSH may explain neoteny, since metamorphosis can be read-

Table 18–2 Plasma Concentrations and Thyroid Content of T_3 and T_4 in Neotenic and Metamorphosed Axolotls[a]

			T_3/T_4	
	T_3 (μg per thyroid)	T_4 (μg per thyroid)	Thyroid	Plasma
Neotenic	0.48 ± 0.04 (12)*	2.17 ± 0.29 (12)*	2.41 ± 0.18 (12)	8.7 ± 0.7 (12)**
Metamorphosed	0.63 ± 0.04 (13)	3.84 ± 0.55 (13)	2.01 ± 0.23 (13)	16.3 ± 1.8 (13)

[a]Values are means ± SEM. Number of animals in parentheses. * $p < 0.05$; ** $p < 0.005$ (ANOVA).

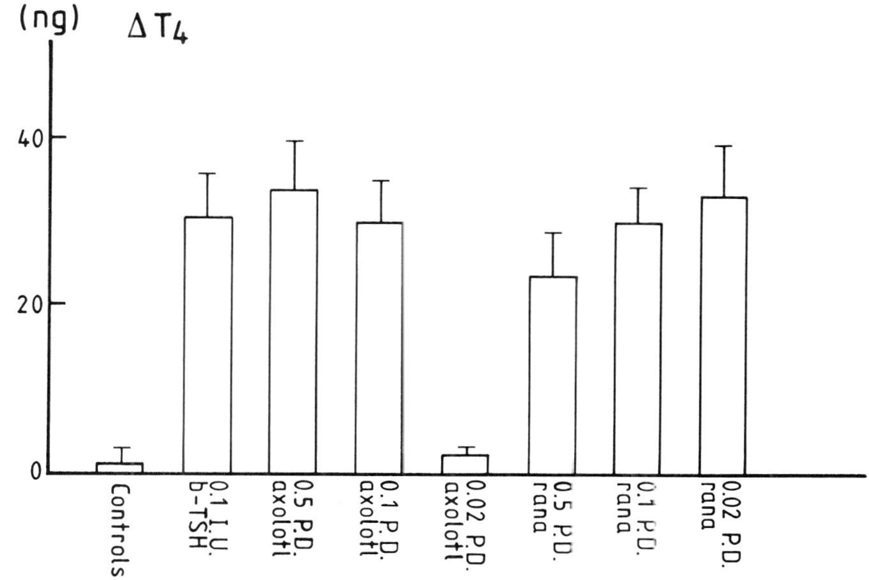

Fig. 18–3. The in vitro activity of β-TSH and pars distalis (P.D.) extracts of *Rana* or axolotl on T_4 release from thyroid glands of *Rana ridibunda*. Each bar represents the mean ± SEM of eight assays (Darras and Kühn, 1983).

ily induced by treatment with mammalian TSH (Norris et al., 1973; Taurog, 1974; Regard and Taurog, 1977; Larras-Regard et al., 1982). Following such stimulation, plasma concentrations of T_3 and T_4 increase (Norris et al., 1981; Larras-Regard, 1985). However, the hypophysis of a neotenic axolotl contains substantial levels of TSH activity, comparable to those found in adult *R. ridibunda*. This was shown in an in vitro assay in which an extract of the pars distalis of the neotenic axolotl caused T_4 to be released from the thyroid of *R. ridibunda* only slightly less effectively than did the hypophysis of the frog itself (Fig. 18–3). Moreover, a 9-fold increase in the plasma concentration of T_4 was seen in neotenic axolotls when the extract of half a pars distalis was injected intravenously (Darras and Kühn, 1983). This result suggests that axolotls remain neotenous because the pituitary fails to release the TSH that it contains.

HYPOTHALAMIC CONTROL OF THYROID FUNCTION

Evidence from pituitary stalk sections, hypothalamic lesions, and pituitary transplantations suggests that the secretion of pituitary gonadotropin is regulated by the hypothalamus in anurans (see Ball, 1981, for a review). In *Triturus*, lesions in the preoptic area, just behind the anterior commissure, delay or arrest spermatogenesis and block ovulation (Mazzi, 1978; Malacarne and Giacoma, 1978). This is also the control center for ovulation in both *Rana* and *Bufo* (Holmes and Ball, 1974).

Therefore, a functional hypothalamus must be present in neotenes, since the gonads mature.

The hypothalamus in both *Ambystoma mexicanum* and *A. tigrinum* contains large quantities of the tripeptide thyrotropin-releasing hormone (TRH) (Taurog et al., 1974; Jackson and Reichlin, 1974; Sawin et al. 1978), as do other species. However, metamorphosis is not induced by injection of TRH or immersion in water containing TRH. Treatment with TRH also has no significant effect on thyroid ^{131}I metabolism. Rapid degradation of TRH cannot be involved, since axolotl plasma, unlike rat plasma, does not degrade TRH (Taurog et al., 1974).

Sawin et al. (1978) observed no change in hypothalamic TRH content during T_3-induced metamorphosis of *A. mexicanum.* In contrast, King and Millar (1981) observed a drastic increase in hypothalamic TRH content during metamorphosis in *Xenopus laevis.* However, it was subsequently shown that the site of TRH injection may be important. An increase in plasma levels of T_4 was obtained when TRH was injected into the vena abdominalis of *R. ridibunda* (Darras and Kühn, 1982). These results prompted us to study the influence of intravenously injected TRH on T_4 release in neotenic *A. mexicanum* (Darras and Kühn, 1983). However, even very high doses of TRH (up to 100 μg) injected intravenously were unable to stimulate the release of T_4 (Fig. 18–4).

In a recent study, a comparison was made between the effect of TRH on neotenic and metamorphosed *A. mexicanum,* in which metamorphosis was induced

Fig. 18–4. The influence of an intravenous injection of TRH, axolotl pars distalis (P.D.) extract, and β-TSH on circulating levels of T_4 in *Ambystoma mexicanum*. Each point represents the mean $\pm$ SEM of nine animals (Darras and Kühn, 1983).

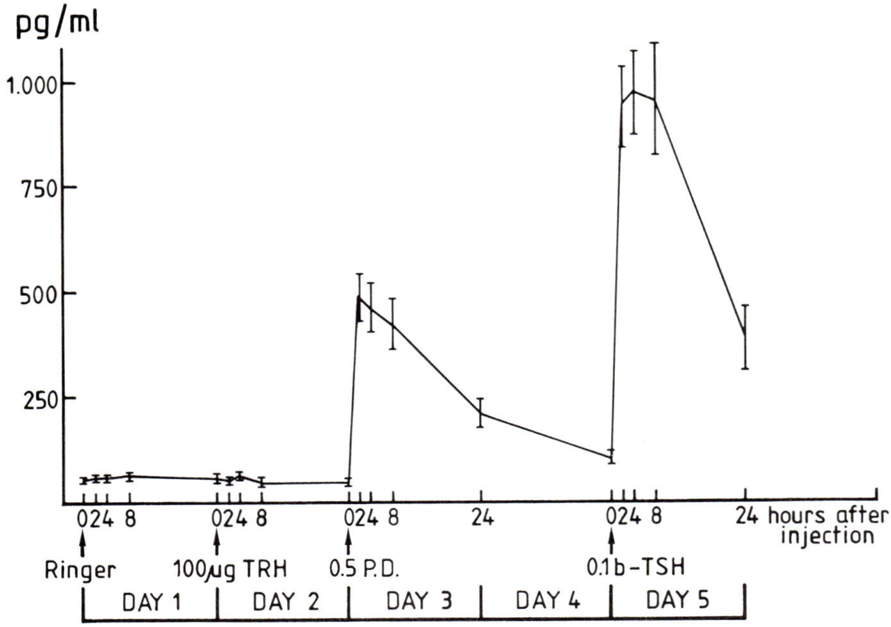

by two injections of 20 μg T_4 (Jacobs and Kühn, 1986). Plasma concentrations of T_4 were lower in neotenic (369 $\pm$ 30 pg/ml) than in metamorphosed animals (2110 $\pm$ 400 pg/ml). An intravenous injection of 10 μg TRH had no influence on T_4 levels in neotenes, but raised T_4 plasma concentrations in metamorphosed animals. These levels remained elevated up to 24 hours (Fig. 18–5). Following subsequent TSH stimulation, plasma T_4 concentrations increased in both neotenic and metamorphosed animals. In a second study, however, TRH was ineffective in neotenic and metamorphosed animals, which had T_4 plasma concentrations of 152 $\pm$ 10.3 and 402 $\pm$ 33 pg/ml, respectively.

Pituitaries from both frogs and axolotls, incubated *in vitro* with TRH, do not show a stimulation of TSH release, suggesting that the amphibian hypophysis is insensitive to TRH (Taurog et al., 1974). Since many authors have been unable to demonstrate a thyrotropin-releasing function for TRH in several anuran species

Fig. 18–5. The influence of 10 μg TRH and of 10 mU β-TSH on T_4 plasma concentrations in neotenic (- - -) and metamorphosed (●—●) axolotls. Each point represents the mean $\pm$ SEM of nine animals. Significant differences compared to pre-TRH injection values (paired *t*-test) are indicated by *5% level), ** (2.5% level), and *** (0.5% level). Significant differences compared to pre-TSH injection values are indicated by *′, *′*′, and *′*′*′.

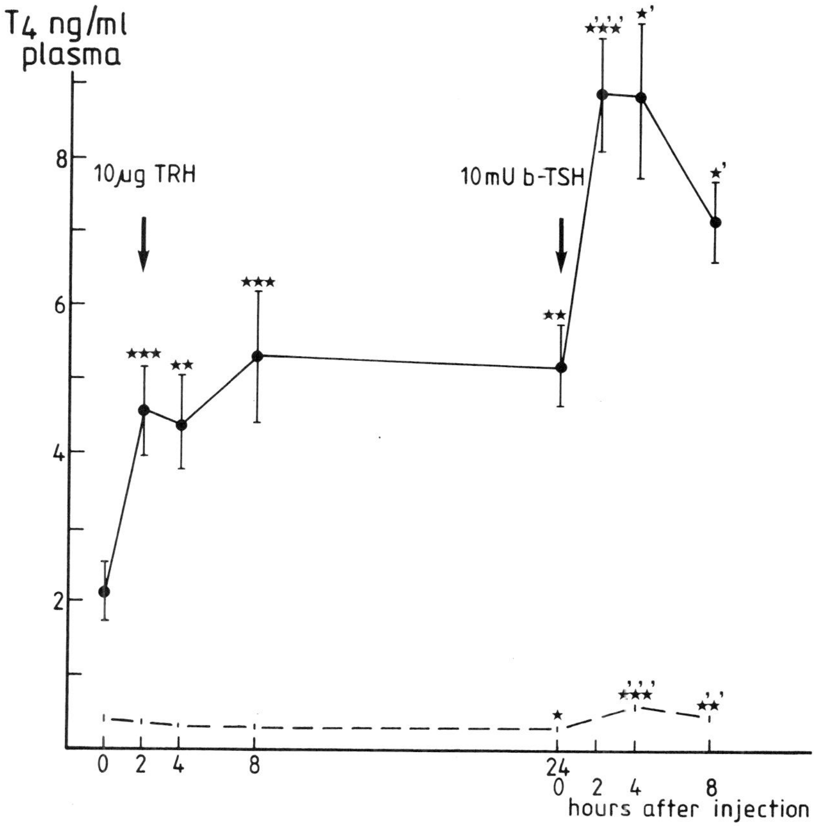

(Rosenkilde, 1979), it has been suggested that TRH may serve some other function in amphibians (Taurog et al., 1974). A role in the release of other hormones, such as growth hormone (Hall and Chadwick, 1984), melanocyte-stimulating hormone (Leroux et al., 1983), and prolactin may not be ruled out. In frogs, TRH is able to promote the secretion of prolactin from the pituitary in vitro (Clemons et al., 1979; Seki and Kikuyama, 1986; Hall and Chadwick, 1984). In *R. ridibunda* the release of prolactin is observed in vivo, together with the release of thyroid hormones, following intravenous injection of TRH (Kühn et al., 1985b).

PROLACTIN AND NEOTENY

An overwhelming amount of data points to prolactin as the growth-promoting hormone in amphibian larvae, whereas an activated thryoid gland is required for metamorphosis (Dodd and Dodd, 1976; Norris, 1978; Rosenkilde, 1979; Kikuyama et al., 1980). An interaction exists between prolactin and the thyroid hormones, which appear to act at either the thyroidal or peripheral level.

Prolactin has been described as capable of counteracting T_4- or TSH-induced metamorphosis in *Ambystoma tigrinum* (Platt, 1976; Larras-Regard, 1985). Although ovine prolactin does not affect iodide uptake by the thyroid in neotenic *A. tigrinum* (Norris and Platt, 1973; Platt, 1976), it does prevent the increase in serum T_4 that follows the administration of TSH (Norris, 1978). We have been unable to confirm these results in neotenic *A. mexicanum*. An injection of 5 mU bovine TSH causes an increase in plasma concentrations of T_4, but this increase was not inhibited by pretreatment with large amounts of ovine prolactin (Fig. 18–2; Darras and Kühn, 1984).

The failure of prolactin to lower basal or TSH-induced plasma concentrations of T_4 in our experiments does not rule out the possibility of a peripheral interaction with thyroid hormones. Prolactin may even turn out to be essential for metamorphosis, together with T_4, since it has been demonstrated that ovine prolactin can stimulate the peripheral conversion of T_4 into T_3, albeit in chicken embryos, by changing the hepatic 5-MDI activity into a 5′-MDI activity (Kühn et al., 1983). Furthermore, prolactin levels are quite high during the advanced stages of *R. catesbeiana* metamorphosis (Clemons and Nicoll, 1977; Yamamoto and Kikuyama, 1982), and in the red-spotted newt, levels of endogenous T_4 can be lowered by injecting 0.2 IU of ovine prolactin (Anderson and Dent, 1982), perhaps by stimulating conversion of T_4 to T_3.

POSITIVE FEEDBACK OF T_4

Another possible explanation for the inability of the pituitary to release TSH may be found in the low endogenous levels of circulating T_4 in the neotenic axolotl. In the kitten, T_4 may have a positive (morphogenic) feedback on TRH-producing neurons (Joseph and Knigge, 1972). In the tadpole, the morphological development of the median eminence and other brain regions is T_4 dependent (Etkin, 1966; Etkin et al., 1966; Pesetsky, 1965; Goos, 1968; Gona, 1973). In the axolotl, T_4 exerts positive feedback on the hypophyseal TSH-producing cells (Gunsbourg, 1971; Hauser-Gunsbourg and Lefebvre, 1976). Notably, Norris and Gern (1976) could induce metamorphosis in *Ambystoma tigrinum* following a single intrahypothalamic T_4 injection, whereas an injection of the same dose intraperitoneally was ineffective. Therefore, the low endogenous level of circulating T_4 may be insufficient to trigger the release of TSH by the pituitary.

It should also be noted that thyroid activity is minimal in hibernating frogs. On several occasions the circulating levels of T_4 start to increase in January before the end of hibernation, while the thyroid glands become virtually depleted of hormone (Kühn et al., 1983). In December, the *in vitro* ability of the thyroid glands to release T_4 following TSH stimulation was investigated. This ability was markedly enhanced when frogs were pretreated with T_4 (Kühn et al., 1984). This means that T_4 may function as a trigger for thyroid activation in hibernating frogs. Therefore, T_4 may also be required to stimulate the thyroid in axolotls. Again, the low circulating level of T_4 would, in most instances, preclude this stimulation, preventing metamorphosis.

SUMMARY

The axolotl is a facultative neotene that rarely metamorphoses in the wild. As with many other amphibians, metamorphosis may be induced in the axolotl by treatment with T_3, T_4, or TSH. In neotenic axolotls, plasma concentrations of T_3 and T_4 are low. The levels of these hormones increase markedly during metamorphosis before dropping back to levels slightly higher than in neotenic animals. Taken together, these results suggest an important role for the thyroid gland in metamorphosis.

Failure to metamorphose in the axolotl cannot be attributed to a lack either of hormones (T_3 and T_4) in the thyroid or of TSH in the pituitary. Rather, it seems to be due to the inability of the pituitary to release the TSH it contains. This appears to have two possible causes. First, the pituitary is insensitive to TRH stimulation. Since the axolotl hypothalamus contains large quantities of TRH, this suggests an alternate role for TRH in the neotenic animal. However, intravenous injections of TRH may stimulate the thyroid gland of metamorphosed axolotls, as it does in adult frogs. Second, T_4 seems to exert a positive-feedback effect on the hypophyseal TSH-producing cells (and possibly on the thyroid gland itself). The low endogenous levels of T_4 in the plasma may therefore be insufficient to allow the pituitary to begin releasing TSH.

III

PRACTICAL INFORMATION ON WORKING WITH AXOLOTLS

19

Developmental-Stage Series of Axolotl Embryos

N. P. BORDZILOVSKAYA, T. A. DETTLAFF,
SUSAN T. DUHON, AND GEORGE M. MALACINSKI

In 1969 Harrison's staging series for *Ambystoma maculatum (Amblystoma punctatum)*, with drawings by Lisbeth Krause, was published. Harrison divided embryonic development into 46 stages based on external morphology and on features visible in dissected embryos. That paper became the model for comparable series for the axolotl published later by Schreckenberg and Jacobson (1975) and by Bordzilovskaya and Dettlaff (1975, 1979). The present chapter represents a revision of the 1979 paper.

Schreckenberg and Jacobson divided the development of the axolotl into 40 stages, based on external morphological features. Timing was measured at 18°C. Their series is illustrated with photographs that are accompanied by brief descriptions.

The staging series originally published by Bordzilovskaya and Dettlaff is also based on external morphology. It furthermore provides a ratio (t_n/t_0) that allows the time and duration of the various stages to be predicted for various temperatures. Such prediction is possible because the duration of different developmental periods in many species changes in a proportional manner in the optimal temperature range (about 16°–25°C) (Dettlaff and Dettlaff, 1960, 1961; Dettlaff, 1964, 1976; Dettlaff and Rudneva, 1973). Such a proportionality has been established for the axolotl as well (Cate, 1956; Skoblina, 1963; Valouch et al., 1971). If the duration of some one period or stage is known at various temperatures, and the durations of the remaining stages are known at one constant temperature (20°C), the duration of each stage at various temperatures can be estimated. The duration of one mitotic cycle during synchronous cleavage division is t_0 (Dettlaff and Dettlaff, 1961; Dettlaff, 1976), corresponding in the axolotl to the interval between the first appearances of the first and second cleavage furrows on the egg surface (Rott, 1973). Skoblina (1963, 1965) determined t_0 at various temperatures, as depicted in Fig. 19–1. At 20°C, t_0 varies from 87 to 93 minutes (M. N. Skoblina, unpublished data). In this chapter, t_0 = 92 minutes is used in all calculations. The time from the appearance of the initial cleavage furrow on the egg surface to the onset of a given devel-

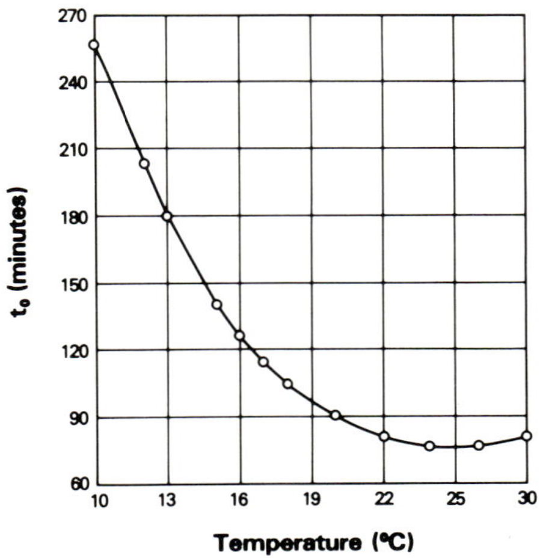

Fig. 19–1. Variation in t_0, the duration (in minutes) of one mitotic cycle in the period of synchronous cleavage divisions, for the axolotl (vertical axis) with the temperature in degrees Celsius (horizontal axis). (After Skoblina, 1965.)

opmental stage is t_n. By multiplying the ratio (t_n/t_0) by the value of t_0 at the temperature of interest, obtained from the graph in Fig. 19–1, the time in minutes to any developmental stage at any temperature in the optimal range can be determined.

PROCEDURES EMPLOYED FOR STAGING

Eggs from natural spawnings were collected within 20–40 minutes of being laid. Timing began with the first appearance of the initial cleavage furrow on the animal pole. The stages have been designated in accordance with the numbering system proposed by Harrison (1969) for *Ambystoma maculatum*. Features common to both species and those that best distinguished successive stages and allowed a precise determination of time of transition were chosen to characterize the stages. Fixed material was used to define the features and timing of the later stages more accurately. Embryos were fixed up to stage 35, at an interval of 0.5–$1t_0$ at first, then at a $5t_0$ interval, and in the latest stages at intervals from 10 to $20t_0$ with reference to precisely definable features. Three to five eggs that began to cleave at different times were fixed for every stage. A greater number of eggs and a greater number of observations were used to define transitional stages such as the beginning of gastrulation, the stage with a slitlike blastopore, and the closure of the neural folds. Size indications are the average of several measurements (Fig. 19–2). The times used in the table were determined on the more advanced embryos in each sample because they have been shown to provide more precise and reproducible results (Dettlaff and Dettlaff, 1961; Zubova, 1962; Ignatieva, 1974).

In some cases, the variation in duration of a certain stage among individual embryos in a sample exceeded the time interval between successive stages. These stages are marked with an asterisk in Table 19–1. In most cases, this variation was less than 10% of the total duration of a given stage.

The descriptions of diagnostic features and the drawings (Plates I–VI) of the

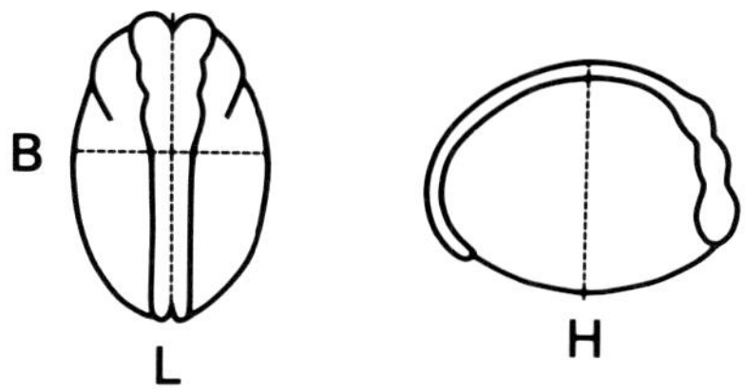

Fig. 19–2. Scheme of measurements. *L,* length; *B,* breadth; *H,* height.

Table 19–1 Stages of Normal Development of *Ambystoma mexicanum* (incubation at 29°C)[a]

Stage Number	Time from First Cleavage: In Hours and Minutes	t_n/t_0	Size (mm)	Features
1			1.85–2.0	Freshly laid fertilized egg in jellycoat.
1½			2.0	Activated egg: a broad perivitelline space has formed.
2–	0.0	0	2.0	First appearance of the first cleavage furrow on the animal pole; timing begins.
2	0.65	0.72	2.0	2 cells.
3	2.40	1.7	2.0	4 cells.
4	4.12	2.7	2.0	8 cells.
5	5.22	3.5	2.0	16 cells.
6	6.45	4.5	2.0	32 cells.
7	8.26	5.5	2.0	64 cells.
8	16.06	10.5	2.0	Early blastula (fall of mitotic index in animal blastomeres: Rott and Beritashvili, 1975).
9	21.28	14	2.0	Late (epithelial) blastula; surface is smooth.
9½	24.32	16	2.0	Morphogenetic function of nuclei begins (Ignatieva, 1972).
10	26.00	17	2.0	Early gastrula I: First signs of dorsal blastopore lip are visible.
10½	32.10	21	2.0	Early gastrula II: Invagination continues; blastopore is a nearly horizontal slit in one quadrant.
10¾	37.00	24	2.0	Middle gastrula I: Dorsal lip forms a semicircle.
11	38.30	25–26	2.0	Middle gastrula II: Blastopore covers three quadrants. Lateral lips are formed; ventral lip is marked only by pigment accumulation. Yolk plug reaches its maximum diameter (1.2 mm).
11½	40–42	27–28	2.0	Late gastrula I: Blastopore is circular; invagination continues; yolk plug diameter, 0.6 mm.
12	47.30	31	2.0	Late gastrula II: Blastopore has an oval or circular shape; size of yolk plug, 0.40 × 0.45 mm.

Table 19–1 *(continued)*

Stage Number	Time from First Cleavage: In Hours and Minutes	Time from First Cleavage: t_n/t_0	Size (mm)	Features
12½	49–51	32–34	2.1	Late gastrula III: Oval blastopore is closing; size of yolk plug, 0.15 × 0.20 mm.
13–	50.30–54.00	33–35	2.1	Stage with slitlike blastopore; boundaries of neural plate are not distinct.
13	55.00–56.30	36–37	2.1	Early neurula I: Blastopore is a narrow vertical slit; groove in the midline of the neural plate; boundaries of neural plate are outlined, but neural folds are not yet raised above surface of embryo; dorsal side is slightly flattened.
14	58.15	38	$L = 2.2$[b] $B = 2.0$	Early neurula II: Neural plate is broad. Neural folds are outlined and begin to rise above the surface in the head region. Embryo is slightly elongated.
15	59.50	39	$L = 2.25$ $B = 2.1$	Early neurula III: Neural plate is shield shaped. Neural folds are raised and bound all regions of the neural plate.
16	63.0	41	$L = 2.25$ $B = 2.1$	Middle neurula: Neural folds become higher, the spinal region of neural plate narrows, and the neural plate becomes sunken.
17	64.30	42	$L = 2.35$ $B = 2.0$	Late neurula I: Neural folds are higher, especially in the head region; further narrowing and deepening of the neural plate occur both in the head and in the spinal regions. Hyomandibular furrow limiting the mandibular arch is slightly outlined. The segmentation of the mesodermal material begins. There are two pairs of somites.
18	66.0–67.30	43–44	$L = 2.4$ $B = 1.9$	Late neurula II: Neural plate is deeply sunken. Neural folds are closing and are especially high in the head region where three slight bulges corresponding to fore-, mid-, and hindbrain vesicles are outlined. The neural folds of the spinal region are almost in contact. Hyomandibular furrow is more marked. There are two pairs of somites.

Table 19–1 *(continued)*

Stage Number	Time from First Cleavage: In Hours and Minutes	t_n/t_0	Size (mm)	Features
19	69.00	45	$L = 2.7$ $B = 1.7$ $H = 2.1$	Late neurula III: Neural folds are in contact throughout, but are not yet fused. Brain curvature is quite distinct in profile; fore-, mid-, and hindbrain vesicles are also distinct. The swelling of optic vesicles is outlined. Hyomandibular furrow is deeper. There are three pairs of somites.
20	70.30	46	$L = 2.7$ $B = 1.5$ $H = 1.7$	Late neurula IV: Neural folds are fused in spinal region; in brain region, they are only in contact. Optic vesicles are distinct and becoming larger. Grooves in ectoderm appear at the level of the hindbrain. A very slight swelling marks the future gill region. Mandibular arch becomes prominent and four pairs of somites are present.
21	72.0	47	$L = 2.7$ $B = 1.5$ $H = 1.8$	Neural folds are completely fused. The posterior boundary of the gill region is more distinct. Pronephros is visible as a very slight swelling, and four pairs of somites are present. The ventral side of the embryo is slightly concave; the head region curves downward from the mandibular arch. The dorsal side forms a semicircle; occipital and parietal brain curvatures are apparent.
22	73.00	48.0	$L = 2.8$ $B = 1.4$ $H = 2.3$	The gill region and the pronephros are now more distinct. The tailbud is very slightly outlined, and five to six pairs of somites are present. The ventral side of the embryo becomes more concave as the downward curvature of the head increases.
23	~74.00	~48.5	$L = 3.5$ $B = 1.35$	The primordium of the ear is outlined as a shallow depression in the ectoderm in the region above the future hyoid arch. In the dorsal region of the gill swelling, the hyobranchial furrow appears, outlining the boundary between the hyoid arch and the first branchial arch.

Table 19–1 *(continued)*

Stage Number	Time from First Cleavage: In Hours and Minutes	Time from First Cleavage: t_n/t_0	Size (mm)	Features
24	80.00	52.0	$L = 3.0$ $B = 1.35$ $H = 1.85$	Ear pit is more distinct. The hyobranchial furrow continues to lengthen ventrally. The pronephric swelling is clearly outlined, and both the pronephros itself and the beginning of the pronephric duct are clearly visible. Eight or nine pairs of somites are present.
25	83.00	54	$L = 3.25$ $B = 1.45$ $H = 2.0$	The gill swelling continues, and the hyobranchial furrow begins to lengthen. The first branchial furrow appears faintly outlined in the dorsal region of gill swelling. The tailbud is still small; nine pairs of somites are present. The body of the embryo continues to lengthen; the ventral side is more concave; downward curvature of the head increases.
26	84.30	55	$L = 3.25$ $B = 1.6$ $H = 1.9$	Ear pit is quite distinct. The first branchial furrow is longer and more marked. The gill region is a large, distinctly outlined swelling, the height of which is accentuated by a deep groove formed between it and the developing pronephros. The pronephric duct is visible along six somites. The primordium of the olfactory organ appears as a tubercle on the anterior part of the head. The tailbud is gradually enlarging. The body of the embryo is stretched; the head is curved downward. There are 10–11 pairs of somites.
27	86.00	56 (54–59)	$L = 3.35$ $B = 1.6$ $H = 1.85$	The second branchial furrow appears faintly in the dorsal part of the gill swelling; 12 pairs of somites are present.
28	92.30	60 (57–63)	$L = 3.55$ $B = 1.6$ $H = 1.75$	Further lengthening of the body of the embryo occurs, and maximum downward curvature of the head is reached. The olfactory pit is distinctly outlined in front of the eye, and 14 pairs of somites are present.
29	97	63.5–64.0	$L = 4.2$ $H = 1.75$	The body of the embryo begins to straighten, visible only by the lesser curvature of the head. The tailbud enlarges; 16 pairs of somites are present.

Table 19–1 *(continued)*

Stage Number	Time from First Cleavage: In Hours and Minutes	Time from First Cleavage: t_n/t_0	Size (mm)	Features
30	102	67 (65–68)	$L = 4.5$ $H = 1.65$	The head continues to straighten and the dorsal curvature is reduced; the body elongates and the tailbud enlarges. The finfold appears for the first time. The dorsal finfold begins at somite 14.
31*[c]	109	71 (69–74)	$L = 4.7$ $H = 1.7$	There are 19 pairs of somites. A groove appears in the region of the lens primordium. The third branchial furrow is apparent in the dorsal part of the gill region. The dorsal finfold begins at somite 12.
32	113	74 (73–75)	$L = 5.0$ $H = 1.7$	There are 20 pairs of somites. The dorsal fin begins at somite 10.
33*	113	74	$L = 5.25$ $H = 1.6$	There are 21–22 pairs of somites. The dorsal fin begins at somite 8.
34	115	75	$L = 5.5$ $H = 1.6$	There are 24–25 pairs of somites.[d] The dorsal fin begins at somite 7.
35	122	80	$L = 6.25$ $H = 1.6$	From this stage on, the body axis from the hindbrain to the tail base is quite straight. Three external gills show as nodules on the surface of the gill swelling. The lateral line reaches to the sixth somite. The dorsal fin begins at the fifth somite. The first chromatophores appear; heart pulsation begins. The somites are now difficult to count.
36	130	85–98	$L = 7.1$ $H = 1.7$	External gills are short sprouts directed to the side from the gill swelling.
37*	177	115	$L = 7.5$ $H = 1.7$	Gills elongate and push ventroposteriorly. No limb buds are yet visible.
38*	178	114–132	$L = 7.9$ $H = 1.8$	Filament sprouts appear as two nodules on each gill. The primordium of the operculum is visible as a fold upon the hyoid arch. Neither of the two rudiments of the operculum reaches the midline. The limb buds are slightly outlined.
39	220	144	$L = 9.0$ $H = 1.9$	The first gill has two pairs of filament sprouts; the second and third have three pairs each. The gills cover the limb buds. Both rudiments of the operculum approach the midline. The angles of the mouth begin to show.

Table 19-1 *(continued)*

Stage Number	Time from First Cleavage: In Hours and Minutes	t_n/t_0	Size (mm)	Features
40	240	157	$L = 9.3$ $H = 2.1$	The gills are longer, and the number of filaments increases. On the first gill are four pairs; on the second and third are six or seven pairs. The rudiments of the operculum join at the midline. The angles of the mouth are marked more distinctly. The limb buds protrude slightly.
41	265	173–177	$L = 10.0$ $H = 2.2$	The gills continue to elongate. The number of filaments increases, and they become longer. The mouth is distinctly outlined. The second lateral line runs along the flank toward the limb bud and bypasses it on the ventral side. The forelimb buds are still small. Hatching begins.
42	296	193	$L = 10.5$ $H = 2.1$	The gills extend far beyond the forelimb buds. The mouth is completely outlined, but is not broken through.
43	342	223	$L = 11.3$ $H = 2.3$	The breaking through of the mouth occurs, or the mouth is already open.

[a]Stage numbers in table correspond to those in Plates I–VI. Plate I (stages 1–13); Plate II (stages 14–22); Plate III (stages 23–30); Plate IV (stages 31–35); Plate V (stages 36–40); Plate VI (stages 41–44).

[b]See Fig. 19–2.

[c]See text for explanation of asterisks.

stages of normal development of axolotl embryos were prepared by N. P. Bordzilovskaya. Numbers correspond to stage numbers (see Table 19–1); drawings without designations are side views.

PRECLEAVAGE EVENTS

As is the case for virtually all amphibian eggs, sperm entrance and bilateral symmetrization precede the first cleavage division in axolotl eggs. Axolotl eggs, in common with other urodele eggs, do not possess a fast (electrical) block to polyspermy; hence they are naturally polyspermic. Figure 19–3 illustrates multiple sperm entrance sites on the surface of an egg shortly after sperm attachment. The entire tail, which consists of an undulating membrane supported by an axial fiber, can be seen penetrating the egg surface in Fig. 19–4. Typically, two to four sperm penetrate

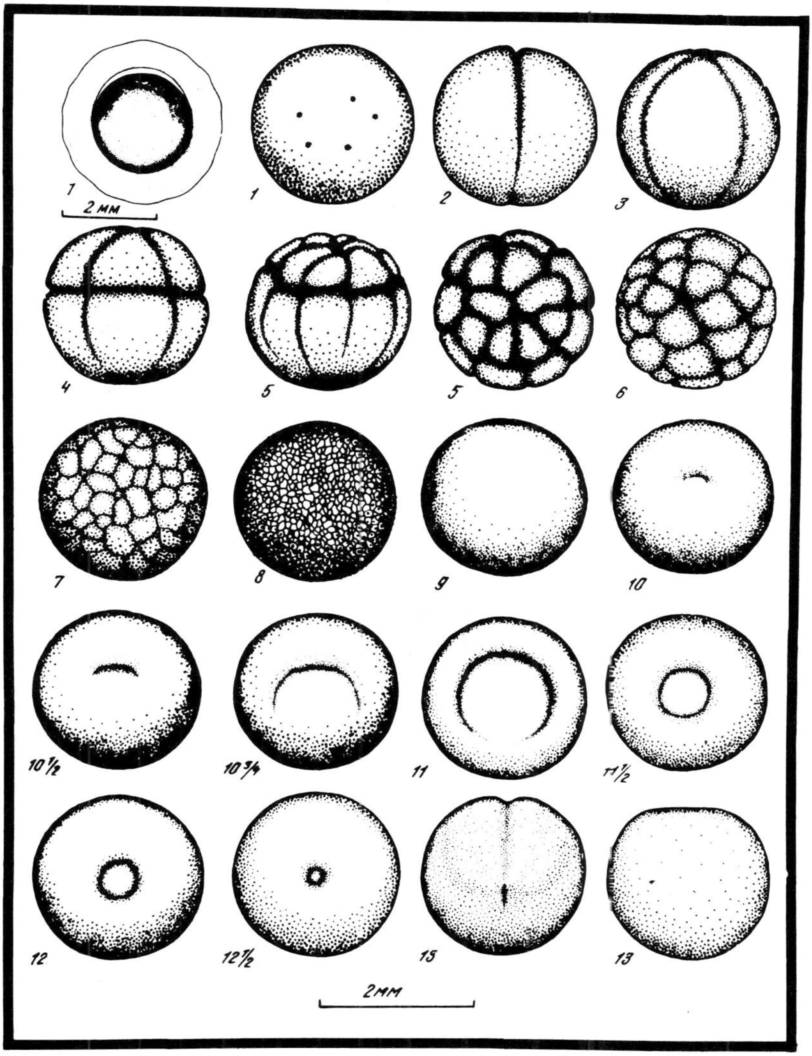

an egg, each penetration being marked by a characteristic and unmistakable local disruption of the egg surface. These penetration points are usually referred to as "sperm pits."

Each of the sperm pronuclei exhibits chromosome condensation and DNA synthesis (Wakimoto, 1979). Eventually, however, the supernumerary sperm nuclei undergo degradation.

Within 3–5 hours postfertilization (18°–20°C) a gray crescent can be observed on one side of the animal hemisphere–equatorial region of some axolotl eggs (Fig. 19–5). Though occasionally easily visualized, usually it is very difficult to observe, since its intensity varies considerably from one spawning to another.

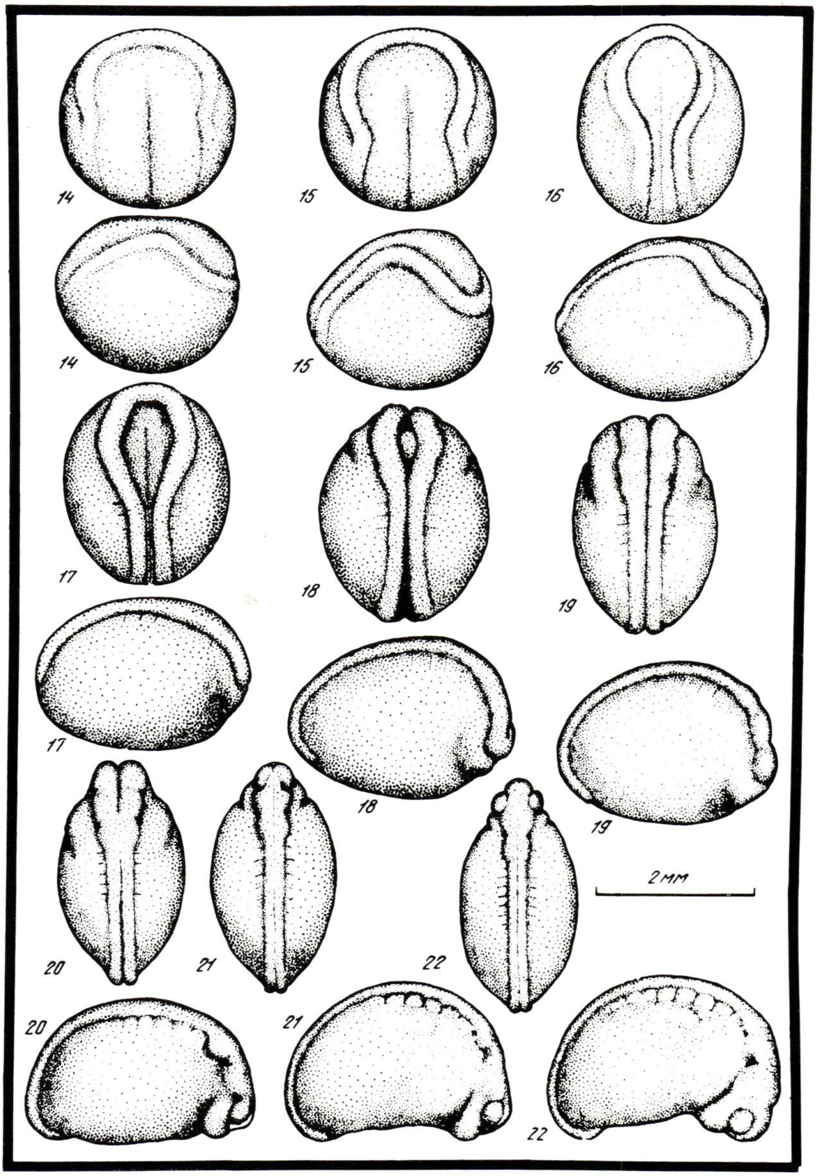

II

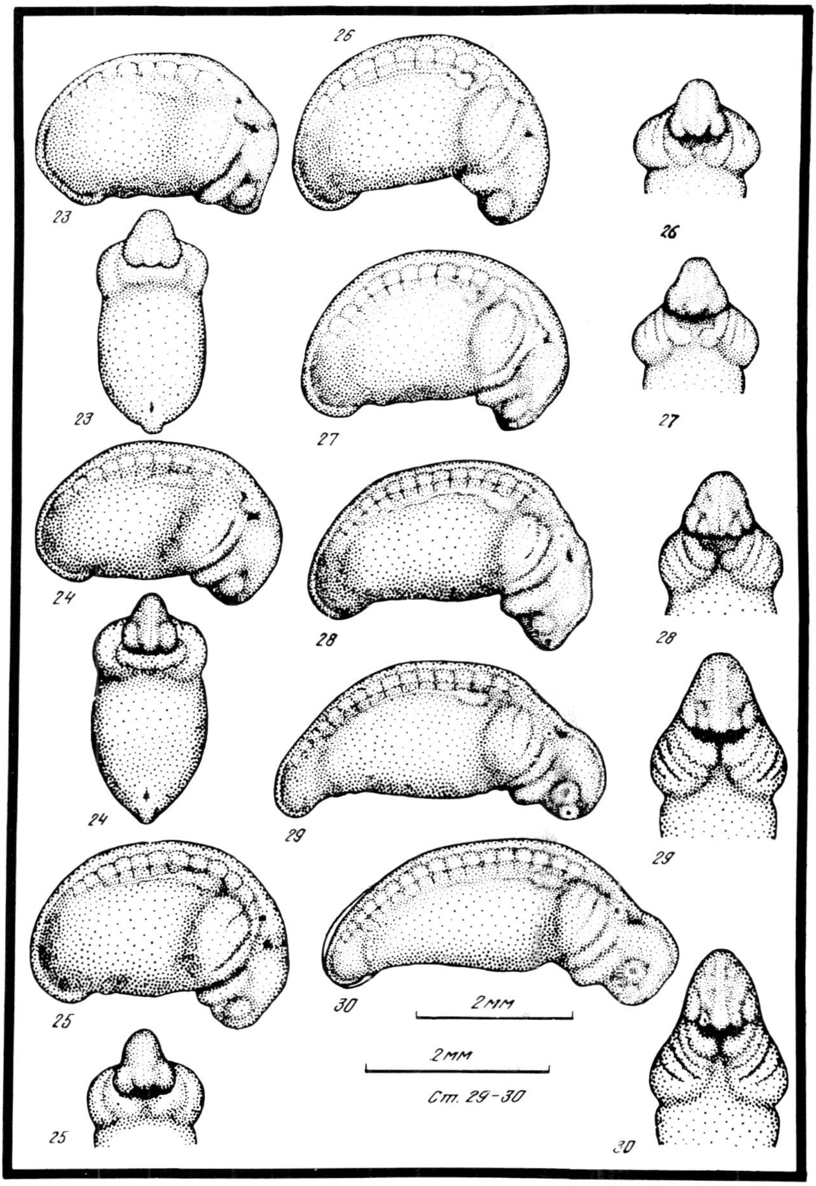

III

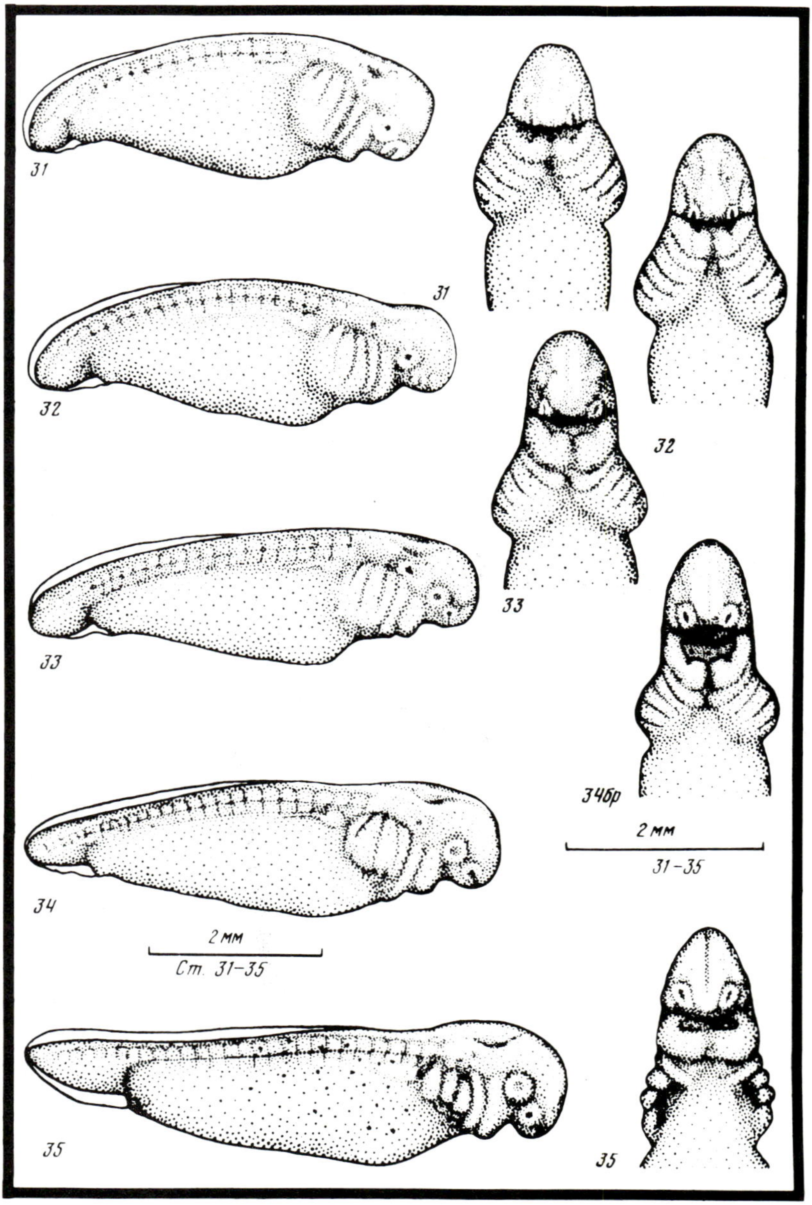

IV

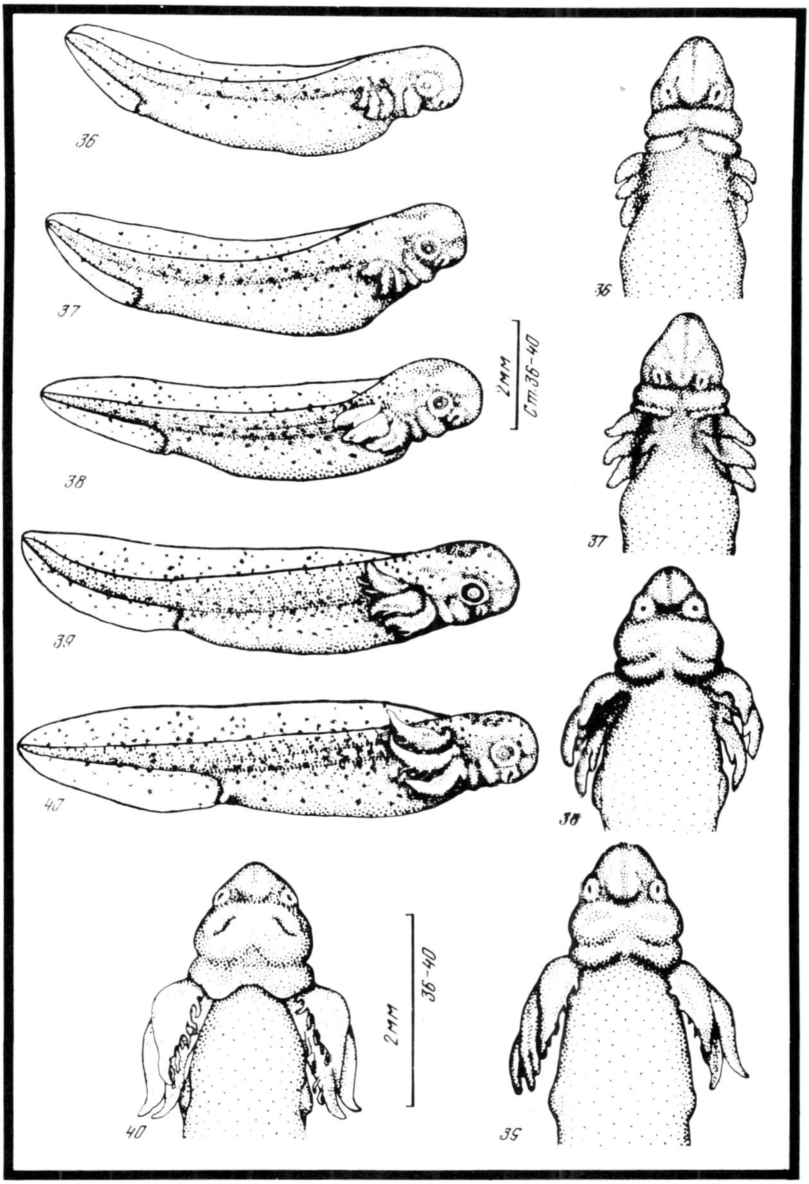

V

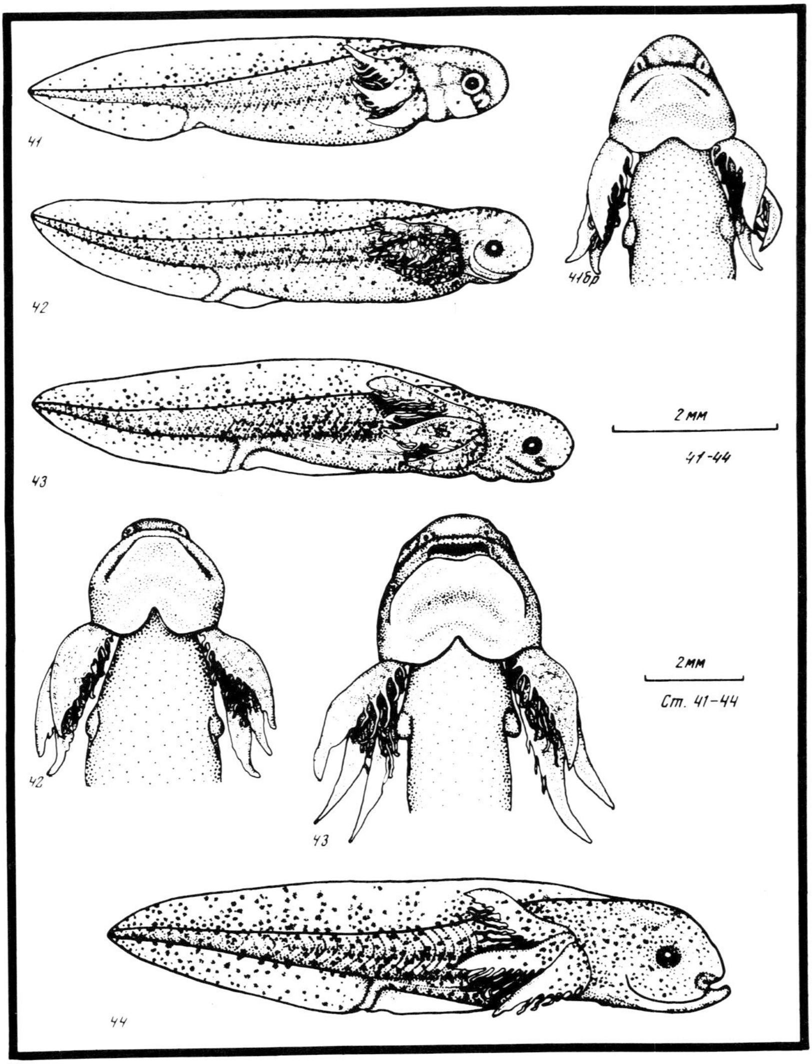

VI

The gray crescent usually serves as a reliable marker to the future dorsal side of the egg. However, no obligatory relationship exists between the original gray-crescent site and the future dorsal side of the egg. Simple 90° rotation, vis-à-vis the gravity vector, of the precleavage egg away from its natural orientation alters the plane of bilateral symmetrization of the egg. Initially, the gray crescent, and later the dorsal lip, form on the side of the 90°-rotated egg that opposed gravity, regardless of where the gray crescent was originally destined to be located (Malacinski and Chung, 1981).

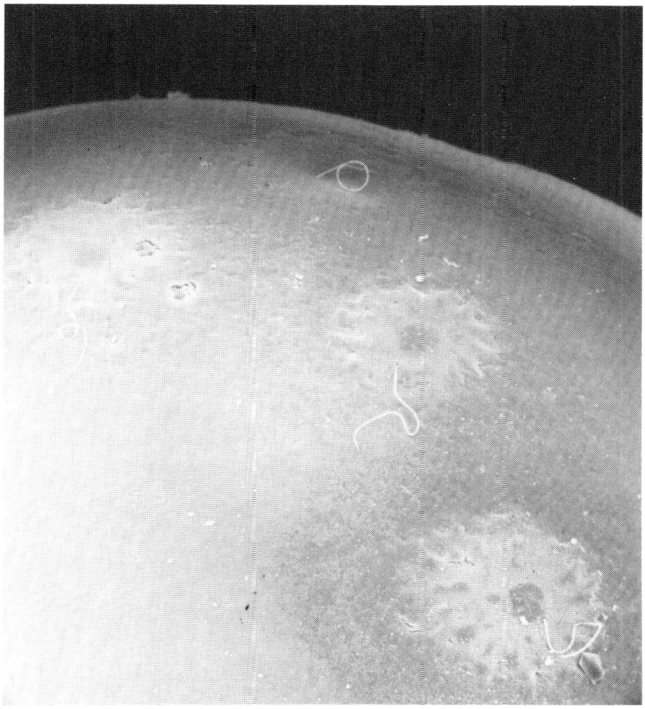

Figure 19–3. Several sperm can be seen penetrating the surface of the axolotl egg. Scanning electron micrograph (×150) of an egg that was fixed approximately 20 minutes after fertilization. (From Chung and Malacinski, 1985.)

COMPARISON TO OTHER STAGING SERIES

Since the drawings and the description of the *Ambystoma maculatum* embryo served as the model for this staging series, it was necessary to examine the degree to which the development of the two species corresponds and what the essential differences are. In general, at least some of the diagnostic features described for each stage are common both to the axolotl and to *A. maculatum*. The principal differences in development are as follows: In axolotl embryos, the gill swelling, hyomandibular, hyobranchial, and branchial furrows, and finfold become apparent at earlier stages; the angles of the mouth also form earlier, but the limb buds develop much later than in *A. maculatum*. The size and shape of the head region differ in the two species at the same stages, and of course the balancer primordia do not develop in the axolotl. Also, Harrison does not indicate at which stage hatching occurs in *A. maculatum*. Details of stage correspondences and differences between the two species are summarized in Table 19–2.

Schreckenberg and Jacobson also based their stages of axolotl development on Harrison's work. Nevertheless, it is difficult at times to compare our series with

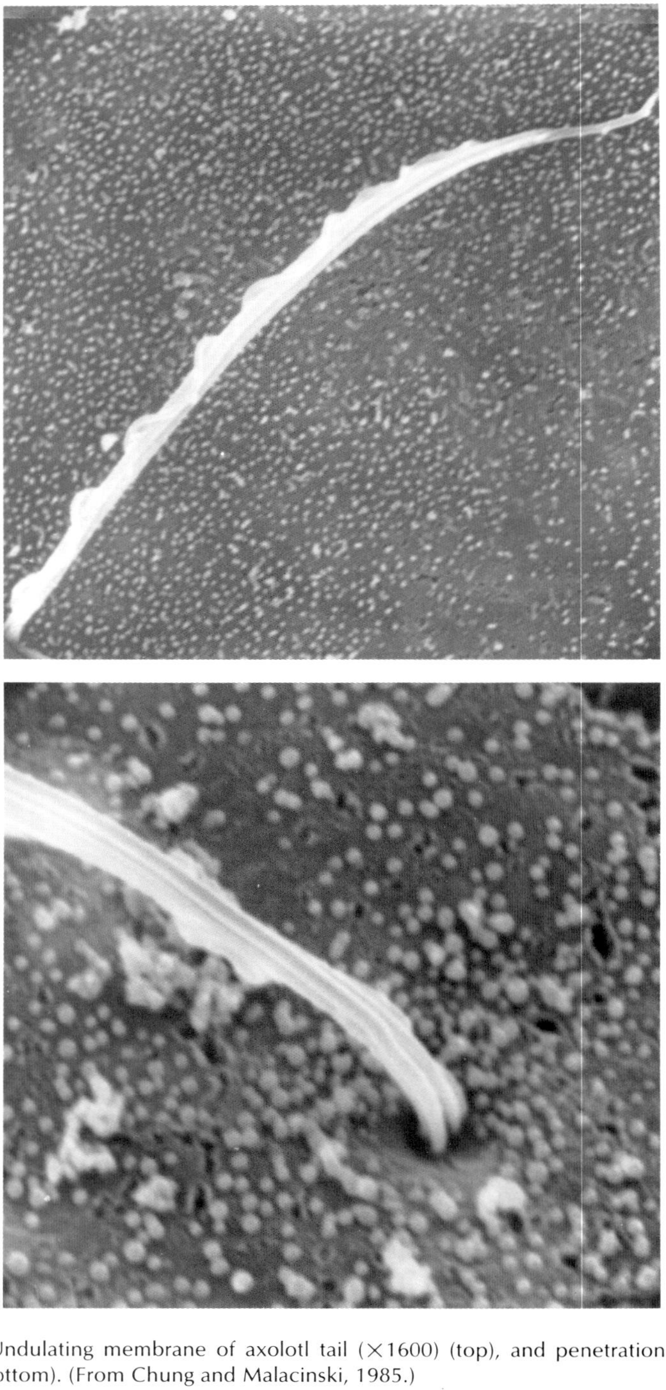

Fig. 19–4. Undulating membrane of axolotl tail (×1600) (top), and penetration of tail region (× 6000) (bottom). (From Chung and Malacinski, 1985.)

Fig. 19–5. Gray crescent of axolotl egg. The gray zone of pigmentation is limited to one side (future dorsal region) of the egg. (From Malacinski et al., 1975.)

theirs, because the photographs do not always show clearly some of our diagnostic features, nor are they always indicated in the descriptions. As far as we can judge, the two series correspond as indicated in Table 19–2.

We also compared our data on the timing of the various stages of development with those of Schreckenberg and Jacobson by recalculating our times to 18°C using t_0 of 105 minutes and correcting for the fact that they began timing from the laying of the egg. The interval between the laying of the egg and first cleavage was about 2.5–$3t_0$. The best agreement was found for the more distinct stages of short duration, such as the beginning of gastrulation, the stage with a slitlike blastopore, and several stages of neurulation. On the other hand, the timing of the stage of hatching differs by about $85t_0$, the embryo of Schreckenberg and Jacobson being much the older.

Timing data obtained for the axolotl by Cate (1956), Skoblina (1963), and Valouch et al. (1971) agree well with ours in the temperature range 16°–23°C. From 26° to 28°C, t_0 ceases to shorten with rising temperature and even lengthens, reflecting the damaging action of these temperatures on the early stages of cleavage, while the duration of the later and less sensitive periods of development continues to shorten, resulting in decreasing values of the ratio (t_n/t_0) (Dettlaff and Dettlaff, 1961; Skoblina, 1963). On the other hand, at 10°C (Cate, 1956) the process of gastrulation lengthens disproportionately, and the fusing of the neural folds is retarded. Thus the ratio t_n/t_0 can only predict the duration of the various stages in the optimal temperature range from about 16° to 25°C.

Table 19–2 Comparison of Stages of Development of Embryos of *Ambystoma maculatum* and *A. mexicanum* in the Tables of Harrison (1969), Bordzilovskaya and Dettlaff (1975), and Schreckenberg and Jacobson (1975)

	A. mexicanum	
A. maculatum Harrison	Bordzilovskaya and Detlaff	Schreckenberg and Jacobson
1	1	1
—	−2	—
2	2	2
3	3	3
4	4	4
5	5	5
6	6	6
7	7	7
8	—	8
—	8	—
9?	9/9.5	8+
10/−10	10	9
—	10.5	10
11	10.75	11
—	12	12
12	—	—
—	12.5	13
13	−13	—
13+	13	14
14	14	14+
15–18	15–18	15–18
19–24	19–24	19–24
25	25	—
26	26	—
27	27	25
28	28	26
29	29	27
—	—	28?
—	—	29?
30–33	30–33	30–33
34	34	—
35	35	34
35	35	35
36	36	36
36	36	37
37	37	38
38	38	39
?	39	—
?	40	—
?	41	40

In order to compare the duration of corresponding developmental stages in different species of amphibians, t_n/t_0 must be determined at an optimum temperature for each species. In closely related species of *Acipenser, Rana* (Dettlaff and Dettlaff, 1961), and *Salmo* (Ignatieva, 1974), the relative duration of the developmental stages of the same name is similar or identical. This may hold true for the various ambystomatids, but no data for comparison are available. On the other hand, the absolute duration of the corresponding developmental stages of even closely relates species (Schreckenberg and Jacobson, 1975) can have only a restricted value, because the optimal temperature ranges may not coincide, and development is strongly temperature dependent.

20

Raising the Axolotl in Captivity

JOHN B. ARMSTRONG, SUSAN T. DUHON,
AND GEORGE M. MALACINSKI

Undoubtedly, the main reason for the popularity of the axolotl among amphibian species employed for studies both in development and in other areas has been the ease with which it can be maintained and bred in the laboratory. The axolotl is normally neotenous, meaning that it achieves sexual maturity without undergoing metamorphosis. It therefore spends its entire life in the water. Consequently, the laboratory scientist need not provide both aquatic and cumbersome terrestrial environments for rearing and maintaining axolotls, as is the case for most amphibians. A single axolotl can be maintained in a small aquarium and fed worms or chopped beef heart. However, there are many variations on this simple theme, some of them involving much more elaborate equipment designed to reduce the labor required to maintain a large colony. There is probably no single setup that is optimal for all laboratory situations. In this chapter we will examine some of the variations used by different laboratories. We will also discuss normal spawning behavior. However, induced spawnings and artificial insemination will be described in Chapter 21.

CONTAINERS

Many laboratories, particularly those that maintain mutant stocks, prefer to keep animals in individual containers for ease in identification. This also has the advantage of minimizing cannibalism and the spread of disease, but is much more labor-intensive than some other systems.

One of the traditional containers is the so-called "squat-style fish globe," often used for goldfish. In the 1-gallon (4.5-liter) size, these containers are 24 cm in diameter. Filled to a depth of about 6 cm, they are ideal for an adult. Smaller size globes are suitable for juveniles. Unfortunately, we have found these containers (particularly the smaller sizes) difficult to obtain from commercial suppliers in recent years. One possible substitute is the rectangular plastic mouse cage sold by many scientific suppliers. The 29 × 19 × 13 cm size is satisfactory for an adult axolotl, but other containers of similar size should be just as good.

W. F. Fox (1980b) has warned that potentially dangerous plasticizers are leached from plastic containers. Polyethylene appears to undergo the least leaching, followed by polystyrene, polycarbonate, and polypropylene in that order. As well as the possibility of direct danger, there may be indirect danger due to increased bacterial growth on the plasticizers. However, we know of several laboratories that have used plastic containers for years without obvious ill effects. As plastic containers are economical, our advice to someone starting out would be that if the animals seem to do poorly in plastic, try a glass container to see if they improve. Another good, though expensive, substitute would be stainless steel. Containers in various sizes are often available from restaurant suppliers.

Animals maintained in individual containers can be changed by carefully pouring them into a plastic colander, which can be set next to the sink while the container is rinsed or washed and the water replaced. A new colander should always be checked carefully to see that none of the holes have rough edges that could injure axolotl skin. The containers may be cleaned with a "scrub mixture" composed of two parts baking soda to one part salt. If necessary, they may also be disinfected with a weak solution of Lysol, after which they must be rinsed thoroughly. The colander should also be rinsed thoroughly between animals to minimize cross-contamination, and should be disinfected periodically. Any sick animals should be isolated, and separate equipment should be used to handle them.

The next step up from the individual container is the aquarium. Some laboratories have reported success maintaining several adults in a 10-gallon (40-liter) aquarium. (The minimum volume per adult axolotl appears to be about 6 liters.) Undergravel filters work well, although dacron wool-type "bubble-up" filters have also been used. The addition of plants, either live or plastic, seems to give the animals a greater sense of security and to reduce aggression (Brun, 1980). Animals may be identified by placing together only those that have clearly distinguishable phenotypes, or by marking them in some way. Brun (1980) has suggested using a collar of colored thread placed behind the gills but anterior to the forelimbs.

Spotte (1979) has suggested the following design criteria for undergravel filters: (1) surface area of the filter bed should equal the surface area of the aquarium; (2) grain size of the gravel should be in the range of 2–5 mm; (3) gravel must be carefully graded to assure uniform grain size; (4) the filter bed should not vary in depth; (5) shape of the gravel should be rough and angular; and (6) minimum depth of the bed should be about 7.6 cm. In addition, for raising axolotls, the bed must be topped with a layer of larger stones of a size (>1 cm diameter) that the animals cannot swallow.

The filter works as an aerobic bacterial bed to convert toxic ammonia and nitrites from decaying food and fecal matter into nitrates that can be used by plants and algae. Algal growth is therefore to be encouraged. Forbes and Forbes (1980) reported that an aquarium with a healthy growth of algae rarely needs cleaning. The bacterial beds should be disturbed as little as possible. However, excess detritus should be siphoned off. Spotte (1979) suggests gently stirring the surface of the bed once every 2 weeks to disturb the detritus and make it easier to siphon off. At least 10% of the water should be changed about every 2 weeks. If dacron wool filters are used, there should be two per aquarium and only one changed at a time.

Raising animals in aquariums minimizes handling of the animals and the labor involved in cleaning containers. However, it does require careful monitoring of water quality. Ammonia and pH should be monitored regularly. Spotte (1979) recommends maintaining the pH of a freshwater aquarium in the range of 7.1–7.8. Ammonia test kits are available from pet shops.

Several laboratories have experimented with larger containers housing a proportionally larger number of axolotls. Justus and Cullum (1971) described a system employing stainless-steel tanks 234 × 50 × 14 cm deep partitioned into 14 equal "cages." Water collected from a spillway at the end of each tank was recirculated through dacron and charcoal filters, sterilized by ultraviolet irradiation, and redistributed through a perforated pipe running the length of the tank. Fresh water was added every few days and uneaten food and fecal matter siphoned out.

Verhoeff-de Fremery (1980) has described keeping animals in 50 × 100 × 50 cm (250-liter) asbestos cement containers, 10 animals to a container. The water runs continuously at a slow rate and is also aerated in the tank. Uneaten food and wastes are siphoned out.

Slack (1983) has described a system employing 38 × 68 × 30 cm polypropylene tanks connected to a water supply that runs at a very slow rate. These tanks, which hold 60 liters and up to 20 adults, are equipped with a drain at the bottom and a siphon-breaker 5 cm up the side. They are flushed and refilled three times a week. The siphon-breaker ensures that the tank does not drain completely. Consequently, the animals do not have to be removed. Slack's report includes a detailed description and design plans.

While maintaining animals together in large numbers may work for adults or even juveniles, it may result in cannibalism among larvae. In the Ottawa colony, eggs and larvae are generally kept in 11-cm-diameter finger bowls (specimen dishes). Prior to hatching, we keep about 50 per bowl. The water need not be changed prior to hatching unless it becomes turbid. Dead embryos are, of course, immediately removed. In the Indiana colony, bowls 20 cm in diameter are used. In excess of 200 embryos can be kept in one bowl, provided that they are kept from clumping. Embryos in the interior of a clump may not receive adequate oxygen as they develop. After hatching, the larvae are gradually thinned out as they grow. Larvae should be well fed and are segregated by size to minimize cannibalism. Some will always grow faster than others and will eat their smaller sibs if permitted to. Larvae that seem particularly aggressive should be separated from the others and, in any event, should be kept one to a bowl by the time they reach 4 cm in length. When they outgrow finger bowls, they are promoted to progressively larger bowls.

WATER

Water quality parameters have not been systematically investigated, but some investigators have reported difficulty in maintaining axolotls on treated municipal water. Distilled water is best avoided. Chlorine is obviously detrimental, and extremes in pH are also to be avoided. To remove chlorine, water can be allowed to stand overnight or passed through a charcoal filter (or both). In some localities, both chlorine and ammonia are added to treated city water. At the Indiana colony, where the tap water is so treated, we have found it necessary to remove the ammonia, chlorine, and the resulting chloramine compounds with a commercially available chloramine remover (Amquel). We also add a water conditioner (Novaqua) to reduce heavy metals and reduce stress on the animals. Even though axolotls are freshwater amphibians, many laboratories find that they thrive best in a dilute saline solution. We use a modified 50% Holtfreter's saline solution containing 1.75 g NaCl, 100 mg $MgSO_4$, 100 g $NaHCO_3$, 50 mg $CaCl_2$, and 25 mg KCl per liter. The saline seems to reduce fungal and bacterial growth, and the animals seem healthier than in straight tap water.

Oxygen levels can be monitored, if necessary, with a portable polarographic meter. However, low oxygen is seldom a problem if the containers are aerated or the water changed regularly. Nitrogenous metabolites can be measured following the procedures described in *Standard Methods for the Examination of Water and Wastewater* (APHA, 1971), or with less (but probably sufficient) accuracy, with test kits for aquariums sold in pet shops. High chloride ($>$3 parts per part nitrite), such as in Holtfreter's saline, reduces nitrite toxicity (Buttner and Nace, 1984).

At the University of Ottawa colony, we noticed an improvement in the general health of our animals after installation of a Millipore filtration system on our city water line. The system consists of a coarse and fine depth filter, the latter removing particles above 0.5 μm. The water is also passed through a charcoal filter and stored overnight in a 600-liter reservoir before use. The surface of the reservoir is irradiated with two 18-watt germicidal lamps to reduce microbial contamination further. This system may be overly elaborate, but there have been no serious disease problems since it was installed.

TEMPERATURE

Most laboratories maintain larval, juvenile, and adult axolotls at 18°–20°C. Embryos have been successfully raised at temperatures ranging from 10° to 28°C (Bordzilovskaya and Dettlaff, 1979; Chapter 19). The developmental rate is temperature dependent, which is often useful for getting embryos to the right stage at a convenient time for observations or operations. However, the developmental rate begins to decrease again about 25°C, and this should probably be considered the practical maximum.

The actual temperature preference of larvae is about 24°C (K. M. Armstrong and J. B. Armstrong, unpublished observations). This result suggests that maintaining larval axolotls at a slightly higher temperature than the usual 18°–20°C would likely accelerate growth without serious detrimental effects.

FOOD

In most laboratories, larvae are fed brine shrimp *(Artemia salina)* on a daily basis until they are 2 or 3 cm in length. Brine shrimp are hatched from eggs, which may be purchased from pet stores or various scientific supply companies; freeze-dried brine shrimp are not satisfactory. Brine shrimp can be hatched in a 2-liter reagent bottle fitted with a two-holed stopper and an air line running to the bottom of the bottle. A suitable salt concentration is 40 g/liter. The amount of brine shrimp added depends on the requirements of the colony. We have successfully hatched as much as 12 g in a single bottle, which is enough to feed about 50 finger bowls. The hatching time depends on the temperature; at 30°C a batch normally hatches in 24 hours.

To harvest the shrimp, the air line is removed, the bottle is tilted up slightly, and a light is placed near the low side. Eggshells float to the surface and, after 10–15 minutes, the shrimp collect near the bottom. Brine shrimp are attracted by light and seem to collect more quickly and in a smaller area than if the light is not used. The shrimp are then siphoned into a shallow glass or plastic dish and the light placed near one corner. This step helps to concentrate the shrimp further and eliminate unhatched eggs that may have been carried over by the siphon. After the shrimp collect in the corner of the dish, they are removed with a wide-mouth pipette and added, as needed, to the bowls of larvae. The collection dish can also be refrigerated and the shrimp will remain viable for a few days.

Once past the brine shrimp stage, most laboratories feed their animals beef heart or liver. This meat should be trimmed of all fat and gristle and is most conveniently handled when frozen as blocks approximately 3–5 cm square by 10–15 cm long. At the time of feeding, fine slivers can be trimmed off with a sharp knife, for younger animals. Larger pieces are cut for adults; a convenient size would be 2 × 5 × 50 mm. Adults are generally fed five or six pieces, three times a week; juveniles are fed daily. At the Ottawa colony, a mature female needed for spawning that does not appear to be in prime spawning condition will also often be fed daily over a 3- to 4-week period. Animals will learn to pick up food from the bottom of their container if it is simply thrown in.

Some laboratories dip the meat in bone meal or vitamins, but at the Ottawa colony we have found that our animals seem healthy on a steady diet of beef heart with only an occasional supplement of surplus larvae! At the Indiana colony, juveniles and adults have been fed for many years on pureed beef liver, usually mixed with Gainesburger brand dog food. We now believe that this diet is not optimal, and may be detrimental due to excessive vitamin A content, low fiber, and poor balance of other nutrients such as calcium and phosphorus. Histopathological examination of the livers of certain of our animals by the International Aquaculture Research Center in Hagerman, Idaho showed ceroid deposits, indicating lipoid liver degeneration, a condition that can be caused by vitamin A toxicity (Halver, 1972). We are now feeding our juveniles soft-moist salmon food pellets, which are available in a wide range of sizes. We wean the young larvae off brine shrimp to ⅛-inch pellets when they are about 6–8 cm long. Juveniles are fed every day, and on the pellet diet they reach an adult size by about 8 months of age. The only drawback we have discovered so far is that the rate of spontaneous metamorphosis appears to be slightly increased, although it is still very low.

Axolotls will accept a variety of live food including tadpoles, small fish, *Daphnia,* and earthworms. Brun (1980) has been feeding his colony live worms for a number of years, but sometimes finds it necessary to cut them in half to prevent regurgitation. Minnows (bait fish) should be avoided because they carry many parasites.

GROWTH AND MATURATION

Embryos hatch after 2.5 weeks and are about 1 cm in length. Thereafter, their growth rate depends on how well they are fed. Under optimal conditions they are mature in a year, but more usually in about 18 months. The females usually mature a few months later than the males. Their useful breeding life is up to 5 or 6 years of age. After that it becomes increasingly difficult to obtain spawnings, and both egg and sperm quality appear to deteriorate. However, axolotls will live 10 years or longer under laboratory conditions.

OBTAINING SPAWNINGS

The general method for obtaining natural spawnings is to place a male and female together in a suitable container 2 days before eggs are required. The container, which may be a small aquarium or a plastic or metal dishpan, the bottom of which is covered with a few large pebbles, is usually covered or left in the dark. The next day the male is removed and, if spermatophores have been deposited on the rocks (see Malacinski, 1978b, for photo), the container with the female is returned to the dark. Oviposition generally occurs about 24 hours after insemination. Some laboratories chill the animals when they are first placed together (Verhoeff-de Fremery, 1984; F. Briggs, 1984). W. F. Fox (1984) noted that this caused the male to deposit spermatophores but did not seem to increase the frequency of successful spawnings.

In most temperate urodeles, the reproductive cycle is closely correlated with the seasons of the year (Lofts, 1974). Although, to our knowledge, the reproductive behavior of the axolotl in the wild has not been reported, Brandon (1970) reported that a closely related neotenic species, *Ambystoma dumerilii,* appears to have a spring breeding season. Most laboratories relying completely on natural spawnings report particular difficulty obtaining spawnings in the summer, suggesting that the axolotl retains some seasonality, even though several generations removed from its natural environment. In many laboratories, the animals are likely exposed to environmental cues such as seasonal fluctuations in temperature or lighting. At the Indiana colony, where all adults are kept in a room with controlled lighting, we have found changes in lighting to be critical for obtaining successful natural spawnings, although the absolute duration of the lighted period is not especially important. In particular, a decrease in "day" length will sharply curtail successful natural spawnings. Such a decrease may be counteracted by gradually increasing the lighted period by 15-minute increments over 2 or 3 weeks, and then holding it constant. Nevertheless, the animals still show marked seasonality, spawning most successfully from January through May. Perhaps they have imprinted the changes they were exposed to as juveniles, when they were kept in a room under natural light conditions.

One way of circumventing this seasonality is to induce ovulation hormonally (Chapter 21). Hormonal stimulation is employed routinely at the Ottawa colony, whereas nearly all the spawnings at Indiana are obtained without the benefit of hormonal induction.

Løvtrup (1981) has reported obtaining natural spawnings year-round. His animals are maintained in 40-liter aquariums at a constant 15°C on a 12 hours on–12 hours off light cycle. He also leaves the animals together constantly, two to three pairs per aquarium, so that spawnings are completely spontaneous. This may be the simplest approach if only wild-type embryos are required, but would be impractical for many purposes.

THE COURTSHIP RITUAL

In most urodeles, chemical secretions of various glands play an important role in mate interactions (Salthe and Mecham, 1974). In the first stage of courtship, the male becomes aware of a potential mate, possibly through substances (pheromones?) originating from glands in the female's cloaca, and approaches her cloacal region. In turn, the female may approach the male from the rear, nuzzling his cloaca with her snout. This further excites the male, and he begins to evert his cloaca, exposing its glands. A male, once excited, will generally approach the female from the side and will push her laterally around the mating pan with his snout. This process may also involve the transfer of chemical stimulants, as many male urodeles have hedonic glands on the side of the head.

In the next stage of courtship, the male tries to position himself ahead of the female with his cloacal glands just in front of her head. He often moves his tail back and forth. The female then nuzzles the cloacal glands again and the male fully exposes them.

When the male is ready to lay a spermatophore, he grasps a rock with his hind legs and deposits the spermatophore on it. The spermatophore consists of a cap of spermatozoa on a conical gelatinous base secreted by the male's cloacal glands.

The excited female should follow the male, bringing her cloacal opening over the spermatophore. She will often grasp the rock between her hind legs. Her secretions dissolve some of the gelatinous matrix, permitting the spermatozoa to make their way into the spermathecal tubules that open into the dorsal portion of the cloaca. The spermatozoa are stored there until spawning begins, at which time a few are expelled against the surface of each egg as it comes into the cloacal chamber from the oviducts. Humphrey (1977a) has shown that sperm stored in the spermatheca remain viable for 5–10 days. Diagrammatic illustrations of the courtship routine can be found in Duellman and Trueb (1986).

If one chooses to observe the courtship, rather than letting it take place in the dark, it is easy to see whether the female picks up the spermatophores. If she does not, she can be encouraged by placing a rock bearing a spermatophore under her just ahead of the cloaca. In our experience at the Ottawa colony, which admittedly is based almost entirely on spawnings where the female is hormonally induced, the animals do not seem bothered by the light.

Sometimes the male continues vigorously to court the female after sufficient spermatophores have been laid. If he makes it difficult for the female to pick up the spermatophores, or even knocks them off the rocks, he should be removed. Otherwise, the male should probably be left with the female, at least until three or four spermatophores have been picked up, as she may stop looking for spermatophores in his absence.

SPERMATOGENESIS

The normal pattern, for a urodele in the temperate zone, is that sperm mature during the summer and are released into the vas deferens in fall and early winter (Lofts, 1974). They are stored in the vas during the winter and are laid as spermatophores in the spring. In the lab, male axolotls still retain much of this cycle but, at least in the Ottawa colony, are not synchronized with the season (Chapter 4; Miltner and Armstrong, 1983). Consequently, two animals examined at the same time may be at widely different places in their cycle, and we find males in good spawning condition during all seasons.

The spermatogenic cycle requires about 60 days from premeiotic S phase to the first mature sperm, and another 60–80 days before sperm are released to the vas deferens. We have not determined how long sperm may remain viable in the vas deferens or what factors, in the absence of environmental cues, trigger a new round of spermatogenesis.

SPAWNING FREQUENCY

Our general policy in the Ottawa colony is not to use a female, after a successful spawning, for 4 months. However, it is often quite possible to use a female again after a shorter interval if she has only produced a few eggs. We have been averaging 660 eggs per spawning (following hormonal stimulation); our largest spawning was 1669 eggs. The most eggs from a single female over a 12-month period was 3457, which probably approaches the maximum number produced by a female during the year.

In the Indiana colony, we will reuse a female for a natural spawning after just 2 months with frequent success and no reduction in the average spawning size (the egg yield depends on the size of the female).

In a series of mutagenesis studies, where males were mated frequently throughout the year, we found that a male could be spawned once every 3–4 weeks, providing he was limited to producing 8–10 spermatophores each time. The maximum number of spermatophores we have obtained in a single spawning was 45, whereas the maximum number obtained from a single male over a year was 113. Ortiz (1977) reported that the average spermatophore contains 3.3×10^5 sperm.

21

Induced Spawnings, Artificial Insemination, and Other Genetic Manipulations

JOHN B. ARMSTRONG AND SUSAN T. DUHON

One of the oft-cited advantages of the axolotl as a subject for experimental and genetic studies is the ease with which it is maintained and bred in captivity (Humphrey, 1975). Yet, as many researchers can attest, it is often difficult to obtain fertile eggs from natural matings year-round. (A natural mating is defined as one in which no stimulants are used.) Even if one chooses not to resort to hormonal stimulation, some knowledge of the normal mating pattern, as described in Chapter 20, can be of use in obtaining successful spawnings.

INDUCED OVULATION

Our experience with several thousand spawnings at the Ottawa colony over more than a decade indicates that eggs can be obtained much more reliably from hormonally stimulated females. There appear to be neither any short-term nor obvious long-term detrimental effects on the animals.

Following the lead of K. M. Newrock and A. J. Brothers [cited in Humphrey (1977a)], we first used follicle-stimulating hormone (FSH), sometimes supplemented with luteinizing hormone (LH), to induce ovulation. However, the success obtained in stimulating *Xenopus* with human chorionic gonadotropin (hCG) prompted us to try this hormone, and we have now used it routinely for several years. Our procedure is to inject the female intramuscularly, somewhat dorsolaterally to the cloaca, in the late afternoon or early evening. The usual dose is 250 IU. The female is left overnight in the mating pan (see Chapter 20) at 12°–18°C. Newrock and Brothers (1973) placed the male and female together at the time of injection and left them covered or in a dark place. We prefer to place the male with the female the following morning so that we can observe courtship behavior. If there is no courtship, or the male produces no spermatophores within the first hour, a second male can be substituted. After the female has picked up a sufficient number of spermatophores, the male is removed and the mating container covered. The female usually starts to lay eggs within a few hours.

Table 21–1 Effect of Hormonal Stimulation of the Female on Spawning Success

Treatment		Male Produced Sperm			No Sperm	
		Fertile Eggs (%)	Infertile Eggs (%)	No Eggs (%)	Infertile Eggs	No Eggs
None	(18)[a]	28.0 (403)[b]	0	33.0	0	39.0
FSH + LH	(199)	70.9 (239)	0	6.0	9.5 (143)	13.6
250 IU hCG	(501)	89.0 (568)	3.9 (261)	3.6	1.6 (313)	2.8
500 IU hCG	(70)	91.4 (639)	4.3 (316)	0	4.3 (124)	0

[a]Number of spawnings.
[b]Average number of eggs laid.

Our procedure does have the disadvantage that a female may sometimes start to lay eggs before morning. This seldom happens if she can be kept in an incubator at 12°–15°C, or if she can be injected late in the evening.

Our results (Table 21–1) indicate that hormonal stimulation definitely affects the probability of ovulation. Although we found no significant difference in this parameter between FSH- plus LH- and hCG-stimulated females, the latter treatment definitely produces more eggs. Although FSH and LH are reportedly the major gonadotropic components of hCG (Gurdon, 1967), minor components may also play a role.

We attempt to maintain animals in an isolated room without windows and with daylight-type fluorescent lighting on a 12 hours on–12 hours off cycle. As far as is practical, no environmental cues concerning seasonality are present. We are usually able to get good spawnings all year with only a trace of residual seasonal variation: both the number of eggs and the frequency of successful spawnings increase slightly in the spring (Armstrong and Fletcher, 1984).

STIMULATION OF MALES

Spermiation—the release of spermatozoa from the testes—is stimulated by LH in a number of species, and by both FSH and LH in some (Lofts, 1974). In *Xenopus*, hCG has been used with both males and females to stimulate spawning. However, *Xenopus* males do not store their sperm after they are released from the testes (Deuchar, 1975). Thus, we are seeing the effects of the hormone on spermiation directly. We have not determined whether gonadotropins affect spermiation in the axolotl, but there is no apparent effect of hCG on spermatophore production (Armstrong and Gillespie, 1981).

ARTIFICIAL INSEMINATION

The earliest procedures described for artificial insemination (Brunst, 1955a; Fankhauser, 1967) involved sacrifice of both male and female in the apparent belief that eggs that had been in contact with water could not be inseminated (Fankhauser, 1967). However, several laboratories have found that eggs laid by hormonally stimulated females can be fertilized successfully if collected at regular intervals and inseminated before the jelly layers have had a chance to swell. Sacrifice of the male is also unnecessary, unless one encounters difficulty getting a male to produce spermatophores. Our procedure, which varies only slightly from that described previously (Gillespie and Armstrong, 1981), is as follows:

1. The female is given an intramuscular injection of 500 IU hCG, 8–12 hours before the planned insemination.
2. When the female begins to lay eggs, or preferably shortly before (as judged by her behavior and/or the production of jelly), a male is placed with her and spermatophores are collected as laid. If it is absolutely essential that normal insemination not occur, a separate "decoy" female can be used to stimulate the male (Trottier and Armstrong, 1975).
3. Each spermatophore is placed in a small tube, on ice, in about 250 μl of 50% Leibovitz L-15 culture medium. Sperm preserved in this manner will remain viable for up to 24 hours.
4. Just before use the spermatophore is gently broken up by pipetting it up and down two or three times in a Pasteur pipette.
5. Eggs are collected as laid and dried immediately on a paper towel for 1–2 minutes to remove the excess water. For maximum fertility, it is important that the jellycoats have not begun to swell and have been well dried (they should be somewhat sticky). The eggs are then placed in a dry petri plate and inseminated with a drop of sperm suspension. After 10 minutes, 25% Holtfreter's saline is added. Eggs are generally collected over a 10-minute period. Thus, some are actually inseminated for closer to 20 minutes.

In case of failure to obtain spermatophores by this procedure, one may have to resort to sacrificing the male and preparing a sperm suspension by maceration of the vas deferens. Brunst (1955a), Humphrey (1962), Fankhauser (1967), and Mohun et al. (1980) all describe variations of the same basic procedure.

SEX REVERSAL

For some experimental purposes, manipulation of the natural reproductive genetics of the axolotl is useful. For example, large numbers of females are desirable for the study of maternal-effect mutations. In order to achieve a sex imbalance favoring females, techniques of sex reversal may be employed.

Sex reversal in *Ambystoma* was first explored by Burns (1925), who discovered that *A. punctatum* parabiotic twins resulted in single-sex pairs to the exclusion of opposite-sex pairs. Burns concluded that in pairs conjoining animals of opposite sex, the gonads of one were reversed in almost every case. Burns's work did not indicate a dominance by either testis or ovary.

Humphrey sought to verify and further Burns's work using orthotopic grafts. In a series of papers he established that *Ambystoma* gonads were initially bisexual in character and potential, and, when a presumptive gonad was grafted into a host of opposite sex, the testis almost invariably dominated, regardless of whether it belonged to the donor or to the host (Humphrey, 1929a,b, 1931a,b,c). He also established that true sex reversal was taking place; he had not obtained only a mosaic of donor and host tissue (Humphrey, 1936, 1938). The fertility of such animals depends on whether correct connections to the ducts are made by the transplanted gonad (Humphrey, 1942, 1944).

An exception to the dominance of the testis occurred in heteroplastic grafts between species of differing size and developmental rate. In that case, the ovary of the more rapidly developing species dominated the testis of the more slowly developing species so that females were obtained (Humphrey, 1935a,b). Humphrey believed that other secondary factors, such as poor nourishment during growth, explained other anomalous data such as that in Burns's original study (Humphrey, 1932).

By studying the progeny of genetic females that had been experimentally converted to males, Humphrey (1945, 1948a, 1957) was able to establish that the male is homozygous for the sex locus (*ZZ*) and the female heterozygous (*WZ*). *WW* individuals arising from such crosses are also female.

The availability of sex hormones in pure form led to sex reversal experiments in amphibians involving either direct administration of hormones via injection or exposure by adding them to the water in which the larvae are reared. In urodeles, the usefulness of the male hormone is limited because of its somewhat paradoxical effects. It leads to severe inhibition of the gonads of both sexes and sometimes even to feminization of the male gonad (reviewed in Gallien, 1955; Burns, 1961).

On the other hand, Gallien succeeded in transforming genetically male newts *(Pleurodeles waltl)* into functional females by prolonged treatment of larvae with estradiol benzoate (E2B) in aqueous solution. The homogametic sex (males in *Pleurodeles* and *Ambystoma*) is the one that appears to be readily reversible by hormone treatment (Burns, 1961), at least in species for which the sex chromosome organization is known.

Bodney (1982) tested whether axolotl males could be sexually reversed by the administration of E2B in the water of developing larvae. He prepared a stock solution of 0.2 g E2B per liter of 95% ethanol and added 0.5 ml of stock solution to each liter of aquarium water for an effective concentration of 100 μg/liter. After keeping newly hatched larvae in this solution (changed daily) for 60 days, he found that 19 of 20 developed into females (the sex of one could not be determined). Sex was determined by histological examination of the gonad, which had been removed at the end of the 60-day period.

Subsequently, this method was used several times in the Indiana colony, but detailed records of the results were not kept. Several groups of treated larvae were raised to sexual maturity, and all proved to be phenotypically female. Several instances have been noted where a treated animal has given rise to only male offspring when spawned with a normal male. We conclude that these females are *ZZ*. The best example is a female (5578.3) from which offspring from three separate spawnings were raised, all of which turned out to be male.

From these experiments, we conclude that functional sex reversal of genetically male axolotls can be achieved through the administration of E2B in the water of developing larvae. However, the method should be used with caution so as to avoid an overabundance of males among the progeny of hormonally treated animals.

MANIPULATION OF PLOIDY

For various types of experimental analyses, it is desirable to have available eggs, embryos, or even adults that display different ploidies. Fortunately, without much effort, haploids, triploids, and even tetraploids can be produced.

Haploids

Although at one time haploid embryos were of interest for creating haploid cell lines (Freed and Mezger-Freed, 1970), haploid embryos generally develop poorly and must be regarded primarily as a laboratory curiosity. In the axolotl, gynogenetic haploids have been produced by activation of eggs with ultraviolet-irradiated sperm (Hronowski et al., 1979). Androgenetic haploids may be produced by inactivating the egg pronucleus with UV light at a time, shortly after insemination, when it still lies near the surface of the egg but the sperm nucleus has penetrated to near the center (Gillespie and Armstrong, 1980).

Triploids

One of the major interests in the production of triploids is for use as a cell lineage marker in developmental experiments (Chapter 14). Practical aspects of using triploids as lineage markers are outlined in Chapter 22. Gene–centromere map distances can also be obtained from the segregation frequencies of genetic markers among triploid progeny of heterozygous females (Lindsley et al., 1956).

Triploids are produced by suppression of second-polar-body release in normally fertilized eggs. Thus, they have two maternal sets and one paternal set of chromosomes. The earliest experiments employed cold or heat shock (Fankhauser and Humphrey, 1942; Sladecek and Lanzová, 1959; Rott and Betina, 1964; Trottier and Armstrong, 1976), but neither survival nor yield was particularly high. Experiments in other species, employing hydrostatic pressure to cause dissolution of the meiotic spindle, led us to try this method in the axolotl (Gillespie and Armstrong, 1979). A triploid yield of better than 90%, with 60–70% survival, was obtained following 6.9×10^4 kPa (10,000 psi) pressure treatment applied 60 minutes after oviposition for a duration of 8 minutes.

Humphrey and Fankhauser (1946) found that triploid females were generally fertile, although they did not reach maturity until 2 years of age and laid fewer eggs. Triploid males, however, are sterile.

Tetraploids

In the newt, several laboratories have used heat shock to suppress first cleavage and create tetraploids. In the axolotl heat shock may also be effective, but our experience with this treatment has been mainly to produce androgenetic diploids from eggs that have had the female pronucleus inactivated (Gillespie and Armstrong, 1981). For this purpose, a 10-minute shock at 36°–37°C, applied 5.5 hours after oviposition, was most effective for suppressing first cleavage.

Hydrostatic pressure is also effective for producing tetraploids. Treatments at

8.2–9.6 × 10^4 kPa (12,000–14,000 psi) for 6–8 minutes, applied 5.75 hours after oviposition, yield over 90% tetraploids, but survival is only 10–15% (Gillespie and Armstrong, 1981).

Tetraploid females are fertile and can be mated with diploids to yield triploids (Fankhauser and Humphrey, 1950), or with each other to produce more tetraploids. Ultimately, for cell lineage experiments, this may be more convenient than making triploids by pressure treatment.

KARYOTYPING

The axolotl's haploid chromosome number is 14. Karyotypes have been prepared from hepatocytes, tailfin epithelial cells, and regenerating-limb blastemas following incubation of either the whole animal or the tissue in 0.1–0.5% colchicine (Callan, 1966; Cuny and Malacinski, 1985). There is an approximately 5-fold difference in size between the largest and smallest chromosome, with the others distributed fairly evenly between the extremes. The largest chromosomes are mostly metacentric; the smallest are acrocentric. None are telocentric.

Callan (1966) distinguished between the chromosomes not only on the basis of size and position of centromere, but also on the basis of secondary constrictions produced by cold shock. Cuny and Malacinski (1985) used Giemsa-banding patterns obtained after incubation with Hoechst 33258, and were able to distinguish 10 of the 14 from the homologous tiger salamander *(Ambystoma tigrinum)* chromosomes. Both groups have identified chromosome 3 as carrying the nucleolar organizer on the basis of its carrying a subterminal constriction. In addition, Cuny and Malacinski (1985) have determined that chromosome 14 carries the sex locus on the basis of sex-correlated differences in its banding pattern.

GENETIC MAPPING

Despite the fact that over 30 mutants have been described (Armstrong, 1985; see also Chapter 10, 11), few linkage relationships have been reported (reviewed by Armstrong, 1984). The greatest progress has been in establishing gene–centromere distances. Both triploids (Lindsley et al., 1956) and gynogenetic diploids (Armstrong, 1984) have been used for this purpose. The latter technique depends on the suppression of second-polar-body release, generally by hydrostatic pressure in eggs that have been activated by heavily irradiated sperm (Gillespie and Armstrong, 1979). If there is no crossing-over between the centromere and the gene in question, only homozygous progeny will be obtained, with wild type and mutant occurring in equal numbers. If crossing-over does occur, heterozygous progeny will also be obtained, and their number will depend on the gene–centromere distance.

The mapping function is the same as that for segregation of tetratypes in fungi and was first adapted to gene–centromere mapping in amphibians by Nace et al. (1970). The equations include a coefficient k "proportional to coincidence" (Barratt et al., 1954). Barratt et al. (1954) maintained that there was a reasonable fit with his *Neurospora* data for $k = 0.2$, and this constant-k assumption was used by

Nace et al. (1970) for *Rana pipiens* and Reinschmidt et al. (1985) for *Xenopus laevis.* However, Weinstein (1958) showed that coincidence increased from near zero for short map distances to near 1.0 for distances of about 50 map units. Consequently, Armstrong (1984) used an empirically derived logistic equation for k based on the *Drosophila* data. However, some caution must be exercised in basing assumptions on data from disparate species. In trout, the percentage heterozygosity approaches 100% for some loci (Thorgaard et al., 1983; Thompson and Scott, 1984), suggesting that only single crossovers occur ($k = 0$).

The genes mapped to date in the axolotl, based on the assumptions of Armstrong (1984), are tabulated in Table 21–2.

The obvious next step should be to develop a method for assigning genes to specific chromosomes. In an amphibian system, the most successful method developed to date is that of Nishioka et al. (1980), for *Rana nigromaculata* and *R. brevipoda.* Hybrid females, which are completely viable and fertile, were backcrossed to males of either species. For any particular chromosome, 50% of the resulting backcross progeny should be hybrid. The chromosomes were identified as *R. nigromaculata* or *R. brevipoda* on the basis of morphological differences in lampbrush chromosomes. If a backcross animal is hybrid for a particular marker, then the chromosome carrying the gene must be hybrid, and the converse must also be true. However, other chromosomes may or may not be hybrid. By determining the karyotypes of backcross progeny of known genotype for various markers, chromosome assignments can be made. Cuny and Malacinski (1985; see also Chapter 10) have suggested using essentially the same approach with axolotl–tiger salamander *(A. tigrinum)* hybrids, except that they planned to work with the F_2 rather than backcrosses.

Kobel and du Pasquier (1979) have explored a different approach, based on the creation of hyperdiploid *Xenopus,* diploid for the chromosomes of one species but carrying supernumerary chromosomes from a second. When hybrids are created between two *Xenopus* species, females frequently produce diploid eggs from allotetraploid oocytes produced by spontaneous endoreduplication in the germ line. When such eggs are fertilized by sperm from one of the parent species, trip-

Table 21–2 Gene–Centromere Map Distances in the Axolotl

Gene	Heterozygosity (%)	Distance from Centromere[a]
Albino (*a*)	60.0	NC[b]
Axanthic (*ax*)	61.2	NC
Cardiac (*c*)	11.7	5.9
White (*d*)	58.2	NC
Esterase (*est-3*)	58.3	NC
Melanoid (*m*)	83.7	59.1
Lethal *p*	53.7	NC
Phosphoglucomutase (*pgm*)	45.6	23.6
Lethal *t*	46.9	24.3
Sex locus	57.1	NC

[a]In centiMorgans (cM).

[b]Not calculated, as percentage heterozygosity is not significantly different from 66.7%.

loids are produced that have two sets of chromosomes from one species and one from the other. The eggs produced by triploid females have a complete haploid set from the backcross species and are expected to have a random distribution of from 0 to 18 chromosomes from the other species. When fertilized by sperm, again from the backcross species, hyperdiploids are produced. These are then analyzed for karyotype and the expression of characters from the second species. Kobel and du Pasquier (1979) found that the hyperdiploids that reached adulthood carried between 8 and 16 chromosomes from the second species, rather than only a few, and suggested that these survive better. This complicates the process of assigning markers to a specific chromosome, and consequently they were unable to make precise assignments.

In principle, this method can be used with a single species. Suppose, for example, that we create triploids by second-polar-body suppression (as described earlier) in wild-type females inseminated by albino (a/a) males. The progeny should be $+/+/a$. Fankhauser and Humphrey (1950) have shown that triploid axolotl females produce eggs with chromosome numbers ranging primarily from haploid to diploid. In our example, the eggs might thus be $+/+$, $+/a$, $+$, or a. If the eggs are fertilized by sperm from an albino male, the only albino progeny that can arise will be a/a, and must carry only two copies of the chromosome bearing the *a* gene. They may, however, carry three copies of any of the other chromosomes. If a number of albinos are analyzed, one should be able to determine the chromosome carrying *a* as the only one present in only two copies in all these animals.

22

Genetic Markers and Their Use in Chimeras

JOHN B. ARMSTRONG AND KEN MUNEOKA

The term *chimera* has been used to describe any animal composed of different cell populations derived from more than one fertilized egg (McLaren, 1976a). Their use has been primarily to allow recognition of the exact fate of different cells or tissues during subsequent development, and therefore requires recognizable differences between the two populations. These differences may arise through application of extrinsic labels (e.g., vital or fluorescent dyes, or radioactive markers), or they may be intrinsic, genetic markers. Only the latter will be discussed here.

Ideally, cell markers should allow us to determine both the proportion of the component cell populations in various tissues, and their distribution. In addition, the best markers should be ubiquitous throughout all tissues, different enough to be recognized as distinct from unmarked cells, and developmentally neutral (McLaren, 1976a; West, 1984). Because many useful markers do not fulfill all these criteria, a choice must be made based on the type of question being asked.

One may also choose between heterospecific and homospecific chimeras. Heterospecific chimeras have been used for many years and generally depend on relatively gross differences in cellular or nuclear size and/or morphology. Thus, Raven (1937) employed the difference in nuclear size between *Triton taeniatus* and the axolotl to study the origin of sheath cells and sympathetic neuroblasts. Much later, Slack and Forman (1980) used the differences in cell size and nuclear staining between *Xenopus laevis* and the axolotl to examine the interaction between explanted dorsal and ventral marginal zones.

The usefulness of heterospecific chimeras depends on the compatibility of the two species with respect to factors such as developmental pattern and rate. Thus, the *Xenopus*–axolotl combination would likely not work very well if the entire embryo were being studied, as the two differ substantially in developmental rate. However, these differences may not be critical to experiments with explants.

Considerations of compatibility can be circumvented by the use of homospecific cell markers, although even these have certain limitations. In this chapter, we will consider three kinds of markers: pigmentation, number of nucleoli, and electrophoretic mobility of isozymes.

PIGMENTATION

Four pigmentation variants have been described in the axolotl: albino, axanthic, melanoid, and white (Chapter 12). Only albino and white have been used as markers. Axanthic and melanoid do not involve reductions in the amount of melanin, and tissue transplanted from such animals to a wild-type background would be difficult to distinguish.

Albino is a useful marker for the study of early developmental stages, since pigmentation differences between wild-type and albino cells are clearly defined. At these early stages one is not restricted to studying cells that synthesize their own pigment, as both wild-type ectoderm and mesoderm contain pigment of maternal origin. Although the mesoderm contains less pigment, individual cells can still be distinguished against an albino background. The epidermis of the albino embryo is sufficiently transparent that cells beneath it can be observed without having to remove the epidermis. Furthermore, melanin granules in the wild-type cells permit them to be distinguished from albino cells in sectioned material. If, for operations, difficulty is encountered because of the lack of contrast in albino embryos, they can be immersed for an hour in a dilute solution of a vital dye such as neutral red until they are lightly stained (Zackson and Steinberg, 1986).

Zackson and Steinberg (1986) made elegant use of albino–wild-type chimeras to study the migration of cranial neural crest and pronephric duct cells. The technique permitted them to trace the movements of small numbers of cells that otherwise would have gone undetected. Migration of the lateral line has also been followed in this way (Fig. 17–3).

The white–wild-type combination has been used in a number of studies on limb regeneration to determine the origin of tissues (reviewed in Muneoka and Bryant, 1984a). As a marker for the epidermis, the color differences are convenient. However, as Muneoka and Bryant (1984a) have shown, boundaries in the underlying dermis cannot be reliably inferred from those in the epidermis. In experiments where supernumerary limbs were stimulated using tissues marked with both ploidy and pigmentation, there was no correlation between the distribution of these two markers (see Fig. 8 in Muneoka and Bryant, 1984a).

NUMBER OF NUCLEOLI

In the axolotl, the diploid–triploid combination has been used to study problems such as cellular commitment during formation of the mesoderm (Forman and Slack, 1980), and development and regeneration of the limb (Chapter 14). The system is based on the correlation between the number of nucleoli and the ploidy of the cell: diploid cells possess two nucleoli and triploid three (Fankhauser and Humphrey, 1943). In studies on limb regeneration, triploidy is an ideal marker, since triploid axolotls regenerate their limbs in a manner indistinguishable from their diploid counterparts. However, the cytological distinction between diploid and triploid cells is based on nucleolar counts on a per-cell basis, and suffers from loss of resolution and the potential introduction of artifacts when analyzed in histological sections.

Triploid axolotls have been generated by exposing newly fertilized eggs to heat, cold, or hydrostatic pressure. These treatments disrupt the spindle apparatus during the second meiotic division of the egg, preventing formation of the second polar body, and thus allowing the egg to contribute an extra set of chromosomes to the embryo. In our experience, hydrostatic pressure results in the highest survival of treated embryos and a high frequency of triploid embryos (Gillespie and Armstrong, 1978, 1979).

Newly laid eggs are collected every 20 minutes, and then pressure-treated 60 minutes after collection. They are held at 6000 psi for 8 minutes, then reared to hatching in 20% modified Steinberg's solution. Pressure-treated larvae are screened for triploidy by microscopic analysis of nucleolar number in live-tissue squashes (usually of the tailtip) viewed with phase-contrast microscopy, or in whole-mount stained tailtips viewed with light microscopy (Muneoka et al., 1984). The trinucleolate frequency has been shown to vary between animals, but not within an individual animal. Hence, animals with a low frequency of trinucleolate cells should be screened out to maximize the resolution of the marker in subsequent experimental analyses.

A number of nucleolus-specific stains have been described (Namenwirth, 1974; Muneoka et al., 1984; Maden and Mustafa, 1984). The one most widely used is the bismuth stain (Fig. 22–1) developed by Locke and Huie (1977) and modified for axolotl tissue by Muneoka et al. (1984). The staining procedure can be used in the analysis of both histological sections and whole-mount tissue preparations.

In whole-mount preparations the frequency of trinucleolate cells is not 100% but varies widely from animal to animal. However, within individual triploid animals the trinucleolate frequency is constant (Muneoka et al., 1984). This has important implications for studies in which cellular contribution in triploid–dip-

Fig. 22–1. Bismuth-stained whole-mount preparation of limb tissues. (a) Triploid dermis; (b) diploid dermis. (From Muneoka et al., 1984.)

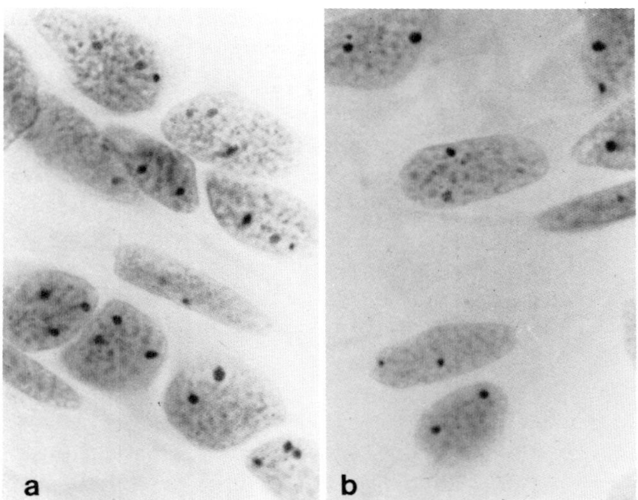

loid chimeras is quantified. In order to convert observed frequencies to real contribution frequencies and thus accurately determine cellular contributions, the trinucleolate cell frequency must be determined for the triploid donor (see, for example, Muneoka and Bryant, 1984a,b; Muneoka et al., 1985, 1986a,b). However, for qualitative studies to determine only whether cellular contribution actually occurred (see, for example, Tank et al., 1985; Rollman-Dinsmore and Bryant, 1984; Gardiner et al., 1986), such control counts are not required.

The resolution of the triploid cell marker is lower in histological sections than in whole-mount preparations. This results from randomly transecting nuclei and analyzing sectioned tissue possessing a population of partial and whole nuclei. Zero, one, two, or three nucleoli may be present in partial triploid nuclei; thus, the frequency of trinucleolate nuclei in a section of triploid tissue is less than the frequency observed in whole-mount preparations. For example, in a study of the cellular contribution to the regeneration blastema in the limb (Muneoka et al., 1986b), we utilized triploid animals with trinucleolate frequencies of greater than 50%, as determined from whole-mount analyses. In histological sections of control triploid blastemas, the trinucleolate frequencies ranged from 13 to 39%, indicating a loss in resolution of between 2- and 4-fold due to sectioning.

Histological sectioning creates scoring artifacts that have been attributed to overlapping of partially transected nuclei, and result in the identification of false-positive trinucleolate cells in sections of pure diploid tissue (Table 22–1). The frequency of false positives varies for different tissue types, suggesting that the plane of section relative to the shape of cells analyzed is an important consideration. No false positives are identified in whole-mount preparations of the limb dermis, where reduced cell density results in clearly distinguishable nuclear boundaries.

Because of the reduced resolution and potential artifacts problems associated with analyzing histological sections, we have developed methods for the analysis of whole-mount tissue preparations whenever possible. Two types of whole-mount preparations have been developed to facilitate the determination of the origin and behavior of cells during regeneration. The first involves the isolation of the limb dermis as a sheet of cells from complete regenerates (dermal preparations; Fig. 22–2). The second permits the cellular origin of the very early blastemal cells to be determined in whole-mount preparations of the distal apical ectodermal cap (AEC). When proximal skin plus AEC is dissected of the stub, blastemal cells adhere to the underside of the AEC as contaminants. The preparation, stained with bismuth, is mounted flat by making longitudinal cuts of the proximal skin, giving the appearance of a flower (Fig. 22–3).

Table 22–1 Diploid Cell Counts

Tissue	Preparation	Total Cells Examined	Frequency of Trinucleolate Cells (%)
Dermis	Whole mount	10,769	0
Cartilage	Section	7,415	0.3
Muscle	Section	7,236	2.4
Blastema	Section	663	0.5

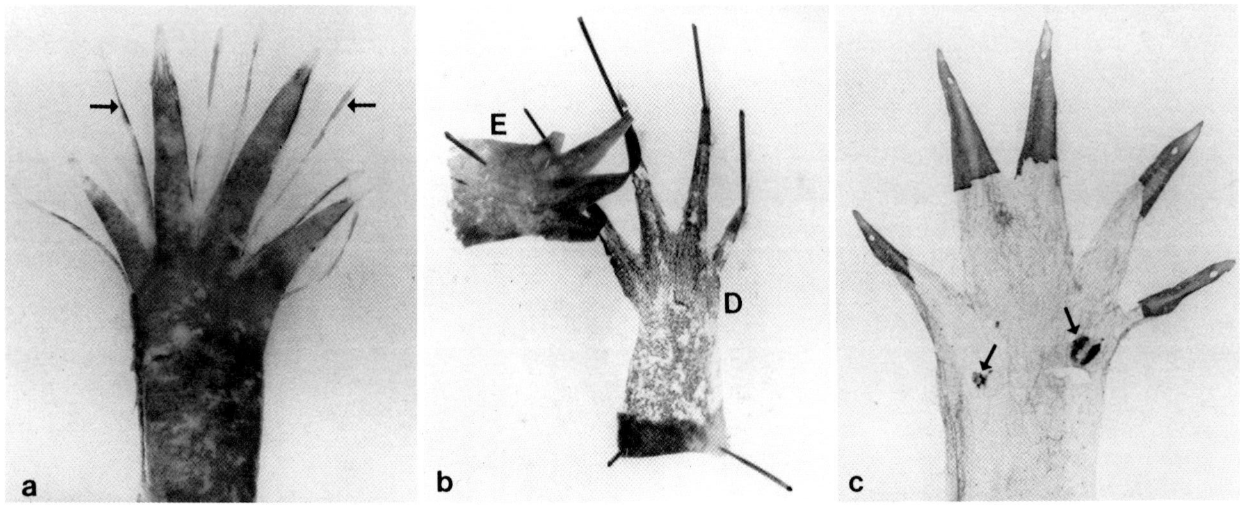

Fig. 22–2. Procedure for isolating the dermis of an axolotl limb. The limb is fixed in Carnoy's fixative for 1–2 days and transferred to 100% ethanol. (a) The skin is cut along the anterior and posterior margin of the limb and the interdigital skin (arrows) is trimmed away. The dorsal and ventral skin is peeled off and transferred to distilled water. (b) The dorsal skin is shown pinned out on wax, and the epidermis (E) has been carefully peeled away from the dermis (D). Epidermis is left at the tips of the digits and at the proximal edge of the skin to prevent the dermis from curling up (arrows). (c) The dermis is mounted on a slide and stained with bismuth. In this example the ventral dermis of a hindlimb is shown. Arrows indicate parts of contaminating epidermis obvious in whole-mount preparation. ×7. (From Muneoka et al., 1984.)

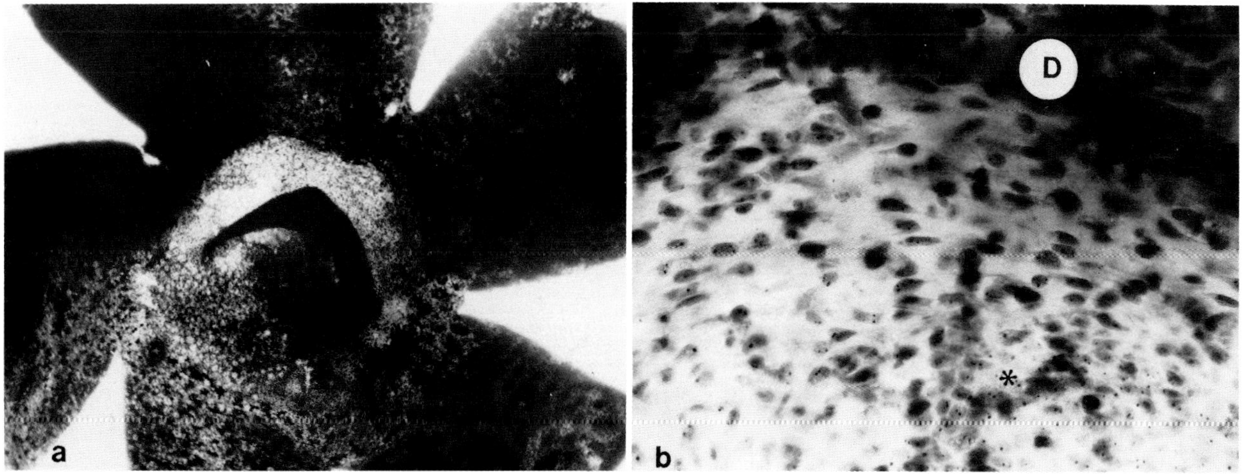

Fig. 22–3. Light micrographs of a whole-mount preparation 15 days postamputation. (a) Low-power photograph of whole-mount preparation to show lighter, central wound area. The dark crescent-shaped region within the wound area is the tip of the blastema that folded over during sample preparation. ×16. (b) Higher magnification of the upper right region of the wound area in (a). The nuclei of blastema cells in the wound area (light area) near the boundary with the mature dermis (D) are oriented parallel to the wound edge (i.e., in an arc from left to right). A patch of epidermal cells (*) is visible in the lower right-hand corner of the micrograph. ×17. (From Gardiner et al., 1986.)

An analysis of these flower preparations has demonstrated that the first blastemal cells to accumulate under the AEC are derived in part from the migration of peripherally located dermal cells (Gardiner et al., 1986). These dermally derived cells subsequently proliferate to give rise to approximately half of all the cells present in the medium-bud stage blastema (Muneoka et al., 1986b).

ISOZYMES

Chimeras constructed from two electrophoretic variants offer the advantage that the relative contributions of two cell populations to a tissue or organ can be quantified, but the disadvantage that spatial relationships cannot be assessed unless a histochemical assay can be developed. In spite of this limitation, isozymes have been used extensively in the analysis of mammalian chimeras (see examples in Le Douarin and McLaren, 1984). In the mouse, an immunohistochemical method has been developed employing antisera specific to one of the glucose phosphate isomerase isozymes (Oster-Granite and Gearhart, 1981), and there seems to be no reason why such a technique could not be adapted to the axolotl. West (1984) has suggested that a similar approach involving an isozyme with a null allele would be even more easily developed.

In the axolotl, a number of enzyme systems have been screened for possible electrophoretic variants. Pavlick (1977) reported genetically controlled variants for seven loci: acid phosphatase, aconitase, β-glucosidase, lactic dehydrogenase (LDH),

Fig. 22–4. Drawing of the patterns seen on polyacrylamide gels stained for phosphoglucomutase (PGM), esterase (Est), and peptidase B (Pep B). The PGM gel shows segregation of alleles *a* and *b* at a proposed single locus to produce phenotypes A *(a/a)*, AB *(a/b)*, and B *(b/b)*. The esterase gels show either a four-zone pattern (A), or a three-zone pattern (O) produced by animals homozygous for a null allele at the *est-3* locus. The Pep B gels show six possible patterns due to the segregation of three alleles (*a*, *b*, and *c*) at a single locus. Homozygotes produce single-band patterns; heterozygotes produce three-band patterns.

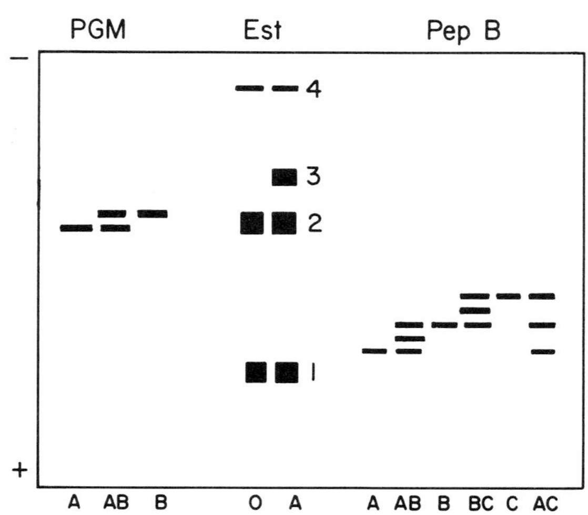

leucyl-glycyl-glycine peptidase (Pep B), and two phosphoglucomutase (PGM) loci. Cooper et al. (1985) were able to confirm variation at one PGM locus, but found none for LDH. In addition, they reported an esterase variant. The PGM results indicate a monomeric enzyme with two alleles that are expressed codominantly in heterozygotes (Fig. 22–4). Esterase gels show four bands, each of which appears to be determined by a separate locus. In some animals, the *Est-3* band is absent. Cooper et al. (1985) proposed that these animals are homozygous for a null allele. In an attempt to devise a histochemical staining procedure, we have searched for a substrate specific to *Est-3,* but without success.

We have recently confirmed Pavlick's report of variation for a leucyl-glycyl-glycine peptidase. This has been named Pep B to be consistent with D. A. Wright and C. M. Richards's nomenclature (1982) for *Rana pipiens* peptidases. The enzyme appears to be a dimer, and three alleles have been found.

The assay procedures for PGM and esterase are outlined in Cooper et al. (1985). Pep B is resolved best by isoelectric focusing and may be run under the same conditions as PGM. The gels are stained for peptidase with a solution of 80 mg leucyl-glycyl-glycine, 10 mg *Crotalus adamanteus* venom, 1500 units peroxidase, and 50 mg *o*-dianisidine dihydrochloride in 50 ml 0.1 *M* Tris buffer pH 8.0 containing 0.1% $MgCl_2$. The gels are incubated overnight at room temperature with shaking.

FUTURE PROSPECTS

The markers described above have all been useful, but each has limitations. Pigmentation can, of course, only be used for pigmented tissue, nucleolar number is difficult to assess in sectioned tissue, and our inability to come up with a suitable histochemical assay for isozyme differences prevents resolution at the single-cell level. However, a novel means of circumventing these problems is suggested by the recent development of recombinant vectors carrying the *Escherichia coli lacZ* gene. When the recombinant vector is introduced into host cells, the β-galactosidase activity in their progeny can readily be detected histochemically. This approach has been successfully used for lineage-labeling in the mouse (Sanes et al., 1986) and rat (Turner and Cepko, 1987). Not only are such vectors useful as lineage labels when injected into individual cells of a developing embryo, but they can also be injected into uncleaved eggs and, if the vector integrates into the germ line, then the progeny of such animals can be used to construct chimeras. Preliminary indications (M. Whiteley, personal communication) suggest that such an approach is equally viable in the axolotl.

23

Primary and Long-Term Culture of Axolotl Cells

SALLY K. FROST, ANTON W. NEFF, AND GEORGE M. MALACINSKI

The first step toward "selection" for a permanent cell line requires the ability to successfully establish and maintain primary cultures of viable cells. The axolotl has, until now, proved to be highly refractory to the established methodologies for long-term cell culture. With the exception of the short-term maintenance of differentiating embryonic cells in vitro (Duprat, 1970; Landström et al., 1976), no reports exist that describe a successful cell culture procedure for the axolotl. This is especially unfortunate, since the Mexican axolotl exhibits a wide variety of interesting and well-described mutant phenotypes (Chapter 10), many of which may lend themselves to in vitro investigation. In fact, one such group of axolotl mutants, tentatively defined as "cell-lethal mutants" (Chapter 11), awaits further phenotypic characterization pending the establishment of successful cell culture methods.

This chapter describes a procedure for generating primary cell cultures from a variety of axolotl tissues. It also describes a procedure for establishing and maintaining long-term cultures of epithelial cells. It appears that the conditions required for the successful dissociation and subsequent growth of axolotl cells in vitro are somewhat different from the methods previously described for other amphibian cells or established cell lines (Rafferty, 1976).

PROCEDURES FOR PRIMARY CULTURES

Source and Preparation of Tissue

Young, sexually immature *Ambystoma mexicanum* larvae (1–5 months old, 2.5–15.0 cm in length) were found to be the most suitable donors of experimental tissue. Animals are sacrificed by an overdose of MS-222 (ethyl *m*-aminobenzoate, methanesulfonic acid salt). The animals are then rinsed with a 5.25% sodium hypochlorite–ethanol (3:7) solution to eliminate surface bacteria and mold. Tissues (including liver, heart, flank muscle, kidney, and gonad) are surgically removed from the animals and placed in sterile wash media consisting of 50% Leibovitz L-15 medium (GIBCO), 1% kanamycin (10,000 μg/ml), and 1% fungizone (250 μg/ml), pH 7.4. When more than one animal is used, appropriate tissues are pooled in individual plates of wash medium.

Cell Dissociation

Each tissue is finely minced, placed in a trypsinizing flask, and dissociated by gentle stirring. Dissociation medium consists of 25% L-15 and 0.25% pronase. The following types of pronase have been utilized: Type VI protease (Pronase P, Sigma Chemical Co.); Type V protease (Pronase AS, Sigma); nonspecific protease (Boehringer-Mannheim Biochemicals); and Pronase, B grade (Calbiochem). According to the manufacturers' catalogs, all were isolated from *Streptomyces griseus.* Their effectiveness in dissociating axolotl tissues is indicated in Table 23–1. At various intervals dissociated cells are decanted into a centrifuge tube and centrifuged at low speed (approximately 200 *g* in a tabletop centrifuge) for 3–5 minutes. The cell pellet is washed once with whole medium (50% L-15, 15% fetal calf serum, 1% kanamy-

Table 23–1 A Summary of Cell Culture Parameters Determined Under Different Dissociation Conditions[a]

Tissue and Exp. No.	Total Dissociation Time (minutes)	Viability (%)	Yield (cells plated/ml)	Attachment (cells/ml)	"Attached Fraction" (%)
1[b]	45	82	4.0×10^6	2.8×10^5	7
2[c]	60	82	3.4×10^6	4.3×10^5	13
3[d]	60	89	4.0×10^6	3.4×10^5	9
4[e]	83	82	2.7×10^6	4.3×10^5	16
1[b]	50	54	9.0×10^5	2.4×10^5	27
2[c]	79	73	1.0×10^6	5.0×10^5	47
3[d]	70	56	1.7×10^6	1.0×10^5	6
4[e]	100	72	1.0×10^6	2.0×10^5	20
1[b]	60	69	8.7×10^5	2.6×10^5	30
2[c]	60	61[f]	1.2×10^6	6.6×10^4	6
3[d]	65	70	1.2×10^6	1.2×10^5	10
4[e]	45	72[f]	3.0×10^6	3.6×10^5	2
1[b]	65	73	7.3×10^5	4.0×10^5	55
2[c]	75	64[f]	1.2×10^6	1.2×10^5	10
3[d]	50	66	1.5×10^6	1.1×10^5	7
4[e]	85	69	1.8×10^6	3.6×10^5	20

[a]Each experiment represents the data from a single set of "representative" experiments for each of the enzymes tested. The results are typical of those obtained during the establishment of many hundreds of primary cell cultures.

[b]Sigma protease Type VI (Pronase P, 3–4 units/mg); 0.25% in 25% L-15 (see text for details).

[c]Sigma protease Type V (Pronase AS, 0.7–1 unit/mg); 0.5% in 25% L-15.

[d]Boehringer-Mannheim "protease" (6 units/mg); 0.125% in 25% L-15.

[e]Calbiochem Pronase (45 units/mg); 0.025% in 25% L-15.

[f]Not reliable data; less than 100 cells counted.

cin, and 1% fungizone, pH 7.4) and then resuspended by gentle pipetting. This suspension is centrifuged again, and the final cell pellet is resuspended in 1–10 ml of whole medium (depending on the size of the pellet). A preliminary cell count is taken using a hemocytometer.

Plating and Maintenance of Cells

After counting, appropriate concentrations of cells (usually $1\text{–}4 \times 10^6$ cells/ml) are plated onto Corning plastic tissue culture dishes. Cells are observed with an inverted light microscope equipped with phase optics and an ocular micrometer. A second cell count is taken within 1–3 hours after plating, and cell viability is determined by trypan blue or the erythrosin B dye exclusion test (Phillips, 1973).

After sufficient numbers of cells attach to the plastic substrate (usually 24–72 hours after plating), the remaining floating cells and debris are removed from the plate by gently washing with whole medium. At this time, another cell count is taken to determine the fraction (i.e., percentage) of attached cells. The "attached fraction" is determined by dividing the number of cells attached at $t = 48$ hours by the number of cells plated ($\times$ 100%).

Fresh medium is added at this point and twice weekly thereafter. Cultures are maintained at room temperature (22°C) at pH 7.6–7.8. After 1–2 weeks, many cultures appear to achieve confluency or to become so dense in certain areas that observation of individual cells becomes difficult. In such cases, cells are subcultured.

OBSERVATIONS OF PRIMARY CULTURES OF VARIOUS TISSUE TYPES

Four groups of axolotl tissues can be established and maintained in culture, including liver, kidney and gonad, heart, and muscle.

Liver

Freshly dissociated liver gives rise to three distinct cell types: small, round cells that attach to the substrate early, but subsequently detach and are washed away; intermediate-sized cells that ultimately flatten and spread to form very large cells with a clear cytoplasm ("clear" cells); and large, round cells that flatten to form sheets of cells with granular cytoplasm ("granular" cells). This latter cell type predominates in all liver cell cultures (see Fig. 23–1).

In general, the liver completely dissociates into single cells or groups of small numbers of cells after three or four changes of dissociation medium (about 40–60 minutes total time). Each change of dissociation medium gives rise to distinctly enriched populations of all cell types (depending to some extent on the original size of the liver and the age of the donor animal). In general, the later medium changes are enriched with the large granular cells.

Granular cells have a very distinct morphology (Fig. 23–1a). Such cells flatten and spread to form a single-layered sheet on the substratum. Sheets of granular cells are often interspersed with occasional large clear cells (Fig. 23–1b).

In comparison with other tissue types, the liver is generally a poor source of

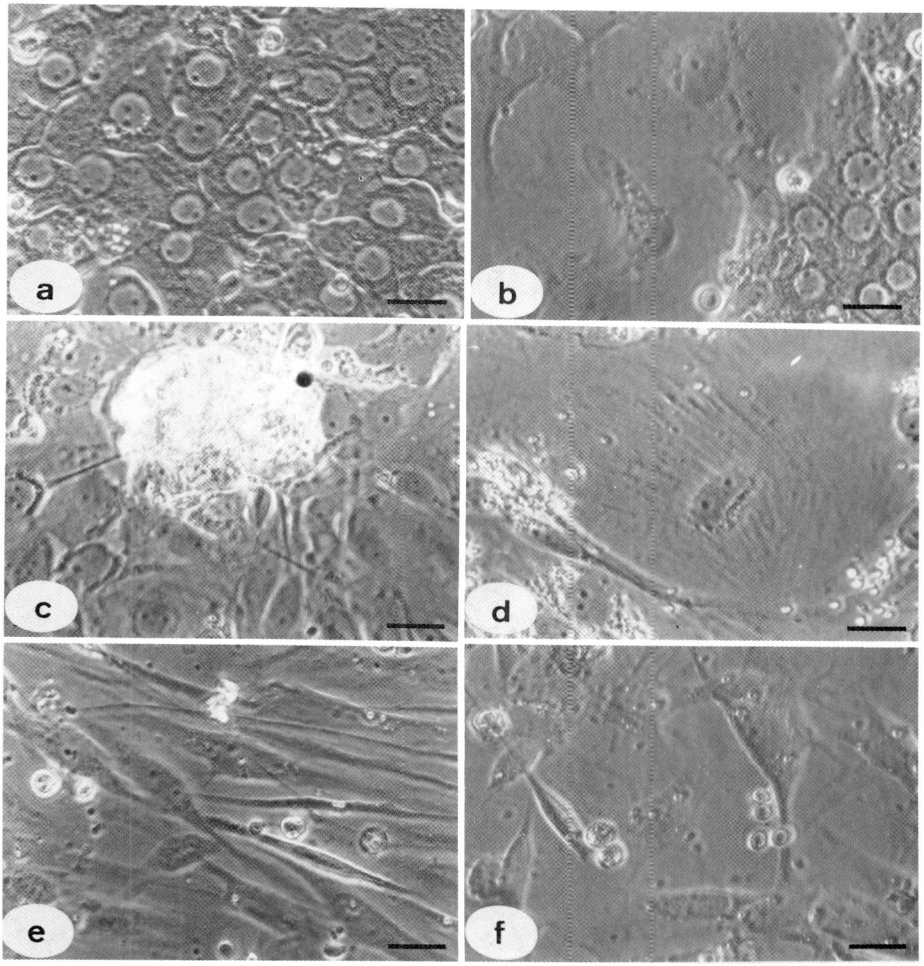

Fig. 23–1. Phase micrographs of cells from various axolotl tissues after short-term culture. (a) Confluent mass of granular liver cells after 1 week. (b) Large clear cells interspersed within a confluent mass of granular liver cells cultured for 1 week. (c) A group of cells forming a confluent sheet around an aggregate of kidney–gonad cells after 1 week. (d) Kidney–gonad cells maintained at a low density for 10 days. Note the characteristic large size of these cells and the striations (stress fibers?) visible in the cytoplasm. (e) Typical heart cells after 1 week in culture. (f) Typical flank muscle cells after 1 week in culture. Bars = 25 μm.

proliferating cells. Only clear cells that maintain a smaller than "normal" size (normal is illustrated in Fig. 23–1b) are seen to undergo mitosis.

The effectiveness of the dissociation enzymes varies somewhat (Table 23–1). In general, the liver from a single larval axolotl yields several million or more viable cells, and plates are therefore seeded with 2–6 $\times$ 10^6 cells. The percentage of cells attached (attached fraction) ranges from 7 to 16% depending on the type of pronase used. The best enzyme for liver dissociation (in terms of viability and percentage of cells attached) appears to be Calbiochem Pronase.

Kidney and Gonad

In sexually immature axolotls, it is difficult to differentiate clearly between kidney and gonad tissue. Moreover, both organs are relatively small in young animals. In order to ensure a sufficient cell yield from the kidney and gonad for cell culture experiments, both tissue types should be dissected out and dissociated together.

Complete dissociation of kidney–gonad tissue occurs within 40–60 minutes after addition of pronase. The yield of viable cells that attach and flatten is relatively high, ranging from approximately 6% to 47% (Table 25–1). Sigma Type V protease was found to be best for kidney–gonad tissue dissociation. After 1 week in culture, kidney–gonad cells frequently attain confluency. These cells can then be subcultured.

Cells typical of kidney–gonad cultures are shown in Fig. 23–1c,d. During the first week or two in culture, kidney–gonad cells maintain a relatively uniform small size (Fig. 23–1c). However, if such cells are transferred to (or maintained at) a low density for periods longer than 1 week, many of the cells increase in size and eventually occupy an area on the substratum more than double their original size (Fig. 23–1d). In some cases, the cytoplasm of these "enlarged" cells appears to be "striated" or folded (stress fibers? Fig. 23–1d). The observed increase in cell size frequently precedes vacuolation. Such cells have never been observed to undergo further proliferation. Of all cultures attempted, the kidney and gonad seem to provide the best source of viable, growing cells (Table 23–1).

Heart

Complete dissociation of heart cells occurs within 40–60 minutes after the addition of pronase. In larval axolotls, the heart is a very small organ and thus dissociates completely in only one aliquot of pronase. Heart cells dissociated in this way suffer little damage (as evidenced by high viability and a reasonable percentage of attached cells: Table 23–1). The best results for heart cell dissociation have been obtained using Sigma Type VI protease.

Two morphologically distinct cell types were observed in heart cultures (Fig. 23–1e): elongated, fibroblastlike cells and small, round cells. Both are found in great abundance in primary heart cultures. Occasionally, after a week or more in culture, heart cells achieved confluency. We have subcultured these cells, but with limited success. Subcultured heart cells usually fail to reattach and flatten. Those that do attach show little or no tendency to proliferate. Primary cultures of heart cells can be maintained for several weeks or months with little or no evidence of degeneration.

Muscle

The entire flank muscle extending across the venter and up each side of the animal to the spinal cord can be used for primary muscle cultures. All of the overlying pigmented skin should be removed and discarded before dissecting out the muscle. In general, flank muscle is difficult to mince into small pieces without considerable shredding of the tissue. A pronase exposure time of 40–60 minutes is required to obtain sufficient numbers of viable dissociated muscle cells.

Both heart and flank muscle cells dissociate best in Sigma Type VI protease. Morphologically, heart and flank muscle cells can be easily distinguished from each other (compare Figs. 23–1e,f). Like heart cells, subcultured muscle cells, in all cases, fail to proliferate further.

PROCEDURES FOR LONG-TERM CULTURE OF EPITHELIAL CELLS

Initial Cultures

Individual stage 36–38 axolotl embryos are cultured as explants for epithelial cell outgrowths in the following manner. Embryos within their jellycoats are preincubated in 20% Steinberg's solution (S) containing antibiotics (penicillin–streptomycin, 50 μg/ml) and fungizone (2.5 μg/ml). They are then *sequentially* immersed in 20% S containing, 1% hypochlorite (Clorox) (approximately 20 seconds); antibiotics; 70% ethanol (approximately 10 seconds); antibiotics (three times). The individual embryos are then mechanically dejellied, anesthetized in 1:5000 MS222 (in 100% S), transferred to 100% S (containing gentamicin, 50 μg/ml), skinned (head, torso, and tail), dissected (yolky abdomen removed), transferred to complete culture medium (55 parts L-15, 20 parts fetal bovine serum, 0.02 units/ml insulin, 1 μg/ml thyroxine, 50 μg/ml kanamycin, 2.5 μg/ml fungizone, pH 7.8), and minced into 1-mm pieces. Next, minced pieces are plated on a prewashed (complete culture medium) tissue culture dish (Corning) containing a minimum volume (about four drops) of culture medium. After 2 hours, 0.5 ml culture medium is added without disturbing the explants, and at 24 hours an additional 1 ml of culture medium is added. The cultures are incubated at room temperature (20°–23°C) in a humid atmosphere, and the medium is changed every third day.

Subculturing

Cultures are washed sequentially in 75% L-15, and then in calcium- and magnesium-free phosphate-buffered saline (PBS), and incubated for 30–60 minutes in dissociation medium containing 1.8 units/ml protease (Type IX, Sigma) in calcium- and magnesium-free PBS on a rotary shaker (60 rpm). Fragments of the epithelial sheets that detach in the dissociation medium are washed in complete medium by centrifugation (200 *g*; 3–5 minutes). The pellet is dissociated into small fragments (50–200 cells per fragment) by repeated pipetting, diluted 4:1, and plated on plastic tissue culture dishes.

OBSERVATIONS ON LONG-TERM CULTIVATION OF EPITHELIAL CELLS

Initial Cultures

Within 24 hours in culture, 85–95% of the explanted pieces attach. By day 5 approximately 90% of those explants display cell outgrowths. By day 25, each culture dish shows at least one area of active cell proliferation of epithelium-like cells

(Fig. 23–2a). The epithelial cell sheets increase in cross-sectional area (epithelial sheets 3–7 mm in diameter are not uncommon) and cell packing density (Neff, 1984). They frequently overgrow surrounding fibroblastic cells (Fig. 23–2b). These epithelial cells are probably *not* epidermal cells because skin had been removed from the embryo, and the morphology and culture behavior of these cells are not consistent with an epidermal cell origin.

The culture conditions utilized also support fibroblastic cell proliferation. The beating of heart cells and ciliated cells is maintained in cultures for up to 2 months.

Subcultivation

Cultures that exhibited actively proliferating, vigorous epithelial sheets by 60–80 days in primary culture (Fig. 23–2a) were subcultured as described. Subculture conditions allow cells of epithelial sheets to detach selectively from the culture dishes, whereas fibroblastic cells are left behind. It is important that the epithelial sheets be dissociated into fragments of about 50–200 cells each. Between 28 and 56% of the epithelial sheet fragments, as well as individual cells, attach within 24 hours. However, all single cells as well as epithelial sheets that cannot maintain confluency eventually degenerate (Figs. 23–2c,d). Three kinds of behavior are observed: (1) In the majority of explants the cells flatten out and the intercellular distance increases (i.e., they lose confluency). These cells have not been observed to proliferate. (2) In a small number of explants there is cell proliferation for a short period of time; then cells lose confluency and the explants degenerate as above. (3) In about 10% of the explants growth is vigorous, with proliferation beginning by day 2 and continuing through day 5. These explants grow with a population-doubling time of about 3.6 days. Subcultured epithelial sheet cells exhibit the same morphology and culture behavior as they did in primary culture.

Further Subcultivation

It is important that the epithelial cells be subcultured as cohorts of cells (50–200 cells). All attempts at subculturing epithelial sheets as single cells failed. Numerous experiments utilizing various dissociation media, including trypsin, protease (Type IX, Sigma), and dispase (Boehringer-Mannheim) at various concentrations and buffers that yielded single-cell suspensions with high cell viabilities (69–91%, try-

Fig. 23–2. Phase micrographs of a long-term axolotl epithelial cell culture. (a) The center of a primary epithelial sheet (diameter 6 mm) cultured for 60 days. Note mitotic figures (arrows). (b) Detached and contracted edge of the same epithelial sheet (esh) as in (a). Note the fibroblastic cells (fc) uncovered by the detached overlying epithelial sheet. (c) Three fragments of an epithelial sheet 13 days after subculturing (186 total days in culture). Note that the confluent central explant exhibits cell proliferation and eventually gave rise to an epithelial sheet. The two flanking explants (less dense) did not exhibit cell proliferation and eventually degenerated. (d) Edge of the same central epithelial sheet shown in (c), at 30 days after subculturing. Substantial growth is obvious. Note the remaining cells of a degenerating nonproliferating epithelial explant on the right (arrows). (a,b) Bars = 100 μm. (c,d) Bars = 200 μm.

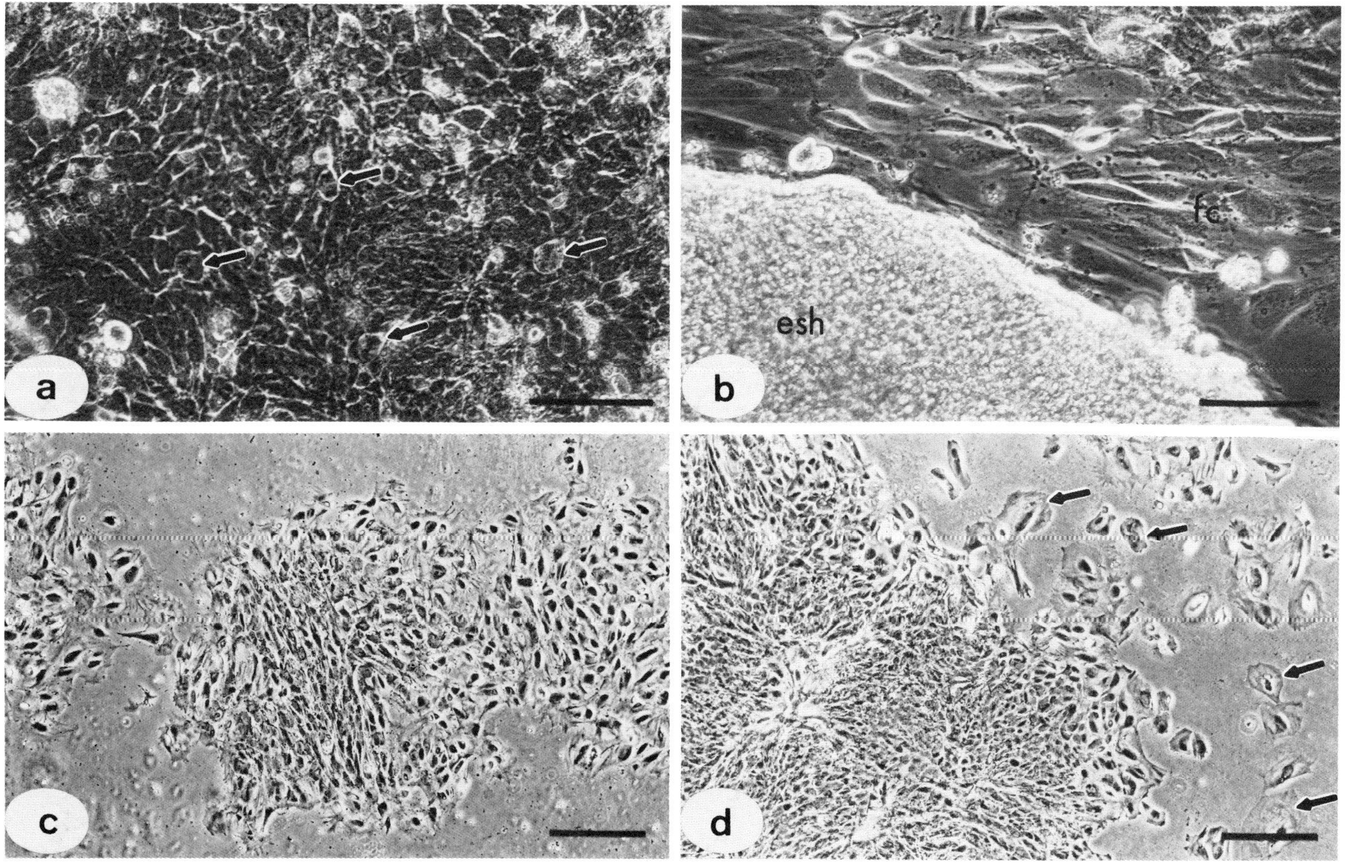
a
b
esh
fc
c
d

pan blue exclusion test) resulted in failure. Although 20–50% of the cells attached by 18 days in culture, few cells remained attached. Proliferation among single cells was rarely observed. The substrate did not appear to make much difference (glass, plastic, with or without rat or axolotl collagen).

Epithelial sheets have been cultured for 305 days, surviving four subcultivations (at 78, 173, 230, and 290 days) and maintaining consistent cell morphology and cell behavior. The duration of mitosis (approximately 80 minutes), as measured by video recording (Neff, 1984), and the mitotic index (1.1–1.2%) and cell cycle time (46–50 hours) remained constant.

The epithelial cells were found to be very sensitive to the type of fetal calf serum used. Different lot numbers of fetal calf serum should therefore be pretested for their capacity to support growth of epithelial sheets.

COMMENTS ON TISSUE DISSOCIATION

Our repeated attempts to develop primary cell cultures by use of methods reported to be successful for *Xenopus* (Solursh and Reiter, 1972) and *Rana* (Freed and Mezger-Freed, 1970) were, at best, minimally successful for axolotl cells. From the results of extensive preliminary experimentation (data not shown), it was concluded that axolotl cells are highly sensitive to the effects of trypsin, trypsin–EDTA, collagenase, and various combinations and commercial preparations of these enzymes. Those common "tissue dissociators" were found to result in very low yields of viable cells, poor attachment, and frequently, considerable production of cellular debris. Thus, the less rigorous methods of tissue dissociation described in this chapter were developed.

To date, the best dissociation results for axolotl cells are obtained with commercial preparations of "pronase" (as illustrated in Table 23–1). Table 23–1 provides representative examples of results from one set of experiments that are considered to be typical of the many hundreds of cell culture experiments that we have performed. For individual types of pronases, the various parameters shown in Table 23–1 differ considerably among the four tissue types employed. The end result (i.e., the "attached fraction") reflects a general tissue-specific preference with respect to the type of enzymatic dissociation employed. In general, Sigma Type VI protease was found to be the most suitable enzyme for the dissociation of all four tissues.

Of all the tissues examined, liver has been the most refractory to enzyme dissociation by pronase. This is evidenced by attached fraction values that rarely exceed 20% (Table 23–1). The reasons for this are not entirely clear. The low percentage is apparently not due to excessively damaged cells, because liver cell viabilities were higher than for any other cell type examined. With respect to the dissociation process, Calbiochem Pronase appears to enhance the plating efficiency of liver cells slightly more than other types of pronase.

Several additional points regarding the data in Table 23–1 also merit discussion. First, enzyme activity varies considerably in each of the commercial enzyme preparations. For example, Type VI protease (used as an arbitrary standard for these experiments), contains 3–4 units/mg of enzyme activity. Type V protease has less than half this activity, and Calbiochem Pronase has more than 10 times the activity of Type VI. Consequently, all enzyme preparations used in these studies were standardized to an approximate equivalent of Type VI activity. Thus, the relative differences in enzyme activity may reflect some critical difference in the purity of each enzyme type, subtle differences in the substrate activities of each enzyme, or some other property intrinsic to the individual enzymes. The facts that Type VI protease is generally suitable for every tissue examined and that Boehringer-Mannheim "protease" was generally unsuitable in every case suggest that substantial differences exist between the qualitative properties of each of the enzyme preparations. Overall, Type VI protease was found to enhance the establishment and maintenance of most axolotl cells in vitro.

Clearly, before any "pronase" is used for axolotl tissue dissociation, it should be tested as outlined in the preceding pages. Sigma no longer supplies Types V and VI protease, but an additional 20 or more proteases are available and may prove to be even more suitable for these types of experiments.

COMMENTS ON INITIATION OF GROWTH

It should also be emphasized that a critical prerequisite to obtaining suitable percentages of attached cells is to obtain initially a high yield of viable cells during the dissociation procedure. In general, the viabilities of all cell types after pronase dissociation are very satisfactory (see Table 23–1). However, consistent success in culturing such cells was not obtained until the average yield (i.e., cells plated) equaled or exceeded 10^6 per milliliter. Even then, it appeared that not every cell type responded to the described culture conditions in the same way. For example, the highest viabilities have been obtained with liver cells and the lowest with kidney–gonad cells. Kidney–gonad cells have been, however, consistently easier to establish and maintain in vitro than the larger number of viable liver cells.

The conditions under which cells were prepared for dissociation were also found to have an influence on whether or not successful primary cultures were obtained. An initial wash of axolotl cells in 50% L-15 (instead of a sterile saline or Ringer's solution) supplemented with antibiotics and an antifungicide proved to be an essential step in the establishment of primary cultures. A report by Seldenrijk et al. (1979) that initial rinsing of tissues in a defined medium (such as dilute L-15) enhanced the subsequent viability and survival of *Xenopus* melanophores in vitro prompted our use of defined medium in these experiments

A variety of amphibian salt solutions including those of Steinberg, Barth, and an amphibian phosphate-buffered saline (APBS) (Arthur and Balls, 1971), as well as others, were found to be unsatisfactory either as initial washes for tissues or as carriers for the dissociating enzymes. The success obtained using L-15 as a wash medium led to the suspicion that a dilute defined medium might also be a better carrier for the dissociating enzyme. In the case of pronase, the use of L-15 as the carrier proved to be essential for the successful and reproducible establishment of primary axolotl cell cultures.

COMMENTS ON LONG-TERM CULTIVATION OF EPITHELIAL CELLS

Success, in this case, depends on (1) establishing primary explant cultures under conditions that select for vigorous epithelial cell outgrowth; and (2) subsequently subcultivating these sheets not as dissociated single cells but rather as fragments of sheets. Utilizing this procedure, axolotl cells have been cultivated for 10 months. An estimated 78 population doublings were achieved with no apparent change in cell morphology or behavior. These observations are consistent with the notion that amphibian cell lines are frequently epithelioid in morphology and display no intrinsic senescence phase (Rafferty, 1976).

It is not known why only 10% of the epithelial sheet fragments that attach grow upon subcultivation. However, there appears to be a correlation with cell–cell contact. Only confluent epithelial cell pieces proliferate. Two hypotheses may be offered to explain these observations: (1) The nonproliferating pieces are derived from nonproliferating areas of epithelial sheets. However, it should be noted that mitotic figures are randomly distributed in epithelial sheets. (2) Cell–cell communication (perhaps via a metabolic, short-lived growth factor, or a nonsoluble cell surface growth factor) among a threshold number of cells may be necessary for cell proliferation. Jones and Elsdale (1963) reported that axolotl blastula and gastrula primary explant cultures require a minimum of 30 cells for subsequent survival and differentiation. This cell–cell contact hypothesis is also supported by the observation that confluent proliferating epithelial sheets appear to have no effect on adjacent single cells or epithelial sheet explants. The apparent requirement for cell–cell contact for proliferation of these epithelial cells may explain why the establishment of long-term cultures of axolotl cells has been difficult to achieve.

It is apparent that the methodologies described for establishing other amphibian cells in vitro are, in general, not applicable to axolotl cells. As additional factors that enhance the ability of axolotl cells to survive in vitro are discovered, it is anticipated that the usefulness of primary cultures as well as long-term cultures will be enhanced. The ultimate goal of studies of the type described in this report is to establish a permanent axolotl cell line.

24
Surgical Manipulation of Embryos

MAKOTO ASASHIMA, GEORGE M. MALACINSKI,
AND STEVEN C. SMITH

Amphibians generally provide excellent material for studies in experimental embryogenesis, and the axolotl is no exception. Axolotl embryos have several advantages when compared with other amphibians. The eggs are relatively large (approximately 2 mm in diameter), as are the cells of older embryos and the chromosomes they contain. Axolotls can produce large numbers of eggs at one spawning (varying from 200 to 300 to over 1000), and the embryos develop relatively slowly, so that the experimenter can operate on large numbers of embryos from a single spawning. As well, their developmental rate can be regulated over an unusually broad range of temperatures, from 4° to 25°C. Therefore, it is relatively easy to bring embryos from different spawnings to exactly the same developmental stage, a prerequisite for most manipulations involving donor and host embryos of different spawnings. Finally, an unusually large number of color (Chapter 12) and developmental (Armstrong, 1985; Chapter 10) mutations have been described in the axolotl, making them ideal for many developmental studies.

Axolotl embryos tolerate microsurgical manipulations well. The experimental procedures employed for extirpation, explantation, grafting, parabiosis, and chimera production are basically the same as those employed for other amphibians. Extensive descriptions of these procedures are included in the laboratory manuals of Hamburger (1960) and Rugh (1962).

Compared with some other species, axolotls have a few disadvantages. Perhaps the most serious of these is the long generation time (about 1 year), which hinders genetic studies. Artifical insemination (Chapter 21) is not as convenient as in anurans, as it is not possible to "strip" large numbers of eggs from gravid females. There is also no suitable chemical method for removing jellycoats and egg membranes, although manual dejelling is quite suitable in most instances.

MEDIA AND EQUIPMENT

The instruments and procedures that are common to most types of surgical manipulation are briefly described below. For detailed descriptions of culture media, instrument making, and procedures, see Hamburger (1960) and Rugh (1962).

Culture Media

Any of a number of media are suitable for axolotl embryos. The most commonly used are variations of the simple salt solutions devised by Steinberg and Holtfreter (see Table 24–1). Addition of glucose, serum, or other nutrients is unnecessary for the culture of embryonic axolotl tissues. The yolk contained in the cells provides sufficient nutrition for them to live for at least a few weeks under normal circumstances.

In fact, the two media listed in Table 24–1 are sufficient for most applications with axolotls. Some groups maintain embryos in 25% and adults in 50% Holtfreter's medium. Full-strength Steinberg's medium is usually the medium of choice for operative procedures on embryos and as a culture medium for explants.

It is important that the salt concentration in the operating medium be optimal for wound healing. However, there is some variation in the exact recipes used by different groups (see Table 24–1). For most applications, 100% Steinberg's medium is quite satisfactory. Some groups have found that extra calcium and magnesium aids in wound healing. Conversely, others find that, for transplants, Ca^{2+}-free medium speeds healing by loosening cell–cell adhesion and thereby retarding curling of the tissues and promoting cell migration. (If this latter procedure is used,

Table 24–1 Culture Media

Medium	Components
Holtfreter's[a,b]	$NaCl$, 3.46 g
	KCl, 0.05 g
	$CaCl_2$, 0.1 g
	$NaHCO_3$, 0.2 g[c]
	+ 1 liter of distilled water
	pH adjusted to 7.4[d]
Steinberg's[e]	$NaCl$, 3.4 g
	KCl, 0.05 g
	$Ca(NO_3)_2 \cdot 4\,H_2O$, 0.08 g[f]
	$MgSO_4 \cdot 7H_2O$, 0.205 g[f]
	Tris, 0.56 g[g]
	+ 1 liter of distilled water
	HCl to pH 7.4[d]

[a]After Holtfreter (1931). This medium is suitable for maintaining adult axolotls (50%) and embryos (25%).

[b]Some workers add 0.2 g $MgSO_4 \cdot 7H_2O$ to the recipe.

[c]Rugh (1960), citing a personal communication from Holtfreter, recommends 0.02 g $NaHCO_3$ as a buffer.

[d]Some workers prefer pH 7.7 for the axolotl (see text).

[e]After Steinberg (1957). See also Hamburger (1960).

[f]Some workers prefer to double the $Ca(NO_3)_2 \cdot 4H_2O$ (to 0.16 g) and $MgSO_4 \cdot 7H_2O$ (to 0.4 g). The nitrate is not necessary, being a historical accident of the medium's development from a plant medium. For regular medium, 0.05 g of $CaCl_2 \cdot 7H_2O$ may be substituted for $Ca(NO_3)_2 \cdot 2H_2O$ (Poole and Steinberg, 1977). For operations involving transplants, some workers use calcium-free medium (i.e., omit $CaCl_2$ or $Ca(NO_3)_2$ from recipe).

[g]Steinberg now uses HEPES buffer (1.10 g/liter) instead of Tris and adjusts the pH to 7.4 with NaOH (Zackson and Steinberg, 1986).

part of the medium in the operating dish is replaced with Ca^{2+}-containing 100% Steinberg's once healing has begun.) It should be noted, however, that Ca^{2+}-free medium is unsuited to operations involving the removal of explants.

The pH of an operating medium is also important, but here again there is some variation in the recipes of individual groups. Most workers find a pH of 7.4 (as originally described by Steinberg) to be quite satisfactory. At least one group, however, uses a pH of 7.7 for axolotl tissues, as this is closer to the pH of axolotl plasma (Hronowski and Armstrong, 1977).

Since infection of embryos by contaminating bacteria is sometimes a problem, sterile solutions are usually supplemented with antibiotics. (Holtfreter's medium, *without* the $NaHCO_3$, and Steinberg's medium may be autoclaved before the addition of sterile antibiotics, or the media with antibiotics may be filter-sterilized.)

The amount of antibiotic used varies among workers. Some find that 50 mg of gentamicin sulfate per liter of solution is quite sufficient. Alternatively, other groups find that combinations of antibiotics work best; up to 400 mg penicillin-G (Na salt), 400 mg streptomycin sulfate, and 25 mg gentamicin may be added together to 1 liter of medium. As an alternative, up to 500 mg of kanamycin sulfate per liter of medium may be used. This latter has the advantage that it is heat stable, so it can be added to the culture medium prior to heat sterilization.

The culture media are used for preoperative washing of embryos, during actual surgery, and for culturing (e.g., 1–10 hours) postoperatively. Once completely healed, embryos are usually transferred to a lower strength (20% Steinberg's or 25% Holtfreter's) culture medium containing antibiotics.

Operating Dishes

The type of operating dish and lining used is largely a matter of personal choice. A lining is used to hold the embryo in place during surgery and to prevent the tissues from adhering to glass or plastic surfaces. A variety of linings were suggested by Hamburger (1960), and the choice depends on the requirements of the worker. Some prefer to use a 60-mm petri dish lined with 2% agar. Small molds may be either cut or melted into the agar surface, before it is flooded, to hold the embryo in place during surgery. This method has several advantages. Cleaner cuts through embryos may be made, when required, since the operating instrument can slice into the substrate. As well, the linings (and the dishes, if plastic ones are used) are disposable and thus may reduce the possibility of contamination. Since they are disposable, the dishes may be used with toxic chemicals (such as fixatives) if necessary.

Some workers have found that modeling clay (that supplied by American Art Clay Co., Indianapolis, IN is satisfactory) makes an excellent lining for operating dishes. Since this is fairly soft, however, it is usually hardened by adding about 20% (w/w) paraffin. Some care is required in the choice of paraffin, however, as some brands tend to be toxic to embryonic tissues. Paraplast (Lancer, St. Louis, MO) *without* DMSO has been found to be satisfactory in this regard. The modeling clay and paraffin are melted together in a water bath, mixed thoroughly, poured into the operating dish (this may be a glass or plastic petri dish or the wells of a three- or nine-well Pyrex spot plate), and molded into a shallow depression in each dish. The lining may be smoothed and sterilized over the heat of an alcohol flame and allowed to cool before use, and rinsed with water and swabbed with 95% ethanol immediately after use.

Dishes lined with modeling clay have several advantages. They are reusable, as long as they do not come into contact with toxic chemicals. (Separate dishes may be kept for procedures involving fixatives, enzymes, and other chemicals.) As well, the lining may be more easily and exactly molded into depressions for holding embryos, it may be packed gently around the embryos to hold them firmly in place during surgery, and it provides a firmer substrate than agar.

Operating Instruments

A variety of specialized operating instruments are required for surgery on amphibian embryos. Fortunately, these are relatively simple to make. The instruments most commonly used include glass or tungsten wire needles for cutting, a variety of hair loops (preferably made with blond baby's hair) for manipulating embryos and pieces of tissue, glass-ball tips for making depressions in operating dishes to hold embryos and explants, sharpened watchmaker's forceps for removing jellycoats and vitelline membranes, and a variety of transfer pipettes. Hamburger (1960) and Rugh (1962) provide excellent descriptions for instrument making. Several things should be noted, however.

First, Pasteur pipettes may be used in place of the glass rods and tubing used by these authors. Pipettes drawn out to thin capillaries and broken off to the appropriate length may be (1) smoothed in a flame and used as handles for hair loops and needles, or (2) rotated in a flame to make smooth glass-ball tips.

Second, satisfactory transfer pipettes for whole embryos may be made by breaking the pipette in the taper, smoothing it in a flame, and placing a rubber bulb over the taper. (Some workers prefer to use disposable plastic transfer pipettes instead.) Similarly, a small transfer pipette for pieces of tissue may be made by cutting the pipette so that 3–4 cm of the narrow end remain, and flaming the cut edge to smooth it. The wide portion may be cut to about one-half the length, and smoothed and narrowed in a flame. This end may then have a mouth-pipette tube attached to it for finer control.

Although many workers find glass needles to be perfectly satisfactory, tungsten needles (made of 0.005-inch tungsten wire) offer advantages in that they are easy to make, are durable, can be resharpened, and can be sterilized in a flame. The involved (and dangerous) procedure described by Hamburger (1960) for the sharpening of tungsten needles should be avoided, however. Instead, the authors recommend that electrolytic sharpening in a 1 *N* NaOH solution be used. By placing one electrode in the solution and attaching the other to the needle, a repeated dipping motion into the solution will sharpen the needle to a fine point.

Larger cutting tools, if required, can be fashioned from razor blades, which are broken into small pieces with pliers. The blade fragment can then be mounted into a wooden rod or a suitable commercial holder.

Finally, we recommend that a fiber-optic light source be used for illuminating operating dishes whenever possible. The heat from an incandescent light source can raise the temperature of the operating medium appreciably.

Sterilization Procedures for Instruments

Glass containers, egg transfer pipettes, and forceps may be heat-sterilized (120°C, 2 hrs). Hair loops and glass needles may be dipped in 70% ethanol, then rinsed in sterile culture medium. Other instruments may be sterilized in the same way, or by dipping in ethanol and passing the instrument through a flame and allowing the ethanol to burn off.

If desired, operations may be performed in an air-flow hood, to reduce the chance of infection by airborne microorganisms. Bench surfaces and the dissecting microscope used for most operations are routinely swabbed with ethanol. In addition, hands should be washed with soap and may be rinsed with ethanol. Some workers prefer to wear a surgical mask as well.

EXPERIMENTAL MANIPULATION OF EMBRYOS

Traditional embryology textbooks such as Hamburger (1960) and Rugh (1962) can be consulted to establish the location of the prospective organ and tissue rudiments. These authors also include several diagrams illustrating the design of various extirpation, explantation, transplantation, vital staining, and parabiosis procedures. Therefore, we will not dwell on these procedures and instead will describe the preparation of embryos for surgery and a few specialized procedures.

Preoperative Treatment of Embryos

The first step in any operative procedure involves preparing the embryos for surgery. If desired, eggs may be immersed in 70% ethanol for 40–50 seconds while still in their jellycoats. They are then quickly transferred to approximately five washes in culture medium containing antibiotics before dejelling. While some authors feel that this is a necessary step, others prefer to omit it and merely dejell the embryos in fresh medium and quickly pass them through several washes of sterile medium.

The first step in embryo preparation involves decapsulation in culture medium. A series of two or three membranes surround most urodele eggs. They may be manually removed by grasping the outer membrane with one pair of watchmaker's forceps and piercing the membranes on the opposite side with a second pair. The forceps are then quickly teased apart, producing a hole in the membranes through which the egg/embryo is gently extruded. Figure 24–1 illustrates this procedure. Alternatively, one tine of the first pair of forceps may be inserted into the hole made by the second, and the two pairs may be used to rip open the membranes.

Usually it is also necessary to remove the vitelline membrane, which lies close to the egg surface. This can be done at early stages by immersing the embryos in a sterile 10% Ficoll solution to dehydrate the perivitelline space (between the membrane and the egg surface). This leaves a slight fold in the membrane that can be grasped and torn with two pairs of forceps. At later stages (e.g., from blastula onward) it is possible to dispense with the Ficoll solution, since if a few surface cells are injured when the membrane is grasped, rapid healing covers the wound.

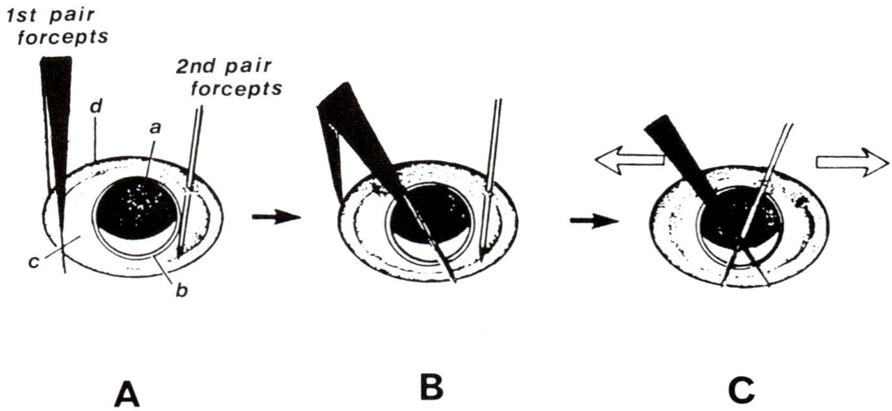

Fig. 24–1. Dejelling procedure. (A) First pair of forceps holds the outer jellycoat while the tips of the second pair pierce the fluid-filled space. (B) One tip of the first pair is used to pierce the fluid-filled space. (C) While gripping firmly with the first pair of forceps, use the second pair to tear a hole quickly in the membranes, through which the egg/embryo is extruded with the first pair. a, Egg; b, vitelline membrane; c, fluid-filled space; d, jellycoats.

Experienced microsurgeons can often manage without the Ficoll by using extremely fine-tipped forceps.

The embryo may now be rinsed with sterile medium and transferred to the operating dish. Two things should be noted. First, it is preferable to allow embryos and pieces of tissue to *fall* from the pipette rather than being expelled. This causes a minimal transfer of (possibly contaminated) medium with the embryo and therefore allows for proper rinsing. Second, all contact between embryos or pieces of tissue and the surface of the medium must be avoided. The surface tension at the medium–air interface is sufficient to rupture the embryo.

Thermal Regulation of Development

The stage of preoperative embryos can be regulated by slowing down (to 4°C) or speeding up (to 25°C) their development. Figure 24–2 contains information that is useful for staging. As mentioned previously, this feature of the axolotl system provides a significant advantage for the experimenter. It should be noted, however, that there is substantial intolerance to either temperature extreme; the degree of intolerance varies from one spawning to another. Some workers have found it a good practice to avoid placing preneurula embryos below 10°C as 4°–5°C temperatures often cause exogastrulation. After stage 13, 4°C usually does not harm embryos if maintained no longer than a few days. Similarly, temperatures over about 22°C should be avoided whenever possible.

Anesthetics

Embryos that have developed muscular response will usually twitch when stimulated by operating instruments. It is therefore necessary to anesthetize such

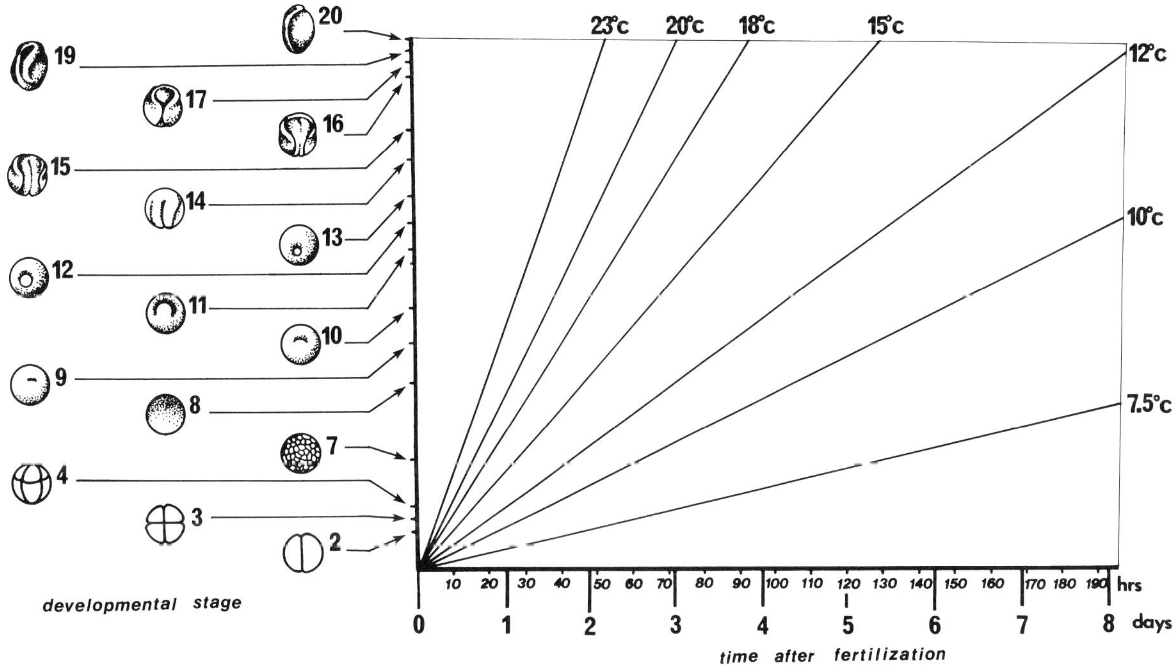

Fig. 24–2. Temperature-dependent development of axolotl embryos.

embryos shortly before operating. MS-222 (also known as Tricaine, 3-aminobenzoic acid ethyl ester, or ethyl *m*-aminobenzoate) at a concentration of approximately 0.01% is effective. It is water soluble, and can be diluted out by washing postoperative embryos in a salt solution. Other authors have found that the less expensive benzocaine (also ethyl *p*-aminobenzoate or 4-aminobenzoic acid ethyl ester) is equally effective, tends to be less toxic, and can be used at a lower dose than MS-222. Benzocaine is dissolved in a minimum volume of 95% ethanol and diluted with culture medium. A concentration of 0.005% benzocaine is recommended for embryos and young larvae (see Vanable, 1985).

Fig. 24–3. Production of a chimeric embryo. (A) Fusion of embryo halves using stainless-steel holders. (B) Adult chimeric axolotl. (Produced by Y. Moriwaki.)

A

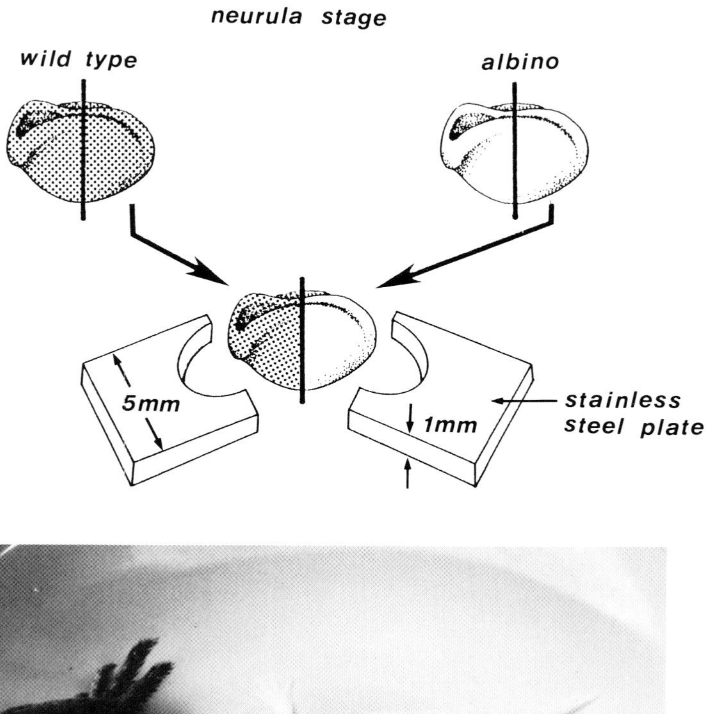

B

Chimeric Embryos

For various types of developmental genetic analyses, chimeric embryos are extremely useful. Neurulae from separate spawnings are dejelled and placed in an operating dish lined with 2% agar. The vitelline membranes are removed and the embryos sliced in half either with needles or a sharp razor blade knife. Using the stainless-steel plates illustrated in Fig. 24–3 (produced by Y. Moriwaki), a single whole is reconstituted. Within 3–4 hours at 20°C, the two halves fuse, and the plates are removed.

Microinjection

For the introduction of isotopes, cytoplasmic components, cloned genes, and so forth, microinjection of volumes as small as a few nanoliters is often necessary. A variety of instruments, including Gordon and Malacinski's precision syringe control and micropipette device (1970), have been described in the literature. The microinjection apparatus of D. L. Stephens et al. (1981), of substantially different design, has also been found to be most satisfactory. Leakage of cytoplasmic contents is frequently encountered, so 5–10% Ficoll is often added to the culture medium to reduce the pressure that fluid in the perivitelline space exerts on the egg. Ficoll shrinks the perivitelline space quickly and effectively.

CONCLUDING REMARKS

The amphibian embryo's large numbers, ease of accessibility, and tolerance for surgical manipulation have always been its main strengths in experimental embryology. The axolotl embryo holds no unpleasant surprises for the novice surgeon. In fact, its relatively large size, broad temperature range, and the large number of mutants available provide notable advantages over other species of amphibian embryo.

As gene transfer techniques, which are already feasible for *Xenopus* (L. D. Etkin and B. Pearman, 1987), are developed for the axolotl, a rebirth in interest in the developmental biology of the axolotl will likely be witnessed.

25
Diseases of Axolotls

SUSAN T. DUHON

Disease is probably the most serious problem that may arise in laboratory colonies of axolotls. A severe outbreak can destroy a colony in the space of a few weeks or months, and less severe, chronic disease may prevent the animals from breeding or alter research data being collected. Despite the potential seriousness of the problem, little is known about causes or treatment, and much of the available information is anecdotal. Some information on disease in other herpetofauna and even in fish may, however, have application to axolotls. This chapter will review some of what is known about disease problems and describe our efforts to combat them.

NONINFECTIOUS DISEASE

Noninfectious causes of disease include poor water quality, inadequate or improper nutrition, poor temperature control, and other environmental stresses.

Water quality may be the single most important consideration. Problems in water quality may stem from the source, if chemicals such as chlorine, ammonia, and chloramines used in water treatment are not removed, or if contaminants such as copper or other heavy metals are present. Toxic leachates from plastic containers can also affect axolotls (National Academy of Sciences, 1974). More often, problems in water quality are due to the buildup of metabolic wastes, especially ammonia, as a result of inadequate housing or infrequent water changes. Published guidelines for acceptable water quality for axolotls are usually based on standards originally developed for fish (National Academy of Sciences, 1974). Axolotls suffering from the toxic buildup of nitrogenous wastes can have increased skin mucus, loss of skin and gill tissue, cloudy or protruding eyes, internal hemorrhaging, and a reduced growth rate. They may gulp air rapidly (Fox, 1982b). A rapid buildup of nitrogenous wastes may result in death without warning. A chronic problem increases the animals' risk of infectious disease. The first step in any disease control program should be to ensure adequate water quality and efficient removal of wastes to eliminate toxic effects.

Nutrition is a neglected area when causes of stress and disease in axolotls are considered, and the nutritional requirements of axolotls are poorly understood.

Gadow (1903) reported that Lakes Xochimilco and Chalco teemed "with fish, larvae of insects, *Daphnia,* [and] worms," all of which probably provided food for axolotls. In the laboratory, axolotls are typically fed beef heart, beef liver, and sometimes earthworms, other small worms, or *Daphnia.* Vitamin supplements are sometimes used. The liver diet may have its origin in early practices and experiments in feeding fish in hatcheries (Hastings and Dickie, 1972). However, possible nutritional complications include rickets (due to calcium–phosphorus depletion) and vitamin A toxicity. Certain animals fed on mostly beef liver show severe liver degeneration and ceroid deposits, indicating lipoid liver degeneration, when examined histologically. Secondary effects of an all-liver diet include anemia, liver enlargement, and colorless gills (Snieszko, 1972).

Neoplasms also occur in axolotl populations, but are rare and unlikely to be a problem in a breeding colony. Humphrey (1969b) reported undefined tumors in the testes of axolotls. These were always considered benign. Brunst (1969) mentioned melanoma, spontaneous teratoma, and olfactory neuroepithelioma in axolotls.

INFECTIOUS DISEASE

Many clinical signs of disease have been reported in axolotls (see, for example, Axolotl Newsletter, nos. 3 and 9). Commonly reported are anorexia, lethargy, cutaneous hemorrhage or ulceration, edema, ascites, anemia, discoloration of the skin, and so forth. Animals that have died often have enlarged livers or spleens, pale or gray livers, inflamed internal organs, and other signs of infection.

There are numerous possible causes for these signs, which reflect—for the most part—a generalized systemic response to toxins, rather than a particular response to a specific infectious disease organism. The responsible agents may be bacterial or viral. Practically nothing is known about viral disease in axolotls, and no treatments are available beyond improvement of husbandry techniques. Most work to date has concentrated on bacterial diseases. Various bacterial pathogens are associated with disease in axolotls and other amphibians. Species of *Aeromonas, Proteus, Pseudomonas, Vibrio, Acinetobacter, Mima,* and *Alcaligenes* are commonly isolated from diseased animals (Anver and Pond, 1984; Marcus, 1981; Reichenbach-Klinke and Elkan, 1965). *Aeromonas hydrophila* has been implicated specifically as a cause of disease in axolotls (Boyer et al., 1971; Fox, 1982a). *Actinomycetes, Salmonella, Pseudomonas, Acinetobacter, Proteus,* and *Alcaligenes* have also been reported in ill axolotls (Brothers, 1977; International Aquaculture Research Center, 1986). The opportunistic nature of these organisms and their ubiquity in aquatic environments, together with the general character of the signs produced, frequently render problematic their positive identification as the disease-causing organism (Amborski and Amborski, 1978).

At the Indiana University axolotl colony we have seen all the signs just described, and more. Usually the first indication that an animal may be ill is that is misses one or more feedings. Healthy animals under 3–4 years of age almost never fail to eat. Another early sign is discoloration of the skin (a grayish cast, or a pallor, or a yellowish hue depending on the original coloration). The best way to detect such discoloration is to compare the animal with siblings having the same pigment phenotype. Small subcutaneous hemorrhages on the tailfin may be visible in white animals at an early stage of disease. Animals with more advanced disease may have a scattering of very small skin ulcerations or sores on the abdomen. They may float in the water, seemingly unable to stay completely submerged. Edema, ascites, severe anemia, and bleeding into the water usually indicate an advanced state of disease. Animals with one or more of these signs seldom recover, regardless of treatment.

The signs that an animal shows before death, or even whether it shows any at all, vary greatly, apparently depending on whether or not the animal possesses any natural immunity (Boyer et al., 1971). Animals may die within a few days of the appearance of signs of disease, or they may linger for weeks.

At the Indiana University axolotl colony we lose a small but variable number of small larvae for unknown reasons. This number is best kept low by keeping the water clean and by generous and frequent meals to promote rapid growth beyond the more vulnerable small sizes. Overfeeding (leading to excessive debris in the water) should be avoided. We seldom see signs of disease in axolotls between about 2 and 8 inches long, and our death rate among such juveniles is very low. As animals approach sexual maturity they again seem to become vulnerable, and our highest death rate—in the absence of treatment—is among young adults. Females sometimes become ill following their first spawning. Some strains are more robust than others. Our white English/Wistar/DeLanney animals show the greatest resistance to disease (and the greatest longevity). On the other hand, animals derived from stock imported from Mexico within the last 20 years succumb more readily.

Ideally, treatment of disease would be based on specific knowledge of the organism responsible and on tests indicating sensitivity to particular antibiotics. Unfortunately, efforts to obtain such information have been mostly unsuccessful. In the face of an unacceptably high death rate, one may choose to treat ill animals blindly with antibiotic. Since the majority of disease-causing bacteria are gram negative, the antibiotic selected should be effective against as many members of this group as possible. The best choices are gentamicin or other aminoglycoside antibiotics, chloramphenicol, and tetracycline (Ross and Marzec, 1984). Another possibility is nitrofurazone. Many antibiotics and treatment regimens are mentioned in informal publications (see Axolotl Newsletter, no. 9), but I will discuss only those with which I have experience at the Indiana University colony.

Tetracycline added directly to the water has been recommended as a treatment for a variety of symptoms, and it is tempting to employ because it is easy to administer and inexpensive (see Axolotl Newsletter, no. 9). It is our experience, however, that when administered in low doses in this fashion, it is useless. Higher doses cause skin damage and are likely to be lethal. Tetracycline administered by intubation (Gibbs, 1963) has been used successfully with axolotls (Boyer et al., 1971). We have also used this method at the Indiana University colony but have found that, although the animals so treated seemed to recover, there was a high disease recurrence rate several months later. Perhaps the dose was sometimes insufficient because the animals frequently regurgitated the medicine. In addition, it was awkward to administer and labor intensive. We have no experience with tetracycline administered by injection.

Gentamicin added to the water has been mentioned as a possible treatment (Fox, 1980a). According to Riviere and co-workers (1979), a concentration of 1 mg/ml is necessary to reach therapeutic levels in the serum in *Rana,* but no toxic effects were observed. The method has never been tested in the Indiana University colony, because it would be very expensive.

The chief disadvantage of gentamicin or other aminoglycoside drugs is reported to be serious kidney damage. Such toxicity might not be apparent until 2–6 months after treatment (Ross and Marzec, 1984, M. K. Stoskopf, personal communication), at which time many animals may die of renal failure. Data on what constitutes a toxic dose in axolotls are unavailable. In *Necturus,* a suitable dose is reported to be 2.0–2.2 mg/kg body weight by intramuscular injection every 3–5 days for a total of two or three doses (Stoskopf et al., 1987, and personal communication).

At the Indiana University colony we have extensive experience with gentamicin administered by intraperitoneal injection, but no histological data. Initial tests with gentamicin in 1984 showed no obvious side effects and resulted in the apparent recovery of many ill animals. Subsequently, a major treatment program was initiated. Between November 1985 and May 1986, 355 adults received gentamicin in an effort to counteract an unacceptably high death rate. Most animals were treated at the first sign of anorexia, but many showed other signs indicating a more advanced state of disease. Although we cannot securely associate the use of gentamicin with the control of the disease, a distinct decrease in the number of deaths among adult axolotls in the colony accompanied the treatment campaign (Table 25–1).

Table 25–1 Deaths by Month at the Indiana Axolotl Colony[a]

Month	No. of Deaths	Total No. Adults	%	Comments
August 1985	46	$<400^b$	>11.5	No treatments in use
September	32	$<400^b$	>8.0	Tetracycline by intubation
October	28	$<400^b$	>7.0	
November	35	$<400^b$	>8.8	Gentamicin begun late in month
December	12	$<400^b$	>3.0	
January 1986	13	$<400^b$	>3.3	
February	7	$<400^b$	>1.8	
March	8	$<400^b$	>2.0	
April	15	389	3.9	Parasites[c]
May	15	394	3.8	Parasites[c]
June	10	389	2.6	Gentamicin still in use, but fewer animals treated
July	9	423	2.1	
August	9	440	2.0	

[a]During the same period, losses at the Ottawa colony were substantially lower, averaging 0.7% per month and ranging from 0 to 1.3%. Since that time, losses have averaged 0.5% per month, and ranged from 0 to 1.7%.—Ed.

[b]Unfortunately, precise figures are unavailable. In spite of deaths and other losses, the total number of adults at this time gradually increased because of maturing juveniles.

[c]Introduced with minnows, which were temporarily employed as food.

The treatment regimen was, for nearly all animals, 0.45 mg/day per animal for 5 consecutive days. A few very small animals received a smaller dose. A few large animals received a larger dose. Our animals weigh on average 80–100 g each, so in most cases they received a dose greater than 4.5 mg/kg body weight, or more than twice the recommended dose (as we learned later) and at a more frequent interval. We determined this dosage on the basis of earlier experience with gentamicin in 1984 and in consultation with the Indiana University veterinarian. An even larger dose (we had used up to twice the dose) resulted in prolonged anorexia after treatment. We reduced the dose until a long recovery time was rare.

Most animals received the treatment only once, but some were treated twice. The second treatment usually came 5–7 months later. Two animals were treated three times. Table 25–2 summarizes the impact of the gentamicin treatment on the treated group of animals.

A good indication of the health of an animal subsequent to treatment is its spawning performance. The 294 animals that received treatment once gave, collectively, 233 spawnings after receiving gentamicin. Of these 233 spawnings, 200 were from those 146 animals still alive and in the colony a year later, and 33 were from animals that were either dead or sent out a year after treatment. The 58 animals that were treated twice gave 21 spawnings after the first treatment but before the second, and 21 spawnings after the second treatment. The two animals that received treatment three times subsequently gave 5 spawnings. All of the spawnings resulted from natural matings. These results led us to conclude that gentamicin—even the large dose used—was not harmful to most animals, and certainly did not discourage spawning. The animals that did not recover may have already been too ill (some died before receiving all doses), or they may have had other problems leading to their death.

Another major class of infectious disease is caused by parasites. Internal parasites identified in certain of our animals after they had been exposed to infected minnows included *Hexamita, Opalina,* and two unidentified ciliates (International Aquaculture Research Center, 1986). Which of these may have been the cause of disease is unknown (cf. Reichenbach-Klinke and Elkan, 1965). Animals that were thought to have died directly as a result of internal parasites were found to have an inflamed stomach and gut. For most animals, however, the parasites were not lethal, but a cause for secondary infection. Internal parasites were successfully treated with metronidazole (500 mg per 100 g of food for three to four feedings).

Other parasites have also been reported in axolotls. Beetschen (1980) reported *Dermocystidium,* a sporozoon. In the Indiana University colony, we have dealt

Table 25–2 Axolotl Survival After Treatment with Gentamicin

Total Treated	No. Alive and in Colony	No. Sent Out[a]	No. Dead[b]
294 (1X)	146 (1 year later)	74 (within 1 year)	74 (within 1 year)
58 (2X)	22 (May 1987)[c]	1818 (May 1987)	18 (May 1987)
2 (3X)	1 (May 1987)[c]	0	1 (May 1987)

[a]Animals sent out could be very ill or apparently healthy, but they were rarely spawning.

[b]The majority of deaths occurred within 3 months.

[c]This was 9–12 months after last treatment.

with *Costia* and *Trichodina,* common skin parasites of fish (van Duijn, 1973). Signs include inflammation of the skin (very visible in white animals), abnormal sloughing of the skin, and a sticky feel to the animal when handled. Identification can easily be made by examining skin smears under a light microscope. Signs can be alleviated by any of several common treatments (discussed in van Duijn, 1973), including temporarily increasing salts in the water, formalin bath, and acriflavine. Particular attention should be paid to water quality. Any treatment should be thoroughly tested on a small number of animals before use on the main group of animals. Complete eradication is, in our experience, very difficult. Prevention—through keeping wild-caught amphibians separate and through feeding live food only if it is aquarium bred and parasite free—is best.

Fungal infections are rare in adult axolotls; their occurrence is probably secondary to other infection and stress, which should be where attention is directed. The best approach to dealing with any fungus observed in larvae is to change the water more frequently and remove affected individuals promptly.

CONCLUSION

All of the infectious disease problems just described are essentially opportunistic. Most axolotls are probably regularly exposed to the various bacteria mentioned, as well as to viruses, fungi, and yeasts (Amborski and Amborski, 1978). Anything that stresses animals in a colony can lead to disease outbreaks; therefore, the first step in any disease control program should be to locate and correct such problems. In some cases, this may be all that is necessary (cf. Boyer et al., 1971). Overall hygiene and water quality should be the first consideration. Handling methods and nutrition are also important. Antibiotics can be, however, a useful measure when disease outbreaks do occur.

Appendix
Directory of Axolotl Colonies

GEORGE M. MALACINSKI AND DAVID ABLE

Approximately two dozen axolotl colonies exist, at various locations throughout the world. At least two of them (Indiana University: G. M. Malacinski, director; University of Ottawa: J. B. Armstrong, director) receive financial support from government agencies for the purpose of serving as genetic stock centers. Those two colonies actively engage in shipping eggs and larvae to research laboratories throughout their own countries, as well as occasionally to foreign countries.

The most sought-after animals are usually either wild type or pigment marked (e.g., white or albino). Both of these colonies therefore maintain large stocks of breeding animals that yield those phenotypes.

In addition, numerous smaller axolotl colonies are often able to supply experimental material. A representative list, in alphabetical order, of several geographically separate colonies follows.

Dr. John B. Armstrong
Department of Biology
University of Ottawa
Ottawa, Ontario K1N 6N5
Canada

Dr. Makoto Asashima
Department of Biology
Yokohama City University
4646, Kawana-Ku
Yokohama 236, Japan

Dr. J.-C. Beetschen
Centre de Biologie du Développement
Université Paul Sabatier
118, route de Narbonne
31062, Toulouse Cedex, France

Dr. J. G. Bluemink
Hubrecht Laboratory
Netherlands Institute for Developmental Biology
Uppsalalaan 8
3584 CT Utrecht, The Netherlands

Dr. Hae-Moon Chung
Seoul National University Axolotl Colony
Department of Biology
College of Education
Seoul, South Korea 151

Dr. T. A. Dettlaff
Institute of Developmental Biology
Academy of Sciences
Vavilov St. 12-2
Moscow, V-113, U.S.S.R.

Dr. James Kaufman
Basel Institute for Immunology
Grenzacherstrasse 487
CH-4005 Basel, Switzerland

Dr. Huguette Løvtrup-Rein
Department of Zoophysiology
University of Umea
S-901-87 Umea, Sweden

Dr. George M. Malacinski
Department of Biology
Indiana University
Bloomington, Indiana 47405

Dr. Pat G. Model
Department of Neuroscience
Albert Einstein College of Medicine
Kennedy Center
Bronx, New York 10461

Dr. J. Signoret
Laboratoire d'Embryologie
Faculté des Sciences
Université de Caen
14-Caen, France

Dr. J. M. W. Slack
ICRF Developmental Biology Unit
Department of Zoology
University of Oxford
South Parks Road
Oxford OX1 3PS, England

In addition to supplying experimental material, these colonies can often provide advice and useful information on axolotl husbandry. For example, procedures for maintaining animals in aquariums (versus individual bowls), instructions for feeding (liver/heart versus fish feed), and advice on increasing the spawning frequency can usually be obtained through communication with colony directors.

It should be pointed out, however, that many of the colonies were begun with small numbers of animals. They therefore contain highly inbred stocks, which may display some of the symptoms of inbreeding depression, such as increased sensitivity to disease or indifferent mating behavior. Often, inbreeding depression is an inevitable by-product of attempts to obtain genetically homogeneous lines of animals, or it arises as sibling matings are made to build a colony.

Wild stocks, recently captured from the canals that represent the remnants of Lake Xochimilco near Mexico City (see Chapter 2), occasionally are obtained by government-authorized collecting expeditions. The Indiana University axolotl colony director often knows about formal collecting trips to Mexico City, and can provide information about the availability of recently imported wild stock.

In obtaining live material from any of the aforementioned sources, it should be kept in mind that *Ambystoma mexicanum* is on the "endangered species" list. It is therefore impossible to transport axolotls legally across international borders without appropriate permits.

References

Adelmann, H. B. (1929). Experimental studies on the development of the eye. I. The effect of the removal of median and lateral areas of the anterior end of the urodelan neural plate on the development of the eyes (*Triton teniatus* and *Amblystoma punctatum*). J. Exp. Zool. *54*:249–290.

Agassiz, J. L. R. (1845). *Nomenclator Zoologicus Continens Nomina Systematica Generum Animalium tam Viventium quam Fossilium,* Fasciculus VII et VIII. Jent and Gassman, Soloduri.

Agassiz, J. L. R. (1847). *Nomenclatoris Zoologici Index Universalis.* Jent and Gassman, Soloduri.

Alberch, P., E. A. Gale, and P. R. Larsen (1986). Plasma T_4 and T_3 levels in naturally metamorphosing *Eurycea bislineata* (Amphibia; Plethodontidae). Gen. Comp. Endocrinol. *61*:153–163.

Amborski, R. L. and G. F. Amborski (1978). Pathogens and diseases of aquatic animals. Lab. Anim. *7*:14–26.

Anderson, S. L. and J. N. Dent (1982). Increased loss of thyroxine from the bloodstreams of newts after injection with ovine prolactin. Gen. Comp. Endocrinol. *47*:276–279.

Anderson, S. and D. A. Ede (1985). The behavior *in vitro* of epidermal cells from normal and *pupoid foetus* mutant mouse embryos. J. Cell Sci. *79*:305–315.

Anver, M. R. and C. L. Pond (1984). Biology and diseases of amphibians. In *Laboratory Animal Medicine* (J. G. Fox, B. J. Cohen, and F. M. Loew, eds.). Academic Press, New York, pp. 427–447.

APHA (American Public Health Association) (1971). *Standard Methods for the Examination of Water and Wastewater,* 13th ed. American Public Health Association, Washington, DC.

Armillas, P. (1971). Gardens on swamps. Science *174*:653–661.

Armstrong, J. B. (1984). Genetic mapping in the Mexican axolotl, *Ambystoma mexicanum.* Can. J. Genet. Cytol. *26*:1–6.

Armstrong, J. B. (1985). The axolotl mutants. Dev. Genet. *6*:1–25.

Armstrong, J. B. and W. S. Fletcher (1984). Residual seasonality in laboratory axolotls. Axolotl Newsletter *13*:32–35.

Armstrong, J. B. and L. L. Gillespie (1981). Induced spawnings and artificial insemination in the axolotl. Axolotl Newsletter *10*:1–4.

Armstrong, J. B., L. L. Gillespie, and G. M. Cooper (1983). Experimental studies on a lethal gene (*t*) in the Mexican axolotl *Ambystoma mexicanum.* J. Exp. Zool. *226*:423–430.

Armstrong, J. B. and A. C. Graveson (1988). Progressive patterning precedes somite segmentation in the Mexican axolotl *(Ambystoma mexicanum).* Dev. Biol., 126:1–6

Arthur, E. and M. Balls (1971). Amphibian cells *in vitro.* I. Growth of *Xenopus* cells in soft agar medium and on an agar surface. Exp. Cell Res. *64*:113–118.

Attardi, D. G. and R. W. Sperry (1963). Preferential selection of central pathways by regenerating optic fibers. Exp. Neurol. *7*:46–64.

Attwell, D., H. Brew, P. Gray, and P. Mobbs (1985a). Single potassium channels in Müller cells of the salamander retina. J. Physiol. *371:*35P.

Attwell, D., P. Mobbs, M. Tessier-Lavigne, and M. Wilson (1987). Neurotransmitter-induced currents in retinal bipolar cells of the axolotl, *Ambystoma mexicanum.* J. Physiol. *387:*125–161.

Attwell, D., M. Wilson, and S. M. Wu (1985b). The effect of light on the spread of signals through the rod network of the salamander retina. Brain Res. *343:*79–88.

Bagnara, J. T., S. K. Frost, and J. Matsumoto (1978). On the development of pigment patterns in amphibians. Am. Zool. *18:*301–312.

Bagnara, J. T. and M. E. Hadley (1973). *Chromatophores and Color Change: The Comparative Physiology of Animal Pigmentation.* Prentice-Hall, Englewood Cliffs, NJ.

Balak, K., M. Jacobson, J. Sunshine, and U. Rutishauser (1987). Neural cell adhesion molecule expression in *Xenopus* embryos. Dev. Biol. *119:*540–550.

Ball, J. N. (1981). Hypothalamic control of the pars distalis in fishes, amphibians, and reptiles. Gen. Comp. Endocrinol. *44:*135–170.

Ballantine, J. E. M., H. R. Woodland, and E. A. Sturgess (1979). Changes in protein synthesis during the development of *Xenopus laevis.* J. Embryol. Exp. Morphol. *51:*137–153.

Barbour, C. D. (1973a). The systematics and evolution of the genus *Chirostoma* Swainson (Pisces; Atherinidae). Tulane Studies in Zoology and Botany *18:*97–141.

Barbour, C. D. (1973b). A biogeographical history of *Chirostoma* (Pisces; Atherinidae): a species flock from the Mexican Plateau. Copeia *1973:*533–556.

Barish, M. E. (1986). Differentiation of voltage-gated potassium current and modulation of excitability in cultured amphibian spinal neurones. J. Physiol. *375:*229–250.

Barratt, R. W., D. Newmeyer, D. D. Perkins, and L. Garnjorst (1954). Map construction in *Neurospora crassa.* Adv. Genet. *6:*1–93.

Baud, C. and M. E. Barish (1984). Changes in membrane hydrogen and sodium conductances during progesterone-induced maturation of *Ambystoma* oocytes. Dev. Biol. *105:*423–434.

Beauchat, H. (1912). *Manuel d'Archéologie Américaine (Amérique Préhistorique-Civilisations Disparues).* Picard, Paris.

Beetschen, J.- C. (1979). Recherches expérimentales sur la symétrisation de l'oocyte et de l'oeuf d'Axolotl: facteurs conditionnant l'apparition précoce du croissant gris à la suite d'un choc thermique. C. R. Acad. Sci. (D), Paris, *288:*643–646.

Beetschen, J.- C. (1980). A lethal sporozoan parasite causing skin pustules. Axolotl Newsletter *9:*13.

Beetschen, J.- C. and J. Gauthier (1987). Heat-shock-induced grey crescent formation in axolotl eggs and oocytes: the role of gravity. Development *100:*599–609.

Bellairs, R., A. S. G. Curtis, and E. J. Sanders (1978). Cell adhesiveness and embryonic differentiation. J. Embryol. Exp. Morphol. *46:*207–213.

Benford, H. H. and M. Namenwirth (1974). Precocious appearance of the gray crescent in heat-shocked axolotl eggs. Dev. Biol. *39:*172–176.

Bennett, M. R. and P. A. McGrath (1980). Segmental innervation of rotated and supernumerary axolotl hindlimbs. J. Exp. Zool. *214:*1–12.

Bennett, M. V. L. (1967). Mechanisms of electroreception. In *Lateral Line Detectors* (P. H. Cahn, ed.). Indiana University Press, Bloomington, IN, pp. 313–393.

Ben-Ze'ev, A. (1985). Cell–cell interaction and cell configuration-related control of cytokeratins and vimentin expression in epithelial cells and in fibroblasts. Ann. N.Y. Acad. Sci. *455:*597–613.

Berg, D. K. (1982). Cell death in neuronal development: regulation by trophic factors. In *Neuronal Development* (N. C. Spitzer, ed.). Plenum Press, New York, pp. 297–331.

Blackshaw, S. E. and A. E. Warner (1976). Low-resistance junctions between mesoderm cells during development of trunk muscles. J. Physiol. *255:*209–230.

Bodney, S. A. (1982). Towards the development of a practical method for reversing the sex (male to female) of large numbers of *Ambystoma mexicanum.* Axolotl Newsletter *11:*12–18.

Bonaparte, C. L. J. L. (1831). Saggio di una distribuzione metodica degli animali vertebrati. Giorn. Arcad. Sci. Lett. Arti, Rome, *49:*3–77.

Bonnanfant-Jaïs, M. L. and P. Mentré (1983). Study of oogenesis in the newt *Pleurodeles waltlii* M. I. Ultrastructural study of the different stages of oocyte development. J. Submicrosc. Cytol. *15:*453–478.

Bordzilovskaya, N. P. and T. A. Dettlaff (1975). Axolotl: *Ambystoma mexicanum* (Cope). In *Objects of the Biology of Development* (T. A. Dettlaff, A. E. Geycinovich, and V. Y. Brodsky, eds.). Series on the Problems of the Biology of Growth, Academy of Sciences, USSR, Moscow, pp. 370–389.

Bordzilovskaya, N. P. and T. A. Dettlaff (1979). Table of stages of the normal development of axolotl embryos and the prognostication of timing of successive developmental stages at various temperatures. Axolotl Newsletter *7:*2–22.

Borgens, R. B. (1984). Are limb development and limb regeneration both initiated by an integumentary wounding? Differentiation *28:*87–93.

Borgens, R. B. (1985). Natural voltage gradients and the generation and regeneration of limbs. In *Regulation of Vertebrate Limb Regeneration* (R. E. Sicard, ed.). Oxford University Press, New York, pp. 6–31.

Borgens, R. B., J. W. Vanable, Jr., and L. F. Jaffe (1977). Bioelectricity and regeneration. I. Initiation of frog limb regeneration by minute currents. J. Exp. Zool. *200:*403–416.

Boterenbrood, E. C. and P. D. Nieuwkoop (1973). The formation of the mesoderm in urodelean amphibians. V. Its regional induction by the endoderm. Wilhelm Roux Arch. EntwMech. Org. *173:*319–332.

Boyer, C. I., Jr., K. Blackler, and L. E. DeLanney (1971). *Aeromonas hydrophila* infection in the Mexican axolotl, *Siredon mexicanum.* Lab. Anim. Sci. *21:*372–375.

Brachet, J. (1974). Observations cytologiques et cytochimiques sur la maturation de l'oocyte chez les Urodèles. Année. Biol. *13:*271–284.

Brachet, J. (1977). An old enigma: the grey crescent of amphibian eggs. Curr. Top. Dev. Biol. *11:*133–186.

Brachet, J. and H. Denis (1963). Effects of actinomycin D on morphogenesis. Nature *198:*205–206.

Brachet, J., E. Hubert, and A. Lievens (1972). The effects of α-amanitin and rifampicins on amphibian egg development. Rev. Suisse Zool. *79:*47–63.

Brandon, R. A. (1970). Size range, size at maturity and reproduction of *Ambystoma (Bathysiredon) dumerilii* (Dugès), a paedogenic Mexican salamander endemic to Lake Patzcuaro, Michoacan. Copeia *1970:*385–388.

Brandon, R. A. (1972). Hybridization between the Mexican salamanders *Ambystoma dumerilii* and *Ambystoma mexicanum* under laboratory conditions. Herpetologica *28:*199–207.

Brandon, R. A. (1976). Spontaneous and induced metamorphosis of *Ambystoma dumerilii* (Dugès), a paedogenetic Mexican salamander, under laboratory conditions. Herpetologica *32:*429–438.

Brandon, R. A. (1977). Interspecific hybridization among Mexican and United States salamanders of the genus *Ambystoma* under laboratory conditions. Herpetologica *33:*133–152.

Brandon, R. A., J. Martan, J. W. E. Wortham, and D. C. Englert (1974). The influence of interspecific hybridization on the morphology of the spermatozoa of *Ambystoma* (Caudata; Ambystomatidae). J. Reprod. Fertil. *41*:275–284.

Brandon, R. A., E. J. Maruska, and W. T. Rumph (1981). A new species of neotenic *Ambystoma* (Amphibia; Caudata) endemic to Laguna Alchichica, Puebla, Mexico. Bull. S. Calif. Acad. Sci. *80*:112–125.

Briggs, F. (1984). A three-year survey of reproductive capacity in the Indiana University axolotl colony. Axolotl Newsletter *13*:21–25.

Briggs, R. (1973). Developmental genetics of the axolotl. In *Genetic Mechanisms of Development* (F. H. Ruddle, ed.). Academic Press, New York, pp. 169–199.

Briggs, R., J. Signoret, and R. R. Humphrey (1964). Transplantation of nuclei of various cell types from neurulae of the Mexican axolotl *(Ambystoma mexicanum)*. Dev. Biol. *10*:233–246.

Brockes, J. P. (1984). Mitogenic growth factors and nerve dependence of limb regeneration. Science *225*:1280–1287.

Brockes, J. P. and C. R. Kintner (1986). Glial growth factor and nerve-dependent proliferation in the regeneration blastema of urodele amphibians. Cell *45*:301–306.

Bronner-Fraser, M. E. and A. M. Cohen (1980). The neural crest: what can it tell us about cell migration and determination? Curr. Top. Dev. Biol. *15*(I):1–25.

Brookes, J. (1828). *A Prodromus of a Synopsis Animalium, Comprising a Catalogue Raisonné of the Zootomical Collection of Joshua Brookes.* Gold and Walton, London.

Brothers, A. J. (1977). Instructions for the care and feeding of axolotls. Axolotl Newsletter *3*:9–16.

Brown, G. M. (1985). Biosynthesis of pterins. In *Folates and Pterins,* vol. 2 (R. L. Blakley and S. J. Benkovic, eds.). Wiley, New York, pp. 115–154.

Brun, R. B. (1978). Experimental analysis of the eyeless mutant in the Mexican axolotl *(Ambystoma mexicanum)*. Am. Zool. *18*:273–279.

Brun, R. B. (1980). Axolotls in the lab. Axolotl Newsletter *9*:21–25.

Brun, R. B. (1981). The movement of the prospective eye vesicles from the neural plate into the neural fold in *Ambystoma mexicanum* and *Xenopus laevis.* Dev. Biol. *88*:192–199.

Brun, R. B. (1983). The eyeless mutant axolotl *(Ambystoma mexicanum)* produces small and delayed optic vesicles. Am. Zool. *23*:998 (Abstract 716).

Brun, R. B. (1985). The eyeless gene *e* in the Mexican salamander *(Ambystoma mexicanum)* might interfere with a head gradient. Cell Differ. *16*(Suppl.):42S (Abstract 100).

Brun, R. B. (1987). The *eyeless* mutant Mexican salamander *(Ambystoma mexicanum)*: evidence for an imbalanced antero-posterior morphogenetic system. J. Neurogenet. *4*:29–46.

Brunst, V. V. (1955a). The axolotl *(Siredon mexicanum)*. I. As material for scientific research. Lab. Invest. *4*:45–64.

Brunst, V. V. (1955b). The axolotl *(Siredon mexicanum)*. II. Morphology and pathology. Lab. Invest. *4*:429–449.

Brunst, V. V. (1969). Structures of spontaneous and transplanted tumors in the axolotl *(Siredon mexicanum)*. In *Biology of Amphibian Tumors* (M. Mizell, ed.). Springer-Verlag, New York, pp. 215–219.

Bryant, S. V. (1976). Regenerative failure of double half limbs in *Notophthalmus viridescens.* Nature *263*:676–679.

Bryant, S. V. and B. A. Baca (1978). Regenerative ability of double-half and half upper arms in the newt, *Notophthalmus viridescens.* J. Exp. Zool. *204*:307–324.

Bryant, S. V., V. French, and P. J. Bryant (1981). Distal regeneration and symmetry. Science *212*:993–1002.

Bryant, S. V., D. M. Gardiner, and K. Muneoka (1987). Limb development and regeneration. Am. Zool. *27*:675–696.

Bryant, S. V., N. Holder, and P. W. Tank (1982). Cell–cell interactions and distal outgrowth in amphibian limbs. Am. Zool. *22:*143–151.

Bullock, T. H., R. G. Northcutt, and D. A. Bodznick (1982). Evolution of electroreception. Trends Neurosci. *5:*50–53.

Bullock, T. H. and T. Szabo (1986). Introduction. In *Electroreception* (T. H. Bullock and W. Heiligenberg, eds.). Wiley, New York, pp. 1–12.

Burns, R. K. (1925). The sex of parabiotic twins in amphibia. J. Exp. Zool. *42:*31–89.

Burns, R. K. (1932). Follicular atresia in the ovaries of hypophysectomized salamanders in relation to yolk formation. J. Exp. Zool. *63:*309–327.

Burns, R. K. (1961). Role of hormones in the differentiation of sex. In *Sex and Internal Secretions,* 3rd ed., vol. 1 (W. C. Young, ed.), Williams & Wilkins, Baltimore, pp. 76–158.

Burnside, B. (1971). Microtubules and microfilaments in newt neurulation. Dev. Biol. *26:*416–441.

Burnside, B. (1973). Microtubules and microfilaments in amphibian neurulation. Am. Zool. *13:*989–1006.

Burnside, N. B. and A. G. Jacobson (1968). Analysis of morphogenetic movements in the neural plate of the newt *Taricha torosa.* Dev. Biol. *18:*537–552.

Buscaglia, M., J. Leloup, and A. De Luze (1985). The role and regulation of monodeiodination of thyroxine to 3,5,3′-triiodothyronine during amphibian metamorphosis. In *Metamorphosis* (M. Balls and M. Bownes, eds.). Clarendon Press, Oxford, pp. 273–293.

Buttner, J. K. and G. W. Nace (1984). Labile water quality parameters in amphibian holding systems. Axolotl Newsletter *13:*38–43.

Callan, H. G. (1966). Chromosomes and nucleoli of the axolotl, *Ambystoma mexicanum.* J. Cell Sci. *1:*85–108.

Callan, H. G. and J. H. Taylor (1968). A radioautographic study of the time course of male meiosis in the newt *Triturus vulgaris.* J. Cell Sci. *3:*615–626.

Cameron, J. A., A. R. Hilgers, and T. J. Hinterberger (1986). Evidence that reserve cells are a source of regenerated adult newt muscle *in vitro.* Nature *321:*607–610.

Carlone, R. L. and A. L. Mescher (1985). Trophic factors from nerves. In *Regulation of Vertebrate Limb Regeneration* (R. E. Sicard, ed.). Oxford University Press, New York, pp. 93–105.

Carlson, B. M. (1974). Morphogenetic interactions between rotated skin cuffs and underlying stump tissues in regenerating axolotl forelimbs. Dev. Biol *39:*263–285.

Carlson, B. M. (1975a). Multiple regeneration from axolotl limb stumps bearing cross-transplanted minced muscle regenerates. Dev. Biol. *45:*203–208.

Carlson, B. M. (1975b). The effects of rotation and positional change of stump tissues upon morphogenesis of the regenerating axolotl limb. Dev. Biol. *47:*269–291.

Carrasco, A. E. and G. M. Malacinski (1987). Localization of *Xenopus* homoeo-box gene transcripts during embryogenesis and in the adult nervous system. Dev. Biol. *121:*69–81.

Carrasco, A. E., W. McGinnis, W. Gehring, and E. M. DeRobertis (1984). Cloning of an *X. laevis* gene expressed during early embryogenesis coding for a peptide region homologous to *Drosophila* homeotic genes. Cell *37:*409–414.

Cate, G. ten (1956). The intrinsic embryonic development. Verh. K. Ned. Akad. Wet. Afd. Natuurk. *51:*1–257.

Chauvin, M. von (1885). Über die Verwandlungsfähigkeit des Mexikanischen Axolotl. Z. Wiss. Zool., Leipzig, *41:*365–389.

Chung, H.-M. and R. Briggs (1975). Experimental studies on a lethal gene (*l*) in the Mexican axolotl, *Ambystoma mexicanum.* J. Exp. Zool. *191:*33–48.

Chung, H.-M. and G. M. Malacinski (1982). Pattern formation during early amphibian development: embryogenesis in inverted anuran and urodele eggs. Dev. Biol. *93*:444–452.

Chung, H.-M. and G. M. Malacinski (1985). SEM study of sperm penetration in polyspermic urodele amphibian *(Ambystoma mexicanum)* eggs. K. J. Zool. *28*:1–8.

Ciment, G. and J. A. Weston (1982). Enteric neurogenesis by neural crest-derived branchial arch mesenchymal cells. Nature *305*:424–427.

Cleine, J. H. and J. M. W. Slack (1985). Normal fates and states of specification of different regions in the axolotl gastrula. J. Embryol. Exp. Morphol. *86*:247–269.

Clemons, G. K. and C. S. Nicoll (1977). Development and preliminary application of a homologous radioimmunoassay for bullfrog prolactin. Gen. Comp. Endocrinol. *32*:531–535.

Clemons, G. K., S. M. Russell, and C. S. Nicoll (1979). Effect of mammalian thyrotropin-releasing hormone on prolactin secretion by bullfrog adenohypophysis *in vitro*. Gen. Comp. Endocrinol. *38*:62–67.

Cochran, D. M. (1961). Type specimens of reptiles and amphibians in the U.S. National Museum. Bulletin 220, U.S. Natl. Museum.

Coe, M. D. (1964). The chinampas of Mexico. Sci. Am. *211*(7):90–98.

Coghill, G. E. (1929). *Anatomy and the Problem of Behaviour*, (reprinted 1964). Hafner, New York.

Collins, J. P., J. B. Mitton, and B. A. Pierce (1980). *Ambystoma tigrinum:* a multispecies conglomerate? Copeia *1980*:938–941.

Conant, R. (1975). *A Field Guide to Reptiles and Amphibians of Eastern and Central North America*. Houghton Mifflin, Boston.

Connelly, T. G. and F. L. Bookstein (1983). Method for 3-dimensional analysis of patterns of thymidine labeling in regenerating and developing limbs. In *Limb Development and Regeneration*, part A (J. F. Fallon and A. I. Caplan, eds.). Alan R. Liss, New York, pp. 525–536.

Cooke, J. (1975). Control of somite number during morphogenesis of a vertebrate, *Xenopus laevis*. Nature *254*:196–199.

Cooke, J. (1977). The control of somite number during amphibian development: models and experiments. In *Vertebrate Limb and Somite Morphogenesis* (D. A. Ede, J. R. Hinchliffe, and M. Balls, eds.). Cambridge University Press, Cambridge, pp. 433–448.

Cooke, J. (1979). Cell number in relation to primary pattern formation in the embryo of *Xenopus laevis*. II. Sequential cell recruitment, and control of the cell cycle, during mesoderm formation. J. Embryol. Exp. Morphol. *53*:269–289.

Cooke, J. (1980). Early organization of the central nervous system: form and pattern. Curr. Top. Dev. Biol. *15*:373–407.

Cooke, J. (1985). Embryonic origins of the nervous system and its pattern. Trends Neurosci. *8*:58–63.

Cooke, J. and E. C. Zeeman (1976). A clock and wavefront model for control of the numbers of repeated structures during animal development. J. Theor. Biol. *58*:455–476.

Cooper, G. M., J. B. Armstrong, and S. Gottlob-McHugh (1985). Allelic isozyme variants in the Mexican axolotl *(Ambystoma mexicanum)* as potential markers for developmental experiments. Dev. Genet. *5*:73–82.

Cope, E. D. (1868). A review of the species of Amblystomidae. Proc. Acad. Nat. Sci., Phila. *19*:166–211.

Couchman, J. R., B. Caterson, J. E. Christner, and J. R. Baker (1984). Mapping by monoclonal antibody detection of glycosaminoglycans in connective tissues. Nature *307*:650–652.

Covell, D. A., Jr. and P. G. Model (1984). Degeneration and regeneration of vestibular axons in axolotl *(Ambystoma mexicanum)* larvae. Soc. Neurosci. Abstr. *10*:1086.

Cuenot, L. (1905). Les races pures et leurs combinaisons chez les souris. Arch. Zool. Exp. Gen. *33:*123–132.

Cuny, R. and G. M. Malacinski (1985). Banding differences between tiger salamander and axolotl chromosomes. Can. J. Genet. Cytol. *27:*510–514.

Cuny, R. and G. M. Malacinski (1986). Axolotl retina and lens development: mutual tissue stimulation and autonomous failure in the eyeless mutant retina. J. Embryol. Exp. Morphol. *96:*151–170.

Cuvier, G. (1828). Le règne animal distribué d'après son organisation, pour servir de l'histoire naturelle des animaux, et d'introduction à l'anatomie comparee. (2nd ed.). Fortin, Paris.

Dale, L. and J. M. W. Slack (1987a). Fate map for the 32-cell stage of *Xenopus laevis.* Development *99:*527–551.

Dale, L. and J. M. W. Slack (1987b). Regional specification within the mesoderm of early embryos of *Xenopus laevis.* Development *100:*279–295.

Dalton, H. C. (1949). Developmental analysis of genetic differences in pigmentation in the axolotl. Proc. Natl. Acad. Sci. USA *35:*277–283.

Dalton, H. C. (1950a). Comparison of white and black axolotl chromatophores *in vitro.* J. Exp. Zool. *115:*17–33.

Dalton, H. C. (1950b). Inhibition of chromatoblast migration as a factor in the development of genetic differences in pigmentation in white and black axolotls. J. Exp. Zool. *115:*151–173.

Darras, V. M. and E. R. Kühn (1982). Increased plasma levels of thyroid hormones in a frog *Rana ridibunda* following intravenous administration of TRH. Gen. Comp. Endocrinol. *48:*469–475.

Darras, V. M. and E. R. Kühn (1983). Effects of TRH, bovine TSH, and pituitary extracts on thyroidal T_4 release in *Ambystoma mexicanum.* Gen. Comp. Endocrinol. *51:*286–291.

Darras, V. M. and E. R. Kühn (1984). Difference of the *in vivo* responsiveness to thyrotropin stimulation between the neotenic and metamorphosed axolotl, *Ambystoma mexicanum:* failure of prolactin to block the thyrotropin-induced thyroxine release. Gen. Comp. Endocrinol. *56:*321–325.

David, J. C., D. Vinson, J. Lefresne, and J. Signoret (1979). Evidence for a DNA ligase change related to early cleavage in axolotl eggs. Cell Differ. *8:*451–459.

David, W. S. and P. G. Model (1982a). Effect of displacing afferent projection on synapse localization on an identified vertebrate central neuron. Anat. Rec. *202:*41A.

David, W. S. and P. G. Model (1982b). Synapse localization on the Mauthner cell is unaffected by displacement of a major afferent projection. Soc. Neurosci. Abstr. *8:*436.

David, W. S. and P. G. Model (1984). Surgically displaced nVIII and nVII follow normal CNS pathways to an identified target neuron. Soc. Neurosci. Abstr. *10:*1160.

Dawid, I. B., S. R. Haynes, M. Jamrich, E. Jones, S. Miyatani, T. D. Sargent, and J. A. Winkles (1985). Gene expression in *Xenopus* embryogenesis. J. Embryol. Exp. Morphol. *89*(Suppl.):113–124.

DeFelice, D. A. and E. M. Rasch (1969). Chronology of spermatogenesis and spermiogenesis in poeciliid fishes. J. Exp. Zool. *171:*191–208.

DeLotto, R. and P. Spierer (1986). A gene required for the specification of dorsal-ventral pattern in *Drosophila* appears to encode a serine protease. Nature *323:*688–692.

Dent, J. N. (1968). Survey of amphibian metamorphosis. In *Metamorphosis: A Problem in Developmental Biology* (W. Etkin and L. I. Gilbert, eds.). Appleton-Century-Crofts, New York, pp. 271–311.

Dent, J. N., J. S. Kirby-Smith, and D. L. Craig (1955). Induction of metamorphosis in *Gyrinophilus palleucus.* Anat. Rec. *121:*429.

Desmond, M. E. and A. G. Jacobson (1977). Embryonic brain enlargement requires cerebrospinal fluid pressure. Dev. Biol. *57:*188–198.

Dettlaff, T. A. (1964). Cell division, duration of interkinetic states and differentiation in early stages of embryonic development. Adv. Morphol. *3:*323–363.

Dettlaff, T. A. (1977). Some thermal-temporal patterns of embryogenesis of poikilothermic animals. In *Problems in Experimental Biology* (D. K. Belyaev, ed.) Nauka, Moscow (in Russian), pp. 269–287.

Dettlaff, T. A. and A. A. Dettlaff (1960). Dimensionless characteristics of the development duration in embryology. Dokl. Akad. Nauk SSSR, *134:*199–202.

Dettlaff, T. A. and A. A. Dettlaff (1961). On relative dimensionless characteristics of the development duration in embryology. Arch. Biol., Liege *72:*1–16.

Dettlaff, T. A. and T. B. Rudneva (1973). Dimensionless characteristic of developmental duration in *Xenopus laevis.* Ontogenez *4:*461–471.

Detwiler, S. R. (1934). An experimental study of spinal nerve segmentation in *Amblystoma* with reference to the plurisegmental contribution to the brachial plexus. J. Exp. Zool. *67:*395–441.

Detwiler, S. R. (1936). *Neuroembryology: an Experimental Study.* Macmillan, New York.

Detwiler, S. R. (1940). Unilateral reversal of the anteroposterior axis of the medulla in *Amblystoma.* J. Exp. Zool. *84:*13–22.

Deuchar, E. M. (1975). *Xenopus: The South African Clawed Frog.* Wiley, London.

Dibble, C. E. and A. J. O. Anderson (1963). *Florentine Codex. General History of Things of New Spain,* in 13 parts, by Fray Bernardino de Sahagún, vol. 12, University of Utah, Salt Lake City.

DiBernardino, M. A. (1980). Genetic stability and modulation of metazoan nuclei transplanted into eggs and oocytes. Differentiation *17:*17–30.

DiBernardino, M. A. and A. J. Brothers (1984). Robert W. Briggs 1911–1983. Differentiation *26:*173–175.

Dijkgraaf, S. (1962). The functioning and significance of the lateral-line organs. Biol. Rev. *38:*51–105.

Dixon, J. R. (1963). A new species of salamander of the genus *Ambystoma* from Jalisco, Mexico. Copeia *1963:*99–101.

Dodd, M. H. I. and J. M. Dodd (1976). The biology of metamorphosis. In *Physiology of the Amphibia* (B. Lofts, ed.), vol. 3. Academic Press, New York, pp. 467–599.

Duellman, W. E. and L. Trueb (1986). *Biology of Amphibians.* McGraw-Hill, New York.

Dugès, A. A. D. (1870). Una nueva especie de ajolote de la Laguna de Patzcuaro. Naturaleza *1:*241–244, pl. 5.

Dugès, A. A. D. (1888). Erpetologia del Valle de México. Naturaleza Ser. 2 *1:*97–146.

Dugès, A. A. D. (1895). Description d'un axolotl des Montagnes de Las Cruces *(Amblystoma cltamirani,* A. Dugès), Imprimèrie de Fomento, Institut Medico-National, México, D. F., 6 pp., 1 pl.

Duméril, A. H. A. (1870a). Quatrième notice sur la ménagerie des reptiles du Muséum d'Histoire Naturelle. Nouv. Arch. Mus. Hist. Nat., Paris, *5:*47–60.

Duméril, A. H. A. (1870b). Création d'une race blanche d'axolotl à la ménagerie des reptiles du Muséum d'Histoire Naturelle, et remarques sur la transformation de ces batraciens. C. R. Acad. Sci., Paris, *70:*782–785.

Duméril, A. H. A. (1872). Notes complémentaires sur les axolotls. Mém. Soc. Linn. N. Fr. *2:*248–251 (1868–1871).

Duméril, A. M. C., G. Bibron, and A. H. A. Dumeril (1854). *Erpétologie Générale ou Histoire Naturelle Complête des Reptiles,* vol. 9. Roret, Paris.

Dundee, H. A. (1957). Partial metamorphosis induced in *Thyplomolge rathbuni.* Copeia *1957:*52–53.

Dunn, E. R. (1928). A new genus of salamander from Mexico. Proc. N. Engl. Zool. Club, *10*:85–86.

Dunn, E. R. (1931). Present status of the axolotl problem. Anat. Rec. *49*(Suppl.):90.

Duprat, A. M. (1970). Action de la puromycine et d'une substance analogue sur la différenciation de cellules embryonnaires d'amphibiens, cultivées *in vitro*. J. Embryol. Exp. Morphol. *24*:119–138.

DuShane, G. P. (1935). An experimental study of the origin of pigment cells in amphibia. J. Exp. Zool. *72*:1–31.

DuShane, G. P. (1939). The role of embryonic ectoderm and mesoderm in pigment production in amphibia. J. Exp. Zool. *82*:193–215.

Dvorak, D. (1984). Off-pathway synaptic transmission in the outer retina of the axolotl is mediated by a kainic acid-preferring receptor. Neurosci. Lett. *50*:7–11.

Eagleson, G. W. and G. M. Malacinski (1983). Development of the preoptic neurosecretory centers in the eyeless mutant axolotl *(Ambystoma mexicanum)*. Am. Zool. *22*:929 (Abstract 443).

Eagleson, G. W. and G. M. Malacinski (1986). A scanning electron microscopy and histological study on the effects of the mutant eyeless (*e/e*) gene upon the hypothalamus in the Mexican axolotl *Ambystoma mexicanum* Shaw. Anat. Rec. *215*:317–327.

Eagleson, G. W. and B. A. McKeown (1978). Changes in thyroid activity of *Ambystoma gracile* (Baird) during different larval, transforming, and postmetamorphic phases. Can. J. Zool. *56*:1377–1381.

Eigenmann, C. H. and W. A. Denny (1900). The eyes of the blind vertebrates of North America. III. The structure and ontogenetic degeneration of the eyes of the Missouri cave salamander. Biol. Bull. *2*:33–41.

Elbert, O. E., C. F. Ide, M. Duda, and R. K. Hunt (1983). *Spastic* mutant axolotl: identification of a phenocopy pathway with implications for the control of axolotl swimming by the vestibulocerebellum. J. Comp. Neurol. *220*:97–105.

Elinson, R. P. (1985). Changes in levels of polymeric tubulin associated with activation and dorsoventral polarization of the frog egg. Dev. Biol. *109*:224–233.

Elsdale, T. and D. Davidson (1983). Somitogenesis in amphibia. IV. The dynamics of tail development. J. Embryol. Exp. Morphol. *76*:157–176.

Elsdale, T., M. Pearson, and M. Whitehead (1976). Abnormalities in somite segmentation following heat shock to *Xenopus* embryos. J. Embryol. Exp. Morphol. *35*:625–635.

Enciclopedia de México (1985). Imagen de la Gran Capital. Departamento del Distrito Federal, México.

Ephrussi, B. (1935). The behavior in vitro of tissues from lethal embryos. J. Exp. Zool. *70*:197–204.

Epp, L. G. (1972). Development of pigmentation in the eyeless mutant of the Mexican axolotl, *Ambystoma mexicanum* Shaw. J. Exp. Zool. *181*:169–180.

Epp, L. G. (1978). A review of the eyeless mutant in the Mexican axolotl. Am. Zool. *18*:267–272.

Epperlein, H. H. and J. Löfberg (1984). Xanthophores in chromatophore groups of the premigratory neural crest initiate the pigment pattern of the axolotl larva. Wilhelm Roux Arch. Dev. Biol. *193*:357–369.

Epperlein, H. H., Perris, R., and J. Löfberg (1986). Environmental control in pigment pattern formation in the axolotl larva. In *Progress in Developmental Biology,* part B (H. C. Slavkin, ed.). Alan R. Liss, New York, pp. 191–194.

Erickson, C. A. (1986). Morphogenesis of the neural crest. In *Developmental Biology,* vol. 2 (L. W. Browder, ed.). Plenum Press, New York, pp. 481–543.

Etkin, L. D. and B. Pearman (1987). Distribution, expression and germ line transmission of exogenous DNA sequences following microinjection into *Xenopus laevis* eggs. Development *99*:15–23.

Etkin, W. (1966). Hypothalamic sensitivity to thyroid feedback in the tadpole. Neuroendocrinology *1*:293–302.

Etkin, W., S. Kikuyama, and J. Rosenbluth (1966). Thyroid feedback to the hypothalamic neurosecretory system in frog larvae. Neuroendocrinology *1*:45–64.

Fankhauser, G. (1967). Urodeles. In *Methods in Developmental Biology* (F. H. Wilt and N. K. Wessells, eds.). Crowell, New York, pp. 85–99.

Fankhauser, G. and R. R. Humphrey (1942). Induction of triploidy and haploidy in axolotl eggs by cold treatment. Biol. Bull. *83*:367–374.

Fankhauser, G. and R. R. Humphrey (1943). The relation between number of nucleoli and number of chromosome sets in animal cells. Proc. Natl. Acad. Sci. USA, *29*:344–350.

Fankhauser, G. and R. R. Humphrey (1950). Chromosome number and development of progeny of triploid axolotl females mated with diploid males. J. Exp. Zool. *115*:207–243.

Fekete, D. M. and J. P. Brockes (1987). A monoclonal antibody detects a difference in the cellular composition of developing and regenerating limbs of newts. Development *99*:589–602.

Fekete, E., D. Fristrom, I. Kiss, and J. W. Fristrom (1975). The mechanism of evagination of imaginal discs of *Drosophila melanogaster*. II. Studies on trypsin-accelerated evagination. Wilhelm Roux Arch. Dev. Biol. *178*:123–138.

Fick, R. (1893). Über die Reifung und Befruchtung des Axolotleies. Z. Wiss. Zool., Leipzig, *56*:529–614.

Flock, Å. (1965). Electron microscopic and electrophysiologic studies on the lateral line canal organ. Acta Otolaryngol. Suppl. *199*:1–90.

Flock, Å. (1967). Ultrastructure and function in the lateral line organs. In *Lateral Line Detectors* (P. H. Cahn, ed.). Indiana University Press, Bloomington, IN, pp. 163–197.

Flock, Å. (1971). The lateral line organ mechanoreceptors. In *Fish Physiology*, vol. 5, *Sensory Systems and Electric Organs* (W. S. Hoar and D. J. Randall, eds.). Academic Press, New York, pp. 241–264.

Flügge, W. (1967). *Stresses in Shells*. Springer-Verlag, New York.

Forbes, W. R. and C. S. Forbes (1980). Notes on maintaining axolotls in aquaria. Axolotl Newsletter *9*:30.

Ford, C. C. and J. B. Gurdon (1977). A method for enucleating oocytes of *Xenopus laevis* J. Embryol. Exp. Morphol. *37*:203–209.

Forman, D. and J. M. W. Slack (1980). Determination and cellular commitment in the embryonic amphibian mesoderm. Nature *286*:492–494.

Fox, R. W. (1984). The world's urban explosion. Natl. Geographic *166*:176–185.

Fox, W. F. (1980a). Raised epidermal ridge disease. Axolotl Newsletter *9*:9.

Fox, W. F. (1980b). Problems with plastics and axolotls. Axolotl Newsletter *9*:32.

Fox, W. F. (1982a). Update on "Raised epidermal ridge disease." Axolotl Newsletter *11*:19–23.

Fox, W. F. (1982b). Ammonia, ammonium, nitrite, and nitrate: causes of disease in axolotls and prevention of their toxic effects. Axolotl Newsletter *11*:24–26.

Fox, W. F. (1984). Factors influencing axolotl spawnings at Developmental Biology Center, University of California, Irvine. Axolotl Newsletter *13*:26–27.

Freed, J. J. and L. Mezger-Freed (1970). Stable haploid cultured cell lines from frog embryos. Proc. Natl. Acad. Sci. USA, *65*:337–344.

Freeman, J. M. and D. F. Davey (1986). The precision of pathway selection by developing peripheral axons in the axolotl. J. Embryol. Exp. Morphol. *91*:117–134.

French, V., P. J. Bryant, and S. V. Bryant (1976). Pattern regulation in epimorphic fields. Science *193:*969–981.

Fritzsch, B. (1981a). The pattern of lateral-line afferents in urodeles. Cell Tissue Res. *218:*581–594.

Fritzsch, B. (1981b). Electroreceptors and direction-specific arrangement in the lateral line system of salamanders? Z. Naturforsch. *360:*493–495.

Fritzsch, B., A. M. Nikundiwe, and U. Will (1984). Projection patterns of lateral-line afferents in anurans: a comparative HRP study. J. Comp. Neurol. *229:*451–469.

Fritzsch, B. and U. Wahnschaffe (1983). The electroreceptive ampullary organs of urodeles. Cell Tissue Res. *229:*483–503.

Fritzsch, B. and M. H. Wake (1986). The distribution of ampullary organs in *Gymnophiona.* J. Herpetol. *20:*90–93.

Frost, D. R., ed. (1985). *Amphibian Species of the World: A Taxonomic and Geographic Reference,* Association of Systematics Collections, Lawrence, KS.

Frost, S. K. and J. T. Bagnara (1979). Allopurinol-induced melanism in the tiger salamander *(Ambystoma tigrinum nebulosum).* J. Exp. Zool. *209:*455–466.

Frost, S. K. and G. M. Malacinski (1980). Developmental genetics of pigment mutants in the Mexican axolotl. Dev. Genet. *1:*271–294.

Frost, S. K. and S. J. Robinson (1984). Pigment cell differentiation in the firebellied toad, *Bombina orientalis.* I. Structural, chemical and physical aspects of the adult pigment pattern. J. Morphol. *179:*229–242.

Frost, S. K., S. J. Robinson, M. E. Borchert, and M. K. Carson (1988). Drug-induced and genetic hypermelanism: effects on pigment cell differentiation. Development, *in press.*

Frost, S. K., S. J. Robinson, M. K. Carson, S. Thorsteinsdottir, and J. Giesler (1987). Effects of exogenous guanosine on chromatophore differentiation in the axolotl. Pigment Cell Res. *1,* 37–43.

Frost, S. K., F. Briggs, and G. M. Malacinski (1984a). A color atlas of pigment genes in the Mexican axolotl *(Ambystoma mexicanum).* Differentiation *26:*182–188.

Frost, S. K., L. G. Epp, and S. J. Robinson (1984b). The pigmentary system of developing axolotls. I. A biochemical and structural analysis of chromatophores in wild-type axolotls. J. Embryol. Exp. Morphol. *81:*105–125.

Frost, S. K., L. G. Epp, and S. J. Robinson (1984c). The pigmentary system of developing axolotls. II. An analysis of the melanoid phenotype. J. Embryol. Exp. Morphol. *81:*127–142.

Frost, S. K., M. E. Borchert, and S. Thorsteinsdottir (1986a). Rapid and sensitive thin-layer chromatographic assay procedure for measuring xanthine dehydrogenase activity from tissue extracts. J. Chromatogr. *382:*314–320.

Frost, S. K., L. G. Epp, and S. J. Robinson (1986b). The pigmentary system of developing axolotls. III. An analysis of the albino phenotype. J. Embryol. Exp. Morphol. *92:*255–268.

Frost, S. K., L. G. Epp, and S. J. Robinson (1986c). The pigmentary system of developing axolotls. IV. An analysis of the axanthic phenotype. J. Embryol. Exp. Morphol. *95:*117–130.

Gadow, H. (1903). The Mexican axolotl. Nature *67:*330–332.

Gallien, L. (1955). The action of sex hormones on the development of sex in amphibia. Mem. Soc. Endocrinol. *4:*188–204.

Gardiner, D. M., K. Muneoka, and S. V. Bryant (1986). The migration of dermal cells during blastema formation in axolotls. Dev. Biol. *118:*488–493.

Gautier, J. (1987). The role of the germinal vesicle for the appearance of maturation-promoting factor activity in the axolotl oocyte. Dev. Biol. *123:*483–486.

Gautier, J. and J. C. Beetschen (1983). Inhibition of protein synthesis elicits early grey crescent formation in the axolotl oocyte. Wilhelm Roux Arch. Dev. Biol. *192:*196–199.

Gautier, J. and J. C. Beetschen (1985). A three-step scheme for grey crescent formation in the rotated axolotl oocyte. Dev. Biol. *110:*192–199.

Gautier, J. and R. Tencer (1986). Changes in patterns of protein synthesis in axolotl oocytes during progesterone-induced maturation. J. Embryol. Exp. Morphol. *92:*103–113.

Gautier, J. and R. Tencer (1987). Involvement of the cytoskeleton in early grey crescent formation in axolotl oocytes. Wilhelm Roux Arch. Dev. Biol. *196:*316–320.

Gehlbach, F. R. (1967). *Ambystoma tigrinum.* In *Catalog of American Amphibians and Reptiles,* American Society of Ichthyologists and Herpetologists, p. 52.

Gerhart, J., M. Wu, M. Cyert, and M. Kirschner (1985). M-phase promoting factors from eggs of *Xenopus laevis.* Cytobios *43:*335–347.

Gibbs, E. L. (1963). An effective treatment for red-leg disease in *Rana pipiens.* Lab. Anim. Care *13:*781–783.

Gillespie, L. L. and J. B. Armstrong (1978). Pressure induction of triploids in the axolotl. Axolotl Newsletter *5:*12–16.

Gillespie, L. L. and J. B. Armstrong (1979). Induction of triploid and gynogenetic diploid axolotls *(Ambystoma mexicanum)* by hydrostatic pressure. J. Exp. Zool. *210:*117–122.

Gillespie, L. L. and J. B. Armstrong (1980). Production of androgenetic diploid axolotls by suppression of first cleavage. J. Exp. Zool. *213:*423–425.

Gillespie, L. L. and J. B. Armstrong (1981). Suppression of first cleavage in the Mexican axolotl *(Ambystoma mexicanum)* by heat shock or hydrostatic pressure. J. Exp. Zool. *218:*441–445.

Gillespie, L. L. and J. B. Armstrong (1985). Formation of the pronephros and pronephric duct rudiment in the Mexican axolotl. J. Morphol. *185:*217–222.

Gillespie, L. L. and J. B. Armstrong (1986). Morphogenetic waves in the development of the lateral mesoderm in the Mexican axolotl *(Ambystoma mexicanum)* and their relationship to pronephric duct migration. J. Exp. Zool. *237:*327–338.

Gillespie, L. L., M. S. Steinberg, and J. B. Armstrong (1985). Experimental evidence for a proteinaceous presegmental wave required for morphogenesis of axolotl mesoderm. Dev. Biol. *107:*220–226.

Gimlich, R. L. and J. Braun (1985). Improved fluorescent compounds for tracing cell lineage. Dev. Biol. *109:*509–514.

Globus, M. (1978). Neurotrophic contribution to a proposed tripartite control of the mitotic cycle in the regeneration blastema of the newt, *Notophthalmus (Triturus) viridescens.* Am. Zool. *18:*855–868.

Globus, M., S. Vethamany-Globus, and Y. C. I. Lee (1980). Effect of apical epidermal cap on mitotic cycle and cartilage differentiation in regeneration blastemata in the newt, *Notophthalmus viridescens.* Dev. Biol. *75:*358–372.

Gluecksohn-Waelsch, S. (1963). Lethal genes and analysis of differentiation. Science *142:*1269–1276.

Gona, A. G. (1973). Effects of thyroxine, thyrotropin, prolactin, and growth hormone on the maturation of the frog cerebellum. Exp. Neurol. *38:*494–501.

González, A., J. L. Camarillo, F. Mendoza, and M. Mancilla (1986). Impact of human expanding populations on the herpetofauna of the Valley of Mexico. Herp. Rev. *17:*30–31.

Goodman, C. S., M. J. Bastiani, C. Q. Doe, S. du Lac, S. L. Helfand, J. W. Kuwada, and J. B. Thomas (1984). Cell recognition during neuronal development. Science *225:*1271–1279.

Goodman, L. A., D. A. Covell, Jr., and P. G. Model (1988). Regulating afferent fibers stimulate the recovery of Mauthner cell dendritic branching in the axolotl. J. Neurosci., *in press.*

Goodman, L. A. and P. G. Model (1988). Superinnervation enhances the dendritic branching pattern of the Mauthner cell in the developing axolotl. J. Neurosci., 8:776–791.

Goos, H. J. T. (1968). Hypothalamic neurosecretion and metamorphosis in *Xenopus laevis*. III. The effect of an interruption of thyroid hormone synthesis. Z. Zellforsch. Mikrosk. Anat. *92:*583–587.

Gordon, G. and G. M. Malacinski (1970). An improved microinjection apparatus for biochemical embryology. Microchem. J. *15:*685–691.

Gordon, R. (1985). A review of the theories of vertebrate neurulation and their relationship to the mechanics of neural tube birth defects. J. Embryol. Exp. Morphol. *89*(Suppl.):229–255.

Gordon, R. and G. W. Brodland (1987). The cytoskeletal mechanics of brain morphogenesis: cell state splitters cause primary neural induction. Cell Biophys. *11:*177–238.

Gordon, R. and A. G. Jacobson (1978). The shaping of tissues in embryos. Sci. Am. *238*(6):106–113.

Goss, R. J. (1956). The relation of bone to the histogenesis of cartilage in regenerating forelimbs and tails of adult *Triturus viridescens*. J. Morphol. *98:*89–123.

Gould, S. J. (1977). *Ontogeny and Phylogeny.* Harvard University Press, Cambridge, MA.

Grabowski, S. R. and W. L. Pak (1975). Intracellular recordings of rod responses during dark adaptation. J. Physiol. *247:*363–391.

Gravenhorst, J. L. C. (1829). *Deliciae Musei Zoologici Vratislaviensis. Fasciculus Primus. Chelonios et Batrachia,* XIV. Voss, Leipzig, 106 pp.

Gray, J. E. (1831). A synopsis of the species of the class Reptilia. In *The Animal Kingdom,* vol. 9 (E. Griffith and E. Pidgeon, eds.). Whittaker, London, pp. 1–110.

Gray, J. E. (1850). *Catalogue of the Specimens of Amphibia in the Collection of the British Museum,* part II, *Batrachia Gradientia, etc.* Taylor & Francis, London.

Green, J. (1818). Descriptions of several species of North American amphibia, accompanied with observations. J. Acad. Nat. Sci. Phila. *1:*348–359.

Green, J. (1825). Description of a new species of salamander. J. Acad. Nat. Sci., Phila. *5:*116–118.

Greenwald, I. (1985). *lin-12,* A nematode homeotic gene, is homologous to a set of mammalian proteins that includes epidermal growth factor. Cell *43:*583–590.

Grimm, L. M. (1971). An evaluation of myotypic respecification in axolotls. J. Exp. Zool. *178:*479–496.

Grinfeld, S. and J. C. Beetschen (1982). Early grey crescent formation experimentally induced by cycloheximide in the axolotl oocyte. Wilhelm Roux Arch. Dev. Biol. *191:*215–221.

Gruberg, E. R. and W. A. Harris (1981). The serotonergic somatosensory projection of the tectum of normal and eyeless salamanders. J. Morphol. *170:*55–69.

Gudernatsch, J. F. (1912). Feeding experiments on tadpoles. 1. The influence of specific organs given as food on growth and differentiation: a contribution to the knowledge of organs with internal secretion. Arch. EntwMech. Org. *35:*457–483.

Gumpel-Pinot, M. (1984). Muscle and skeleton of limbs and body wall. In *Chimeras in Developmental Biology* (N. Le Douarin and A. McLaren, eds.). Academic Press, London, pp. 281–310.

Gunsbourg, N. (1971). Etude de la variation de charge de péricaryons neurosécréteurs au cours de la métamorphose provoquée chez *Ambystoma mexicanum* (Shaw). Ann. Endocrinol. *32:*705–708.

Gurdon, J. B. (1967). African clawed frogs. In *Methods in Developmental Biology* (F. H. Wilt and N. K. Wessells, eds.). Crowell, New York, pp. 75–84.

Häcker, V. (1907). Über Mendellsche Vererbung bei Axolotln. Verh. Ges. Dtsch. Naturforsch. Ärzte *31:*99–102.

Hadorn, E. (1961). *Developmental Genetics and Lethal Factors.* Methuen, London.

Hall, B. K. (1984). Developmental processes underlying heterochrony as an evolutionary mechanism. Can. J. Zool. *62:*1–7.

Hall, T. R. and A. Chadwick (1984). Effects of synthetic mammalian thyrotrophin-releasing hormone, somatostatin and dopamine on the secretion of prolactin and growth hormone from amphibian and reptilian pituitary glands incubated *in vitro.* J. Endocrinol. *102:*175–180.

Halver, J. E., ed. (1972). *Fish Nutrition.* Academic Press, New York.

Hamburgr, V. (1960). *A Manual of Experimental Embryology.* University of Chicago Press, Chicago.

Hamburger, V. (1985). Hans Spemann, Nobel Laureate 1935. Trends Neurosci. *8:*385–387.

Harlan, R. (1835). Genera of North American Reptilia, and a synopsis of the species. Med. Phys. Res. *1835:*84–163.

Harris, W. A. (1980). The effects of eliminating impulse activity on the development of the retinotectal projection in salamanders. J. Comp. Neurol. *194:*303–317.

Harris, W. A. (1982). The transplantation of eyes to genetically eyeless salamanders: visual projections and somatosensory interactions. J. Neurosci. *2:*339–353.

Harris, W. A. (1983). The *eyeless* axolotl: experimental embryogenetics and the development of the nervous system. Trends Neurosci. *6:*505–510.

Harris, W. A. (1984). Axonal pathfinding in the absence of normal pathways and impulse activity. J. Neurosci. *4:*1153–1162.

Harris, W. A. and J. Cole (1984). Common mechanisms in vertebrate axonal navigation: retinal transplants between distantly related amphibia. J. Neurogenet. *1:*127–140.

Harrison, R. G. (1904). Experimentelle Untersuchungen über die Entwicklung der Sinnesorgane der Seitenlinie bei den Amphibien. Arch. Mikrosk. Anat. Entw. *63:*35–149.

Harrison, R. G. (1921). On relations of symmetry in transplanted limbs. J. Exp. Zool. *32:*1–136.

Harrison, R. G. (1969). Harrison stages and description of the normal development of the spotted salamander, *Amblystoma punctatum* (Linn.). In *Organization and Development of the Embryo* (S. Wilens, ed.). Yale University Press, New Haven, CT, pp. 44–66.

Harsa-King, M. L. (1978). Experimental studies on a mutant gene, *a,* causing albinism in the axolotl, *Ambystoma mexicanum.* Dev. Biol. *62:*370–388.

Harsa-King, M. L. (1980). Melanogenesis in oocytes of wild-type and mutant albino axolotls. Dev. Biol. *74:*251–262.

Hastings, W. H. and L. M. Dickie (1972). Feed formulation and evaluation. In *Fish Nutrition* (J. E. Halver, ed.). Academic Press, New York, pp. 327–374.

Hauser-Gunsbourg, N. and V. Lefebvre (1976). Action de la thyroxine sur les cellules thyréotropes de l'axolotl mexicain (*Ambystoma mexicanum,* Shaw) avant et après la maturation sexuelle. Gen. Comp. Endocrinol. *29:*260 (Abstract 54).

Hay, E. D. and D. A. Fischman (1961). Origin of the blastema in regenerating limbs of the newt *Triturus viridescens.* An autoradiographic study using tritiated thymidine to follow cell proliferation and migration. Dev. Biol. *3:*26–59.

Hedrick, S. M., D. I. Cohen, E. A. Nielson, and M. M. Davis (1984). Isolation of cDNA clones encoding T cell-specific membrane-associated proteins. Nature *308:*149–158.

Heidemann, S. R., M. A. Hamborg, J. E. Balasz, and S. Lindley (1985). Microtubules in immature oocytes of *Xenopus laevis.* J. Cell Sci. *77:*129–141.

Herrick, C. J. (1948). *The Brain of the Tiger Salamander, Ambystoma tigrinum,* University of Chicago Press, Chicago.

Hetherington, T. E. and M. H. Wake (1979). The lateral line system in larval *Ichthyophis* (Amphibia; Gymnophiona). Zoomorphologie *93*:209–226.

Hibbard, E. and R. L. Ornberg (1976). Restoration of vision in genetically eyeless axolotls (*Ambystoma mexicanum*). Exp. Neurol. *50*:113–123.

Hill, T. L. and M. W. Kirschner (1982). Bioenergetics and kinetics of microtubule and actin filament assembly–disassembly. Int. Rev. Cytol. *78*:1–125.

Hiramoto, Y. (1979). Mechanical properties of the dividing sea urchin egg. In *Cell Motility: Molecules and Organization* (S. Hatano, H. Ishikawa, and H. Sato, eds.). University Park Press, Baltimore, pp. 653–663.

Holder, N. and R. Glade (1984). Skin glands in the axolotl: the creation and maintenance of a spacing pattern. J. Embryol. Exp. Morphol. *79*:97–112.

Holder, N., J. Mills, and D. A. Tonge (1982). Selective reinnervation of skeletal muscle in the newt *Triturus cristatus*. J. Physiol. *326*:371–384.

Holder, N. and P. W. Tank (1979). Morphogenetic interactions occurring between blastemas and stumps after exchanging blastemas between normal and double-half forelimbs in the axolotl, *Ambystoma mexicanum*. Dev. Biol. *68*:271–279.

Holmes, R. L. and J. N. Ball (1974). *The Pituitary Gland: A Comparative Account*. Cambridge, London.

Holmgren, N. and T. Pehrson (1949). Some remarks on the ontogenetical development of the sensory lines on the cheek in fishes and amphibians. Acta Zool. *30*:249–314.

Holtfreter, J. (1931). Über die Aufzucht isolierter Teile des Amphibienkeimes. Wilhelm Roux Arch. EntwMech. Org. *124*:404–466.

Holtfreter, J. (1938). Differenzierungspotenzen isolierter Teile der Urodelengastrula. Wilhelm Roux Arch. EntwMech. Org. *138*:522–656.

Hörstadius, S. (1950). *The Neural Crest: Its Properties and Derivatives in the Light of Experimental Research*. Oxford University Press, London.

Hörstadius, S. and S. Sellman (1946). Experimentelle Untersuchungen über die Determination des knorpeligen Kopfskelettes bei Urodelen. Nova Acta Regiae Soc. Sci. Ups., Series IV, *13*:1–170.

Hronowski, L. and J. B. Armstrong (1977). Ionic composition of the plasma of *Ambystoma mexicanum*. Comp. Biochem. Physiol. *58A*:181–183.

Hronowski, L., L. L. Gillespie, and J. B. Armstrong (1979). Development and survival of haploids of the Mexican axolotl, *Ambystoma mexicanum*. J. Exp. Zool. *209*:41–48.

Huchon, D., C. Jessus, and R. Ozon (1985). Redistribution des microtubules au cours de la première division méiotique de l'ovocyte de *Xenopus laevis:* étude en immunofluorescence. C. R. Acad. Sci. (III), Paris, *300*:463–465.

Humboldt, A., and A. Boupland (1811). *Recueil d'Observations de Zoologie et d'Anatomie Comparée,* vol. 1. Schoell et Dufour, Paris.

Humphrey, R. R. (1929a). Studies on sex reversal in *Amblystoma*. I. Bisexuality and sex reversal in larval males uninfluenced by ovarian hormones. Anat. Rec. *42*:119–155.

Humphrey, R. R. (1929b). Studies on sex reversal in *Amblystoma*. II. Sex differentiation and modification following orthotopic implantation of a gonad preprimordium. J. Exp. Zool. *53*:171–219.

Humphrey, R. R. (1931a). Studies on sex reversal in *Amblystoma*. III. Transformation of the ovary of *A. tigrinum* into a functional testis through the influence of a testis resident in the same animal. J. Exp. Zool. *58*:333–365.

Humphrey, R. R. (1931b). Studies on sex reversal in *Amblystoma*. IV. The developmental potencies exhibited by the modified ("freemartin") ovary of *Amblystoma tigrinum* following removal of the testis which had induced its modification. J. Exp. Zool. *58*:367–399.

Humphrey, R. R. (1931c). Studies on sex reversal in *Amblystoma*. V. The structure of ovaries of *A. tigrinum* subjected for long periods to the influence of a testis resident in the same animal. Anat. Rec. *51:*135–153.

Humphrey, R. R. (1932). Sex of parabiotic twins in *Amblystoma maculatum.* Proc. Soc. Exp. Biol. Med. *29:*713–717.

Humphrey, R. R. (1935a). Studies on sex reversal in *Amblystoma.* VII. Reversal of sex type in gonadic preprimordia of *A. punctatum* males implanted in females of more rapidly growing species. Anat. Rec. *62:*223–245.

Humphrey, R. R. (1935b). Sex reversal in *Amblystoma.* VIII. Sex type of gonads developed from gonadic preprimordia of *A. punctatum* implanted in axolotl females. Proc. Soc. Exp. Biol. Med. *33:*102–104.

Humphrey, R. R. (1936). Studies on sex reversal in *Amblystoma.* IX. Reversal of ovaries to testes in parabiotic *A. tigrinum.* J. Exp. Zool. *73:*1–21.

Humphrey, R. R. (1938). Studies on sex reversal in *Amblystoma.* XI. Reversal in *A. tigrinum* females induced by ectopic implantation of testis preprimordia. Anat. Rec. *72:*451–467.

Humphrey, R. R. (1942). Studies on sex reversal in *Amblystoma.* XII. Sterility after reversal of ovary to testis in the axolotl. Growth *6:*185–201.

Humphrey, R. R. (1944). The functional capacities of heteroplastic gonadal grafts in the Mexican axolotl, and some hybrid offspring of grafted animals. Am. J. Anat. *75:*263–282.

Humphrey, R. R. (1945). Sex determination in ambystomid salamanders: a study of the progeny of females experimentally converted into males. Am. J. Anat. *76:*33–66.

Humphrey, R. R. (1946). The Mexican axolotl as a laboratory animal. Ward's Nat. Sci. Bull. *20:*19–23.

Humphrey, R. R. (1948a). Reversal of sex in females of genotype *WW* in the axolotl (*Siredon* or *Amblystoma mexicanum*) and its bearing upon the role of the Z chromosomes in the development of the testis. J. Exp. Zool. *109:*171–185.

Humphrey, R. R. (1948b). A lethal fluid imbalance in the Mexican axolotl: inherited as a simple Mendelian recessive. J. Hered. *39:*255–261.

Humphrey, R. R. (1957). Male homogamety in the Mexican axolotl: a study of the progeny obtained when germ cells of a genetic male are incorporated in a developing ovary. J. Exp. Zool. *134:*91–101.

Humphrey, R. R. (1960). A linked gene determining the lethality usually accompanying a hereditary fluid imbalance in the Mexican axolotl. J. Hered. *50:*279–286.

Humphrey, R. R. (1962). Mexican axolotls, dark and mutant white strains: care of experimental animals. Bull. Phila. Herpetol. Soc. *10:*21–25.

Humphrey, R. R. (1964). Genetic and experimental studies on a lethal factor (*r*) in the axolotl which induces abnormalities in the renal system and other organs. J. Exp. Zool. *155:*139–150.

Humphrey, R. R. (1966). A recessive factor (*o,* for ova deficient) determining a complex of abnormalities in the Mexican axolotl *(Amblystoma mexicanum).* Dev. Biol. *13:*57–76.

Humphrey, R. R. (1967a). Albino axolotls from an albino tiger salamander through hybridization. J. Hered. *58:*95–101 (and cover).

Humphrey, R. R. (1967b). Genetic and experimental studies on a lethal trait ("short toes") in the Mexican axolotl *(Ambystoma mexicanum).* J. Exp. Zool. *164:*281–296.

Humphrey, R. R. (1969a). A recently discovered mutant, "eyeless," in the Mexican axolotl *(Ambystoma mexicanum).* Anat. Rec. *163:*306 (Abstract).

Humphrey, R. R. (1969b). Tumors of the testis in the Mexican axolotl (*Ambystoma,* or *Siredon, mexicanum*). In *Biology of Amphibian Tumors* (M. Mizell, ed.). Springer-Verlag, New York, pp. 220–228.

Humphrey, R. R. (1973). Experimental studies on a new lethal trait in Mexican axolotls of the Wistar Institute white strain. J. Exp. Zool. *183:*201–208.

Humphrey, R. R. (1975). The axolotl, *Ambystoma mexicanum*. In *Handbook of Genetics*, vol. 4 (R. C. King, ed.). Plenum Press, New York, pp. 3–17.

Humphrey, R. R. (1977a). Factors influencing ovulation in the Mexican axolotl as revealed by induced spawings. J. Exp. Zool. *199:*209–214.

Humphrey, R. R. (1977b). A lethal mutant gene in Mexican axolotl. J. Hered. *68:*407–408.

Humphrey, R. R. and J. T. Bagnara (1967). A color variant in the Mexican axolotl. J. Hered. *58:*251–256.

Humphrey, R. R. and R. K. Burns, Jr. (1939). An incompatibility manifested in heteroplastic parabiosis or grafting in *Amblystoma* due to a toxin of cutaneous origin. J. Exp. Zool. *81:*1–42.

Humphrey, R. R. and H.-M. Chung (1977). Genetic and experimental studies on three associated mutant genes in the Mexican axolotl: *st* (for stasis), *mi* (for microphthalmic) and *h* (for hand lethal). J. Exp. Zool. *202:*195–202.

Humphrey, R. R. and H.-M. Chung (1978). Experimental studies on two mutant genes, *r* and *x*, in the Mexican axolotl *(Ambystoma mexicanum)*. Dev. Biol. *62:*34–43.

Humphrey, R. R. and G. Fankhauser (1946). The development, structure and functional capacity of the ovaries in triploid ambystomid salamanders. J. Morphol. *79:*467–510.

Humphrey, R. R., G. M. Malacinski, and H.-M. Chung (1978). Developmental studies on an apparent cell-lethal mutant gene—*ut*—in the Mexican axolotl, *Ambystoma mexicanum*. Cell Differ. *7:*47–59.

Ide, C. F. (1977). Neurophysiology of *spastic*, a behavior mutant of the Mexican axolotl: altered vestibular projection to cerebellar auricle and area acoustico-lateralis. J. Comp. Neurol. *176:*359–371.

Ide, C. F., N. Miszkowski, C. B. Kimmel, E. Schabtach, R. Tompkins, O. Elbert, and M. Duda (1977). Analysis of *spastic:* a neurological mutant of the Mexican axolotl. In *Cellular Neurobiology* (Z. W. Hall, C. F. Fox, and R. Kelley, eds.). Alan R. Liss, New York, pp. 267–289.

Ide, C. F. and R. Tompkins (1975). Development of locomotor behavior in wild-type and *spastic* (*sp/sp*) axolotls, *Ambystoma mexicanum*. J. Exp. Zool. *194:*467–478.

Ignatieva, G. M. (1972). Time relationships between the onset of RNA synthesis in nuclei and the expression of their morphogenetic function in the axolotl embryos. Ontogenez *3:*626–629.

Ignatieva, G. M. (1974). Relative duration of the same periods of early embryogenesis in teleosts. Ontogenez *5:*427–436.

Imoh, H. and Y. Miyazaki (1984). Distribution of the germinal vesicle material during progesterone-induced oocyte maturation in *Xenopus* and in *Cynops*. Dev. Growth Differ. *26:*157–165.

Ingham, C. A. and F. H. Güldner (1980). Constant occurrence of an ipsilateral retino-tectal projection in the axolotl *(Ambystoma mexicanum)* revealed by horseradish peroxidase tracing. Neurosci. Lett. *17:*17–22.

International Aquaculture Research Center (1986). Clinical diagnosis: final summary report. Unpublished report on three axolotls from the Indiana University colony.

International Commission on Zoological Nomenclature (1963). *Ambystoma* Tschudi, 1838 (Amphibia): validation under the plenary powers. Bull. Zool. Nom. *20:*102–104.

International Commission on Zoological Nomenclature (1974). *Salamandra tigrina* Green, 1825 (Amphibia): grant of priority under the plenary powers over *Gyrinus mexicanus* Shaw, 1789. Bull. Zool. Nom. *31:*133–134.

Iten, L. E. and S. V. Bryant (1973). Forelimb regeneration from different levels of amputation in the newt, *Notophthalmus viridescens:* length, rate and stages. Wilhelm Roux Arch. Dev. Biol. *173:*263–282.

Jäckle, H. and G. W. Eagleson (1980). Spatial distribution of abundant proteins in oocytes and fertilized eggs of the Mexican axolotl *(Ambystoma mexicanum)*. Dev. Biol. *75:*492–499.

Jackson, I. M. D. and S. Reichlin (1974). Thyrotropin-releasing hormone (TRH): distribution in hypothalamic and extrahypothalamic brain tissues of mammalian and submammalian chordates. Endocrinology *95:*854–862.

Jacobs, G. F. M. and E. R. Kühn (1986). TRH injection induces thyroxine release in the metamorphosed but not in the neotenic axolotl *Ambystoma mexicanum.* 13th conference of European comparative endocrinologists, Belgrade (Abstract 8).

Jacobs, G. F. M., R. P. A. Michielsen and E. R. Kühn (1988). Thyroxine and triiodothyronine in plasma and thyroids of the neotenic and metamorphosed axolotl *Ambystoma mexicanum;* influence of TRH injections. Gen. Comp. Endocrinol., 20:145–151.

Jacobson, A. G. (1978). Some forces that shape the nervous system. Zoon *6:*13–21.

Jacobson, A. G. and R. Gordon (1976). Changes in the shape of the developing vertebrate nervous system analyzed experimentally, mathematically and by computer simulation. J. Exp. Zool. *197:*191–246.

Jacobson, A. G. and S. Meier (1984). Morphogenesis of the head of a newt: mesodermal segments, neuromeres, and the distribution of neural crest. Dev. Biol. *106:*181–193.

Jacobson, A. G., G. F. Oster, G. M. Odell, and L. Y. Cheng (1986). Neurulation and the cortical tractor model for epithelial folding. J. Embryol. Exp. Morphol. *96:*19–49.

Jacobson, C. O. (1959). The localization of the presumptive cerebral regions in the neural plate of the axolotl larva. J. Embryol. Exp. Morphol. *7:*1–21.

Jacobson, C. O. (1964). Motor nuclei, cranial nerve roots, and fibre pattern in the medulla oblongata after reversal experiments on the neural plate of the axolotl larvae. I. Bilateral operations. Zool. Bidr. Upps. *36:*73–160.

Jacobson, C. O. (1976). Motor nuclei, cranial nerve roots, and fibre pattern in the medulla oblongata after reversal experiments on the neural plate of axolotl larvae. II. Unilateral operations. Zoon *4:*87–100.

Jacobson, M. (1968). Development of neuronal specificity in retinal ganglion cells of *Xenopus.* Dev. Biol. *17:*202–218.

Jacobson, M. (1983). Clonal organization of the central nervous system of the frog. III. Clones stemming from individual blastomeres of the 128-, 256-, and 512-cell stages. J. Neurosci. *3:*1019–1038.

Jacobson, M. and G. Hirose (1978). Origin of the retina from both sides of the embryonic brain: a contribution to the problem of crossing at the optic chiasma. Science *202:*637–639.

Jacobson, M. and U. Rutishauser (1986). Induction of neural cell adhesion molecule (NCAM) in *Xenopus* embryos. Dev. Biol. *116:*524–531.

Jacoby, J. and C. B. Kimmel (1982). Synaptogenesis and its relation to growth of the postsynaptic cell: a quantitative study of the developing Mauthner neuron of the axolotl. J. Comp. Neurol. *204:*364–376.

Jarocki, F. P. (1822). *Zoology,* vol. 3. *Reptiles and Amphibians.* Latkiewicz, Warsaw.

Jasch, L. G. (1979). Separation of the epithelial and mesenchymal components of the newt limb regenerate with salt. J. Exp. Zool. *209:*443–454.

Jenkin, P. H. (1970). *Control of Growth and Metamorphosis.* Pergamon, Oxford.

Jenkinson, J. W. (1904). Observations on the maturation and fertilization of the egg of the axolotl. Q. J. Microsc. Sci. *48:*407–482.

Jessus, C., D. Huchon, and R. Ozon (1986). Distribution of microtubules during the breakdown of the nuclear envelope of the *Xenopus* oocyte: an immunocytochemical study. Biol. Cell *56:*113–120.

Johnen, A. G. (1956). Experimental studies about temporal relationships in the induction process. I. Experiments on *Amblystoma mexicanum.* Proc. K. Ned. Akad. Wet. *59C:*554–660.

Jones, K. W. and T. R. Elsdale (1963). The culture of small aggregates of amphibian embryonic cells *in vitro.* J. Embryol. Exp. Morphol. *11:*135–154.

Joseph, S. A. and K. M. Knigge (1972). Neural regulation of TSH secretion: thyroxine feedback in the newborn. Neuroendocrinology *10:*197–206.

Justus, J. T. and L. Cullum (1971). A new method of housing axolotls and other aquatic amphibians. Lab. Anim. Sci. *21:*110–111.

Kalt, M. R. (1976). Morphology and kinetics of spermatogenesis in *Xenopus laevis.* J. Exp. Zool. *195:*393–408.

Karfunkel, P. (1974). The mechanisms of neural tube formation. Int. Rev. Cytol. *38:*245–271.

Keller, R. E., M. Danilchik, R. L. Gimlich, and J. Shih (1985). The function and mechanism of convergent extension during gastrulation of *Xenopus laevis.* J. Embryol. Exp. Morphol. *89* (Suppl.):185–209.

Keller, R. E., J. Löfberg, and J. Spieth (1982). Neural crest cell behavior in white and dark embryos of *Ambystoma mexicanum:* epidermal inhibition of pigment cell migration in the white axolotl. Dev. Biol. *89:*179–195.

Keller, R. E. and J. Spieth (1984). Neural crest cell behavior in white and dark larvae of *Ambystoma mexicanum:* time-lapse cinemicrographic analysis of pigment cell movement *in vivo* and in culture. J. Exp. Zool. *229:*109–126.

Kelly, D. J. and R. A. Tassava (1973). Cell division and ribonucleic acid synthesis during the initiation of limb regeneration in larval axolotls *(Ambystoma mexicanum.)* J. Exp. Zool. *185:*45–54.

Kezer, J. (1952). Thyroxin-induced metamorphosis of the neotenic salamanders *Eurycea tynerensis* and *Eurycea neotenes.* Copeia *1952:*234–237.

Kikuyama, S., K. Yamamoto, and T. Seki (1980). Prolactin and its role in growth, metamorphosis and reproduction in amphibians. Gunma Symp. Endocrinol. *17:*3–13.

Kimmel, C. B. and P. G. Model (1978). Developmental studies of the Mauthner cell. In *Neurobiology of the Mauthner Cell* (D. S. Faber and H. Korn, eds.,) Raven Press, New York, pp. 183–220.

Kimmel, C. B. and E. Schabtach (1974). Patterning in synaptic knobs which connect with Mauthner's cell *(Ambystoma mexicanum).* J. Comp. Neurol. *156:*49–80.

Kimmel, C. B., E. Schabtach, and R. J. Kimmel (1977). Developmental interactions in the growth and branching of the lateral dendrite of Mauthner's cell *(Ambystoma mexicanum).*Dev. Biol. *55:*244–259.

King, J. A. and R. P. Millar (1981). TRH, GH-RIH and LH-RH in metamorphosing *Xenopus laevis.* Gen. Comp. Endocrinol. *44:*20–27.

Kinney, H. C., G. K. Klinworth, J. Lesiewicz, L. A. Goldsmith, and B. Wilkening (1982). Congenital cystic microphthalmia and consequent anophthalmia in the rat: a study in abnormal ocular morphogenesis. Teratology *26:*203–212.

Kintner, C. R. and J. P. Brockes (1984). Monoclonal antibodies identify blastemal cells derived from dedifferentiating muscle in newt limb regeneration. Nature *308:*67–69.

Kintner, C. R. and J. P. Brockes (1985). Monoclonal antibodies to the cells of a regenerating limb. J. Embryol. Exp. Morphol. *89.*37–55.

Kirsche, K. and W. Kirsche (1964). Experimental study on the influence of olfactory nerve regeneration of forebrain regeneration of *Ambystoma mexicanum.* J. Hirnforsch. *7:*315–333.

Kirsche, K. and W. Kirsche (1968). Über Homotransplantation eines Endhirndrittels von *Ambystoma mexicanum.* Z. Mikrosk.-anat. Forsch. *79:*223–243.

Knouff, R. A. (1935). The developmental pattern of ectodermal placodes in *Rana pipiens*. J. Comp. Neurol. *62*:17–65.

Kobel, H. R. and L. du Pasquier (1979). Hyperdiploid species hybrids for gene mapping in *Xenopus*. Nature *279*:157–158.

Koebke, J. (1977). Über das Differenzierungsverhalten des mesodermalen Keimbezirks verschieden alter Prägastrulationsstadien von *Ambystoma mexicanum*. Z. Mikrosk. anat. Forsch. *91*:215–228.

Kokoros, J. J. and R. G. Northcutt (1977). Telencephalic efferents of the tiger salamander *Ambystoma tigrinum tigrinum* (Green). J. Comp. Neurol. *173*:613–627.

Krebs, S. L. and R. A. Brandon (1984). A new species of salamander (family Ambystomatidae) from Michoacan, Mexico. Herpetologica *40*:238–245.

Kreig, P. A. and D. A. Melton (1985). Developmental regulation of a gastrula-specific gene injected into fertilized *Xenopus* eggs. EMBO J. *4*:3463–3471.

Kühn, E. R., V. M. Darras, and T. M. Verlinden (1985a). Annual variations of thyroid reactivity following thyrotropin stimulation and circulating levels of thryoid hormones in the frog *Rana ridibunda*. Gen. Comp. Endocrinol. *57*:266–273.

Kühn, E. R., E. Decuypere, K. Hemschoote, L. Berghman, and J. Paulussen (1983). Antagonism of serum triiodothyronine changes after injections of prolactin in the domestic fowl before and after hatching. J. Endocrinol. *99*:401–407.

Kühn, E. R., H. Gevaerts, G. Jacobs, and G. Vandorpe (1987). Reproductive cycle, thyroxine and corticosterone in females of the giant swamp frog *Dicroglossus occipitalis* at the equator. Gen. Comp. Endocrinol. *66*:137–144.

Kühn, E. R., G. F. M. Jacobs, and R. P. A. Michielsen (1984). Thyroxine as the trigger for thyroid activation in the hibernating frog *(Rana ridibunda)*. IRCS Med. Sci. *12*:168.

Kühn, E. R., S. Kikuyama, K. Yamamoto, and V. M. Darras (1985b). *In vivo* release of prolactin in *Rana ridibunda* following an intravenous injection of thyrotropin-releasing hormone. Gen. Comp. Endocrinol. *60*:86–89.

Lamoreux, M. L., L. Gerrity, and R. E. Boissy (1987). Vitiligo: a comparison of several animal models. Am. J. Pathol., *in press*.

Landacre, F. L. (1921). The fate of the neural crest in the head of the urodeles. J. Comp. Neurol. *33*:1–43.

Landmesser, L. T. (1980). The generation of neuromuscular specificity. Ann. Rev. Neurosci. *3*:279–302.

Landois, H. (1874). *Thierstimmen*. Herder, Freiburg.

Landström, U., H. Løvtrup-Rein, and S. Løvtrup (1976). On the determination of the dorsoventral polarity in the amphibian embryo: suppression by lactate of the formation of Ruffini's flask-cells. J. Embryol. Exp. Morphol. *36*:343–354.

Lannoo, M. J. (1985). Neuromast topography in *Ambystoma* larvae. Copeia *1985*:535–539.

Lannoo, M. J. (1987a). Neuromast topography in anuran amphibians. J. Morphol. *191*:115–129.

Lannoo, M. J. (1987b). Neuromast topography in urodele amphibians. J. Morphol. *191*:247–263.

Lannoo, M. J. (1988). The evolution of the lateral line system in amphibians and its bearing on amphibian phylogeny. Z. Systematik Evolutionforsch., *26*:128–134.

Larras-Regard, E. (1985). Hormonal determination of neoteny in facultative neotonic urodeles. In *Metamorphosis* (M. Balls and M. Bownes, eds.). Clarendon Press, Oxford, pp. 294–312.

Larras-Regard, E., A. Taurog, and M. Dorris (1981). Plasma T_4 and T_3 levels in *Ambystoma tigrinum* at various stages of metamorphosis. Gen. Comp. Endocrinol. *43*:443–450.

Larras-Regard, E., A. Taurog, and M. Dorris (1982). Effets de la TSH exogène sur l'activité thyroidienne des Amphibiens adultes ou néoténiques. C. R. Soc. Biol. *176*:795–803.

Lawrence, P. A. and G. Morata (1979). Pattern formation and compartments in the tarsus of *Drosophila,* In *Determinants of Spatial Organization* (S. Subtelny and I. R. Konigsberg, eds.). Academic Press, New York, pp. 317–323.

Lazard, L. (1979). Steroidogenesis in axolotl testis: histochemistry of two major enzymes related to cell type, spermatogenesis and substrate. Gen. Comp. Endocrinol. *39:*381–387.

Leber, S. M. and P. G. Model (1981). Effect of delayed afferent arrival on synapse localization on the amphibian Mauthner cell. Soc. Neurosci. Abstr. *7:*177.

Leber, S. M. and P. G. Model (1982). Normal localization of synapses on the amphibian Mauthner cell despite precocious synaptogenesis. Soc. Neurosci. Abstr.*8:*436.

Le Douarin, N. M. (1980). Migration and differentiation of neural crest cells. Curr. Top. Dev. Biol. *16*(2):31–85.

Le Douarin, N. M. (1982). *The Neural Crest.* Cambridge University Press, Cambridge.

Le Douarin, N. M. (1986). Cell line segregation during peripheral nervous system ontogeny. Science *231:*1515–1522.

Le Douarin, N. M. and A. McLaren (1984). *Chimeras in Developmental Biology.* Academic Press, London.

Le Douarin, N. M. and M. A. M. Teillet (1974). Experimental analysis of the migration and differentiation of neuroblasts of the autonomic nervous system and of neuroectodermal mesenchymal derivatives, using a biological cell marking technique. Dev. Biol. *41:*162–184.

Lefresne, J., J. C. David, and J. Signoret (1983). DNA ligase in axolotl egg: a model for study of gene activity control. Dev. Biol. *96:*324–330.

Leghissa, S. (1941). Sviluppo dell apparato di Mauthner in larve di *Ambystoma mexicanum.* Arch. Zool. Ital. *29:*231–253.

Lehman, H. E. (1957). The developmental mechanics of pigment pattern formation in the black axolotl, *Amblystoma mexicanum.* The formation of yellow and black bars in young larvae. J. Exp. Zool. *135:*355–386.

Leroux, P., M. C. Tonon, M. E. Stoeckel, S. Jégou, F. Leboulenger, C. Delarue, I. Perroteau, P. Netchitailo, G. Kupryszewski, and H. Vaudry (1983). Role of TRH in the control of melanotropin release in amphibia. In *Thyrotropin-Releasing Hormone* (E. C. Griffitsh, and G. W. Bennett, eds.). Raven Press, New York, pp. 229–240.

Lessman, C. A., W. S. Marshall, and H. R. Habibi (1986). Movement and dissolution of the nucleus (germinal vesicle) during *Rana* oocyte meiosis: effect of demecolcine (colcemid) and centrifugation. Gamete Res. *14:*11–23.

Lewis, W. H. (1947). Mechanics of invagination. Anat. Rec. *97:*139–156.

Licht, P., B. R. McCreery, R. Barnes, and R. Pang (1983). Seasonal and stress-related changes in plasma gonadotropins, sex steroids, and corticosterone in the bullfrog, *Rana catesbeiana.* Gen. Comp. Endocrinol. *50:*124–145.

Lindsley, D. L., G. Frankhauser, and R. R. Humphrey (1956). Mapping centromeres in the axolotl. Genetics *41:*58–64. *(13-4)*

Lipsett, J. C. (1941). Disproportionate dwarfism in *Ambystoma.* J. Exp. Zool. *86:*441–460.

Liversage, R. A., D. S. McLaughlin, and H. M. G. McLaughlin (1985). The hormonal milieu in amphibian appendage regeneration. In *Regulation of Vertebrate Limb Regeneration* (R. E. Sicard, ed.). Oxford University Press, New York, pp. 54–80.

Locke, M. and P. Huie (1977). Bismuth staining for light and electron microscopy. Tissue Cell *9:*347–371.

Löfberg, J. and K. Ahlfors (1978). Extracellular matrix organization and early neural crest cell migration in the axolotl embryo. In *Form-shaping Movements in Neurogenesis* (C. O. Jacobson and T. Ebendal, eds.). Almqvist & Wiksell, Stockholm, pp. 87–101.

Löfberg, J., K. Alhfors, and C. Fällström (1980). Neural crest cell migration in relation to extracellular matrix organization in the embryonic axolotl trunk. Dev. Biol. *75:*148–167.

Löfberg, J., A. Nynäs-McCoy, C. Olson, L. Jönsson, and R. Perris (1985). Stimulation of initial neural crest cell migration in the axolotl embryo by tissue grafts and extracellular matrix transplanted on microcarriers. Dev. Biol. *107:*442–459.

Löfberg, J., R. Perris, & H. H. Epperlein (1988). Timing in the regulation of neural crest cell migration: Retarded "maturation" of regional extracellular matrix inhibits pigment cell migration in embryos of the white axolotl mutant, submitted.

Lofts, B. (1974). Reproduction. In *Physiology of the Amphibia,* vol 2 (B. Lofts, ed.). Academic Press, New York, pp. 107–218.

Lofts, B. (1984). Amphibians. In *Marshall's Physiology of Reproduction,* vol 1, *Reproductive Cycles of Vertebrates* (G. E. Lamming, ed.). Churchill Livingstone, Edinburgh, pp. 127–205.

Løvtrup, S. (1981). Manipulation of the sex life of the Mexican axolotl. Axolotl Newsletter *10:*5.

Luft, J. H. (1971a). Ruthenium red and violet. I. Chemistry, purification, methods of use for electron microscopy and mechanisms of action. Anat. Rec. *171:*347–368.

Luft, J. H. (1971b). Ruthenium red and violet. II. Fine structural localization in animal tissues. Anat. Rec. *171:*369–416.

Lynn, W. G. (1961). Types of amphibian metamorphosis. Am. Zool. *1:*151–161.

Maden, M. (1976). Blastemal kinetics and pattern formation during amphibian limb regeneration. J. Embryol. Exp. Morphol. *36:*561–574.

Maden, M. (1978). Neurotrophic control of the cell cycle during amphibian limb regeneration. J. Embryol. Exp.Morphol. *48:*169–175.

Maden, M. (1982). Vitamin A and pattern formation in the regenerating limb. Nature *295:*672–675.

Maden, M. (1983). Vitamin A and the control of pattern in regenerating limbs. In *Limb Development and Regeneration,* part A (J. F. Fallon and A. I. Caplan, eds.). Alan R. Liss, New York, pp. 445–454.

Maden, M. (1985). Retinoids and the control of pattern in limb development and regeneration. Trends Genet. *1:*103–107.

Maden, M. and K. Mustafa (1984). The cellular contributions of blastema and stump to 180° supernumerary limbs in the axolotl. J. Embryol. Exp. Morphol. *84:*233–253.

Malacinski, G. M. (1978a). Rufus R. Humphrey (1892–1977). Am. Zool. *18:*191–193.

Malacinski, G. M. (1978b). The Mexican axolotl, *Ambystoma mexicanum:* its biology and developmental genetics, and its autonomous cell lethal genes. Am. Zool. *18:*195–206.

Malacinski, G. M. (1988). Amphibian developmental genetics. In *Developmental Genetics of Higher Organisms,* (G. M. Malacinski, ed.). Macmillan, New York, pp. 393–411.

Malacinski, G. M. and D. Barone (1985). Towards understanding paternal extragenic contributions to early amphibian pattern specification: the axolotl *ts-1* gene as a model system. J. Embryol. Exp. Morphol. *89*(Suppl.):53–68.

Malacinski, G. M., H. Benford, and H.-M. Chung (1975). Association of an ultraviolet irradiation-sensitive cytoplasmic localization with the future dorsal side of the amphibian egg. J. Exp. Zool. *191:*97–110.

Malacinski, G. M. and A. J. Brothers (1974). Mutant genes in the Mexican axolotl. Science *184:*1142–1147.

Malacinski, G. M. and H.-M. Chung (1981). Establishment of the site of involution at novel locations on the amphibian embryo. J. Morphol. *169:*149–159.

Malacinski, G. M., H.-M. Chung, and M. Asashima (1980). The association of primary embryonic organizer activity with the future dorsal side of amphibian eggs and early embryos. Dev. Biol. *77:*449–462.

Malbranc, M. (1876). Von der Seitenlinie und ihren Sinnesorganen bei Amphibien. Z. Wiss. Zool., Leipzig *26*:24–86.

Malcarne, G. and C. Giacoma (1978). Lesions to the CNS and ovariectomy in *Triturus cristatus carnifex* Laur. Effects on sexual behaviour. Atti. Acad. Sci. Torino Cl. Sci. Mat. Fis. Natur. *112*:49–61.

Maldonado-Koerdell, M. (1948). Las colecciones de anfibios del Museo Alfredo Duges en la Universidad de Guanajuato, I-Urodelos. Mem. Rev. Acad. Nac. Cienc. *56*:185–226.

Maller, J. L. and D. S. Smith (1985). Two-dimensional polyacrylamide gel analysis of changes in protein phosphorylation during maturation of *Xenopus* oocytes. Dev. Biol. *109*:150–156.

Maller, J., M. Wu, and J. C. Gerhart (1977). Changes in protein phosphorylation accompanying maturation of *Xenopus laevis* oocytes. Dev. Biol. *58*:295–312.

Mangold, O. (1931). Versuche zur Analyse der Entwicklung des Haftfadens bei Urodelen: ein Beispiel für die Induktion artfremder Organe. Naturwissenschaften *19*:905–911.

Marcus, L. C. (1981). *Veterinary Biology and Medicine of Captive Amphibians and Reptiles.*. Lea & Febiger, Philadelphia.

Mark, R. F. (1980). Synaptic repression at neuromuscular junctions. Physiol. Rev. *60*:355–395.

Marlin, J. T., I. Ness, and S. T. Collins (1986). *Book of World City Rankings.* Free Press, New York.

Martínez, U. V. (1961). *La Ciudad de México (1325–1960).* Departamento del Distrito Federal, Mexico City.

Masui, Y. and C. L. Markert (1971). Cytoplasmic control of nuclear behavior during meiotic maturation of frog oocytes. J. Exp. Zool. *177*:129–146.

Mazzi, V. (1978). Effects on spermatogenesis of permanent lesions to the rostral preoptic area in the crested newt (*Triturus cristatus carnifex* Laur.) Gen. Comp. Endocrinol. *34*:247–250.

McDowell, B. (1980). The Aztecs. Natl. Geographic *158*:704–751.

McDowell, B. (1984). Mexico City: an alarming giant. Natl. Geographic *166*:138–175.

McLaren, A. (1976a). *Mammalian Chimaeras.* Cambridge University Press, London.

McLaren, A. (1976b). Genetics of the early mouse embryo. Ann. Rev. Genet. *10*:361–388.

Meier, S. (1979). Development of the chick embryo mesoblast: formation of the embryonic axis and the establishment of metameric pattern. Dev. Biol. *73*:25–45.

Meijer, L. and P. Guerrier (1984). Maturation and fertilization in starfish oocytes. Int. Rev. Cytol. *86*:129–196.

Mellinger, M. W. (1975). Reconstruction of the cerebral hemispheres in *Ambystoma mexicanum* following bilateral brain ablation and ablation/transplantation. Anat. Rec. *181*:424.

Merrem, B. (1820). *Versuch eines Systems der Amphibien. Tentamen systematis amphibiorum.* Marburg.

Merriam, R. W. (1971). Progesterone-induced maturational events in oocytes of *Xenopus laevis.* II. Change in intracellular calcium and magnesium distribution at germinal vesicle breakdown. Exp. Cell Res. *68*:81–87.

Mescher, A. L. and D. Gospodarowicz (1979). Mitogenic effect of a growth factor derived from myelin on denervated regenerates of newt forelimbs. J. Exp. Zool. *207*:497–503.

Mes-Hartree, M. and J. B. Armstrong (1980). Evidence that the premature death mutation (*p*) in the Mexican axolotl *(Ambystoma mexicanum)* is not an autonomous cell lethal. J. Embryol. Exp. Morphol. *60*:295–302.

Messenger, E. A. and A. E. Warner (1979). The function of the sodium pump during differentiation of amphibian embryonic neurones. J. Physiol. *292*:85–105.

Metcalfe, W. K. (1985). Sensory neuron growth cones comigrate with posterior lateral line primordial cells in zebrafish. J. Comp. Neurol. *238:*218–224.

Metcalfe, W. K., C. B. Kimmel, and E. Schabtach (1985). Anatomy of the posterior lateral line system in young larvae of the zebrafish. J. Comp. Neurol. *233:*377–389.

Meuler, D. C. and G. M. Malacinski (1985). An analysis of protein synthesis patterns during early embryogenesis of the urodele *Ambystoma mexicanum.* J. Embryol. Exp. Morphol. *89:*71–92.

Miltner, M. J. and J. B. Armstrong (1983). Spermatogenesis in the Mexican axolotl, *Ambystoma mexicanum.* J. Exp. Zool. *227:*255–263.

Minuth, W. W. (1978). Mesodermalization of amphibian gastrula ectoderm in transfilter experiments. Med. Biol. *56:*349–354.

Mivart, S. G. (1870). On the axial skeleton of *Urodela.* Proc. Zool. Soc. Lond. *1870:*260–278.

Model, P. G. (1978a). Aspects of Mauthner cell differentiation in the axolotl, *Ambystoma mexicanum.* Am. Zool. *18:*253–265.

Model, P. G. (1978b). Regulation of the Mauthner cell following unilateral rotation of the prospective hindbrain in axolotl *(Ambystoma mexicanum)* neurulae. Brain Res. *153:*135–143.

Model, P. G. (1982). Prospective forebrain-midbrain from axolotl neurulae can be reprogrammed to differentiate as Mauthner cell-containing medulla. Dev. Brain Res. *3:*109–121.

Model, P. G. and S. Wurzelmann (1982). Vestibular axons form synapses on abnormally derived Mauthner cells. Dev. Brain Res. *3:*123–129.

Mohun, T. J., R. Tilly, R. Mohun, and J. M. W. Slack (1980). Cell commitment and gene expression in the axolotl embryo. Cell *22:*9–15.

Morgan, G. T. (1979). The time course of male meiosis in the red-backed salamander, *Plethodon cinereus.* J. Cell Sci. *38:*345–356.

Muneoka, K. and S. V. Bryant (1982). Evidence that patterning mechanisms in developing and regenerating limbs are the same. Nature *298:*369–371.

Muneoka, K. and S. V. Bryant (1984a). Cellular contribution to supernumerary limbs in the axolotl, *Ambystoma mexicanum.* Dev. Biol. *105:*166–178.

Muneoka, K. and S. V. Bryant (1984b). Cellular contribution to supernumerary limbs resulting from the interaction between developing and regenerating tissues in the axolotl. Dev. Biol. *105:*179–187.

Muneoka, K., G. V. Holler-Dinsmore, and S. V. Bryant (1985). A quantitative analysis of regeneration from chimeric limb stumps in the axolotl. J. Embryol. Exp. Morphol. *90:*1–12.

Muneoka, K., G. V. Holler-Dinsmore, and S. V. Bryant (1986a). Pattern discontinuity, polarity and directional intercalation in axolotl limbs. J. Embryol. Exp. Morphol. *93:*51–72.

Muneoka, K., W. F. Fox, and S. V. Bryant (1986b). Cellular contribution from dermis and cartilage to the regenerating limb blastema in axolotls. Dev. Biol. *116:*256–260.

Muneoka, K., L. D. Wise, W. F. Fox, and S. V. Bryant (1984). Improved techniques for use of the triploid cell marker in the axolotl, *Ambystoma mexicanum.* Dev. Biol. *105:*240–245.

Münz, H., B. Claas, and B. Fritzsch (1984). Electroreceptive and mechanoreceptive units in the lateral line of the axolotl, *Ambystoma mexicanum.* J. Comp. Physiol. A *154:*33–44.

Murphy, C. (1974). Cell death and autonomous gene action in lethals affecting imaginal discs in *Drosophila melanogaster.* Dev. Biol. *39:*23–36.

Nace, G. W., C. M. Richards, J. H. Asher (1970). Parthenogenesis and genetic variability. I. Linkage and inbreeding estimations in the frog, *Rana pipiens.* Genetics *66:*349–368.

Nakamura, O. and S. Toivonen, eds. (1978). *Organiser—A Milestone of a Half-Century from Spemann.* Elsevier North-Holland, Amsterdam.

Namenwirth, M. (1974). The inheritance of cell differentiation during limb regeneration in the axolotl. Dev. Biol. *41:*42–56.

National Academy of Sciences (1974). *Amphibans: Guidelines for the Breeding, Care, and Managment of Laboratory Animals.* A report of the Subcommittee on Amphibian Standards, Committee on Standards, Institute of Laboratory Animal Resources, National Research Council, Washington, DC, 153 pp.

Needham, J. (1936). *Order and Life* (reprinted 1968). M.I.T. Press, Cambridge, MA.

Neff, A. W. (1984). Long-term culture of epithelial cells derived from axolotl *(Ambystoma mexicanum)* embryos. Indiana Acad. Sci. *93:*131–135.

Neff, A. W., F. Briggs, and H.-M. Chung (1987). Craniofacial development mutant *pi* (pin-head) in the axolotl *(Ambystoma mexicanum)* which exhibits reduced interocular distance. J. Exp. Zool. *241:*309–316.

Neff, A. W., W. G. Mcturnan, S. K. Frost, and G. M. Malacinski (1980). An *in vitro* model for the study of axolotl cell-lethal genes. J. Cell Biol. *87:*23a.

Nei, M. (1978). Estimation of average heterozygosity and genetic distance from a small number of individuals. Genetics *89:*583–590.

Nelkin, B., C. Nichols, and B. Vogelstein (1980). Protein factor(s) from mitotic CHO cells induce meiotic maturation in *Xenopus laevis* oocytes. FEBS Lett. *109:*233–238.

Nelson, C. E. and R. R. Humphrey (1972). Artificial interspecific hybridization among *Ambystoma.* Herpetologica *28:*27–32.

Newgreen, D. F. and C. A. Erickson (1986). The migration of neural crest cells. Int. Rev. Cytol. *13:*89–145.

Newth, D. R. (1960). Black axolotl and white. Am. Sci. *48:*300–310.

Nichol, C. A., G. K. Smith, and D. S. Duch (1985). Biosynthesis and metabolism of tetrahydrobiopterin and molybdopterin. Ann. Rev. Biochem. *54:*729–764.

Nieuwkoop, P. D. (1969). The formation of the mesoderm in urodelean amphibians. I. Induction by endoderm. Wilhelm Roux Arch. EntwMech. Org. *162:*341–373.

Nieuwkoop, P. D.,R. de Fremery, T. W. Glenister, K. Hara, D. Harari, O. Hassa, R. Hori, A. G. Jonen, D. J. McCallim, V. M. McMurray, and M. Sala (1955). Origin and establishment of organization patterns in embryonic fields during early development in amphibians and birds. Proc. K. Ned. Akad. Wet. *C58:*356–367.

Nieuwkoop, P. D., A. G. Johnen, and B. Albers (1985). *The Epigenetic Nature of Early Chordate Development: Inductive Interaction and Competence.* Cambridge University Press, Cambridge.

Nieuwkoop, P. D. and G. A. Ubbels (1972). The formation of the mesoderm in urodelean amphibians. IV. Quantitative evidence for the purely "ectodermal" origin of the entire mesoderm and of the pharyngeal endoderm. Wilhelm Roux Arch. EntwMech. Org. *169:*185–199.

Nishioka, M., H. Ohtani, and M. Sumida (1980). Detection of chromosomes bearing the loci for seven kinds of proteins in Japanese pond frogs. Sci. Rep. Lab. Amphibian Biol., Hiroshima Univ. *4:*127–184.

Noden, D. M. (1975). An analysis of the migratory behaviour of avian cephalic neural crest cells. Dev. Biol. *42:*106–130.

Noden, D. M. (1978). Interactions directing the migration and cytodifferentiation of avian neural crest cells. In *Specificity of Embryological Interactions* (D. R. Garrod, ed.). Chapman & Hall, London, pp. 5–49.

Noden, D. M. (1984). The use of chimeras in analyses of craniofacial development. In *Chimeras in Developmental Biology* (N. M. Le Douarin and A. McLaren, eds.). Academic Press, London, pp. 241–280.

Norris, D. O. (1978). Hormonal and environmental factors involved in the determination of neoteny in urodeles. In *Comparative Endocrinology* (P. J. Gaillard and H. H. Boer, eds.). Elsevier North-Holland, Amsterdam, pp. 109–112.

Norris, D. O. (1985). *Vertebrate Endocrinology*, 2nd ed. Lea & Febiger, Philadelphia.

Norris, D. O., D. Duvall, K. Greendale, and W. A. Gern (1977). Thyroid function in pre- and postspawning neotenic tiger salamanders *(Ambystoma tigrinum).* Gen. Comp. Endocrinol. *33:*512–517.

Norris, D. O. and W. A. Gern (1976). Thyroxine-induced activation of hypothalamo-hypophyseal axis in neotonic salamander larvae. Science *194:*525–527.

Norris, D. O., W. A. Gern, and K. Greendale (1981). Diurnal and seasonal variations in thyroid function of neotenic tiger salamanders *(Ambystoma tigrinum).* Gen. Comp. Endocrinol. *45:*134–137.

Norris, D. O., R. E. Jones, and B. B. Criley (1973). Pituitary prolactin levels in larval, neotenic and metamorphosed salamanders *(Ambystoma tigrinum).* Gen. Comp. Endocrinol. *20:*437–442.

Norris, D. O., M. F. Norman, M. K. Pancak, and D. Duvall (1985). Seasonal variations in spermatogenesis, testicular weights, vasa deferentia, and androgen levels in neotenic male tiger salamanders, *Ambystoma tigrinum.* Gen. Comp. Endocrinol. *60:*51–57.

Norris, D. O. and J. E. Platt (1973). Effects of pituitary hormones, melatonin, and thyroidal inhibitors on radioiodide uptake by the thyroid glands of larval and adult tiger salamanders, *Ambystoma tigrinum* (Amphibia; Caudata). Gen. Comp. Endocrinol. *21:*368–376.

Norris, D. O. and J. E. Platt (1974). T_3 and T_4 induced rates of metamorphosis in immature and sexually mature larvae of *Ambystoma tigrinum* (Amphibia; Caudata). J. Exp. Zool. *189:*303–310.

Northcutt, R. G. (1986a). Electroreception in nonteleost bony fishes. In *Electroreception* (T. H. Bullock and W. Heiligenberg, eds.). Wiley Series in Neurobiology (R. G. Northcutt, ed.). Wiley, New York, pp. 257–285.

Northcutt, R. G. (1986b). Embryonic origin of amphibian electroreceptors. Soc. Neurosci. Abstr. *12:*103.

Northcutt, R. G. and C. Gans (1983). The genesis of neural crest and epidermal placodes: a reinterpretation of vertebrate origins. Q. Rev. Biol. *58:*1–28.

Northcutt, R. G. and E. Kicliter (1980). Organization of the amphibian telencephalon. In *Comparative Neurology of the Telencephalon* (S. O. E. Ebbesson, ed.). Plenum Press, New York, pp. 203–255.

Olsen, C. L., P. M. Barger, and R. A. Tassava (1984). Rescue of blocked cells by reinnervation in denervated forelimb stumps of larval *Ambystoma.* Dev. Biol. *106:*399–405.

Opdam, P. and R. Nieuwenhuys (1976). Topological analysis of the brain stem of the axolotl *Ambystoma mexicanum.* J. Comp. Neurol. *165:*285–306.

Ortiz, M. F. (1977). *Alkylation Treatment of the Mexican Axolotl: An Approach to the Induction of New Mutations.* Master's thesis, University of Ottawa.

O'Steen, W. K. and B. E. Walker (1961). Radioautographic studies of regeneration in the common newt. II. Regeneration of the forelimb. Anat. Rec. *139:*547–555.

Oster-Granite, M. L. and J. Gearhart (1981). Cell lineage analysis of cerebellar Purkinje cells in mouse chimeras. Dev. Biol. *85:*199–208.

Owen, R. (1844). Characters of a new species of axolotl. Ann. Mag. Nat. Hist. *1:*23.

Pan American Union (1964). *Motoring in Mexico.* Pan American Union, Washington, DC.

Pasteels, J. (1942). New observations concerning the maps of presumptive areas of the young amphibian gastrula *(Amblystoma and Discoglossus).* J. Exp. Zool. *89:*255–281.

Pavlick, F. (1977). *Variability and Inheritance of Isozyme Patterns in the Mexican Axolotl.* Master's thesis, University of Texas, Houston.

Pearson, M. and T. Elsdale (1979). Somitogenesis in amphibian embryos. I. Experimental evidence for an interaction between two temporal factors in the specification of somite pattern. J. Embroyl. Exp. Morphol. *51:*27–50.

Peltonen, J. (1978). *Elmo.* Söderström, Helsinki.

Perris, R. (1987). *Cell–Matrix Interactions in Neural Crest Development.* Doctoral thesis, Department of Zoology, Uppsala University. Summary also published separately, *Acta Universitatis Upsaliensis* No. 94, Almqvist & Wiksell International, Stockholm.

Perris, R., von Boxberg, Y. and J. Löfberg (1987). Local embryonic matrices determine region-specific phenotypes in neural crest cells. In *Cell–Matrix Interactions in Neural Crest Development* (doctoral thesis, R. Perris). Department of Zoology, Uppsala University.

Perris, R. and S. Johansson (1987). Amphibian neural crest cell migration on purified extracellular matrix components: a chondroitin sulphate proteoglycan inhibits locomotion on fibronectin substrates. J. Cell Biol. *105:*2511–2521.

Perris, R. and J. Löfberg (1986). Promotion of chromatophore differentiation in isolated premigratory neural crest cells by extracellular material explanted on microcarriers. Dev. Biol. *113:*327–341.

Pesetsky, I. (1965). Thyroxine-stimulated oxidative enzyme activity associated with precocious brain maturation in anurans. A histochemical study. Gen. Comp. Endocrinol. *5:*411–417.

Peters, W. C. H. (1863). Fernere Mittheilungen über neue Batrachier. Mber. Dtsch. Akad. Wiss. Berl. *1863:*445–471.

Phillips, H. J. (1973). Dye exclusion tests for cell viability. In *Tissue Culture Methods and Applications* (P. F. Kruse, Jr. and M. K. Patterson, Jr., eds.). Academic Press, New York, pp. 406–408.

Piatt, J. (1957). Studies on the problem of nerve pattern. III. Innervation of the regenerated forelimb in *Amblystoma.* J. Exp. Zool. *136:*229–247.

Platt, J. E. (1976). The effects of ergocornine on tail height, spontaneous and T_4-induced metamorphosis and thyroidal uptake of radioiodine in neotenic *Amblystoma tigrinum.* Gen. Comp. Endocrinol. *28:*71–81.

Poole, T. J. and M. S. Steinberg (1977). SEM-aided analysis of morphogenetic movements: development of the amphibian pronephric duct. Scanning Electron Microscopy *1977* (II):43–52.

Poole, T. J. and M. S. Steinberg (1981). Amphibian pronephric duct morphogenesis: segregation, cell rearrangement and directed migration of the *Ambystoma* duct rudiment. J. Embryol. Exp. Morphol. *63:*1–16.

Poole, T. J. and M. S. Steinberg (1982). Evidence for the guidance of pronephric duct migration by a craniocaudally travelling adhesion gradient. Dev. Biol. *92:*144–158.

Poole, T. J. and M. S. Steinberg (1984). Different modes of pronephric duct origins among vertebrates. Scanning Electron Microscopy *1984*(I):475–482.

Pringle, J. R. and L. H. Hartwell (1981). The *Saccharomyces cerevisiae* cell cycle. In *Molecular Biology of the Yeast Saccharomyces, Life Cycle and Inheritance* (J. N. Strathern, E. W. Jones, and J. R. Broach, eds.). Cold Spring Harbor Laboratory, pp. 97–142.

Raff, R. A. and T. C. Kaufman (1983). *Embryos, Genes and Evolution.* Macmillan, New York.

Rafferty, K. A., Jr. (1976). The physiology of amphibian cell in cultures. In *Physiology of the Amphibia*, vol. 3 (B. Lofts, ed.). Academic Press, New York, pp. 111–162.

Raven, C. P. (1931). Zur Entwicklung der Ganglienleiste. I. Die Kinematik der Ganglienleistenentwicklung bei den Urodelen. Wilhelm Roux Arch. EntwMech. Org. *125:*210–292.

Raven, C. P. (1937). Experiments on the origin of the sheath cells and sympathetic neuroblasts in amphibia. J. Comp. Neurol. *67:*221–240.

Raven, C. P. and J. Kloos (1945). Induction by medial and lateral pieces of the archenteron roof, with special reference to the determination of the neural crest. Acta Neerl. Morphol. *5*:348–362.

Regard, E. and A. Taurog (1977). A comparison of T_4 (RIA) and T_3 (RIA) levels in spontaneous and TSH-induced metamorphosis in amphibians. Ann. Endocrinol. *38*:20A (Abstract 35).

Regard, E., A. Taurog, and T. Nakashima (1978). Plasma thyroxine and triiodothyronine levels in spontaneously metamorphosing *Rana catesbeiana* tadpoles and in adult anuran amphibia. Endocrinology *102*:674–684.

Reichenbach-Klinke, H. and E. Elkan (1965). *Diseases of Amphibians.* T. F. H. Publications, Hong Kong.

Reinschmidt, D., J. Friedman, J. Hauth, E. Ratner, M. Cohen, M. Miller, D. Krotoski, and R. Tompkins (1985). Gene–centromere mapping in *Xenopus laevis.* J. Hered. *76*:345–347.

Reynhout, J. K. and L. D. Smith (1974). Studies on the appearance and nature of a maturation-inducing factor in the cytoplasm of amphibian oocytes exposed to progesterone. Dev. Biol. *38*:394–400.

Richter, W. (1968). Regerative Vorgänge nach einseitiger Entfernung des caudalen Endhirnabschnittes einschliesslich des telo-diencephalen Grenzbereiches bei *Ambystoma mexicanum.* J. Hirnforsch. 10:515–534.

Richter, W. and D. Kranz (1981). Autoradiographic investigations on postnatal proliferative activity of the telencephalic and diencephalic matrix zones in the axolotl *(Ambystoma mexicanum),* with special references to the olfactory organ. Z. Mikrosk.-anat. Forsch. *95*:883–904.

Riddiford, L. M. (1960). Autoradiographic studies of tritiated thymidine infused into the blastema of the early regenerate in the adult newt, *Triturus.* J. Exp. Zool. *144*:25–31.

Riviere, J. E., D. P. Shapiro, and G. L. Coppoc (1979). Percutaneous absorption of gentamicin by the leopard frog, *Rana pipiens.* Journal Paper no. 7237, Agricultural Experiment Station, Purdue University, West Lafayette, IN.

Roach, F. C. (1945). Differentiation of the central nervous system after axial reversals of the medullary plate of *Amblystoma.* J. Exp. Zool.*99*:53–77.

Robles, V. A. (1941). *Alejandro de Humboldt, Ensayo Politico Sobre el Reino de la Nueva Espana,* vol. 2, 6th ed. Editorial Pedro Robredo, Mexico.

Rollman-Dinsmore, C. and S. V. Bryant (1982). Pattern regulation between hind- and forelimbs after blastema exchanges and skin grafts in *Notophthalmus viridescens.* J. Exp. Zool. *223*:51–56.

Rollman-Dinsmore, C. and S. V. Bryant (1984). The distribution of marked dermal cells from small localized implants in limb regenerates. Dev. Biol. *106*:275–281.

Romer, A. S. (1966). *Vertebrate Paleontology,* 3rd ed. University of Chicago Press, Chicago.

Rosenkilde, P. (1976). Biological half-life of thyroxine in amphibians. Gen. Comp. Endocrinol. *29*:260–261.

Rosenkilde, P. (1979). The thyroid hormones in amphibia. In *Hormones and Evolution,* vol. 1 (E. J. W. Barrington, ed). Academic Press, New York, pp. 437–491.

Rosenkilde, P., E. Mogensen, G. Centervall, and O. S. Jorgensen (1982). Peaks of neuronal membrane antigen and thyroxine in larval development of the Mexican axolotl. Gen. Comp. Endocrinol. *48*:504–514.

Rosin, S. (1943). Experimente zur Entwicklungsphysiologie der Pigmentierung bei Amphibian. Rev. Suisse Zool. *50*:485–578.

Ross, R. A. and G. Marzec (1984). *The Bacterial Diseases of Reptiles: Their Epidemiology, Control, Diagnosis and Treatment.* Institute for Herpetological Research, Stanford University.

Rott, N. N. (1973). The correlation between karyo- and cytokinesis during the first cleavage divisions in the axolotl (*Ambystoma mexicanum* Cope). Ontogenez *4*:190–192.

Rott, N. N. and D. R. Beritashvili (1975). Changes in potassium and sodium content in early embryogenesis of the axolotl. Ontogenez *6:*93–95.

Rott, N. N. and M. I. Betina (1964). Obtaining triploid axolotl larvae by heat treatments (in Russian). Tsitologiia *6:*65–68.

Rugh, R. (1962). *Experimental Embryology.* Burgess, Minneapolis.

Russell, I. J. (1976). Amphibian Lateral Line Receptors. In *Frog Neurobiology, A Handbook* (R. Llinas and W. Precht, eds.). Springer-Verlag, Berlin, pp. 513–550.

Sadler, T. W., K. Burridge, and J. Yonker (1986). A potential role for spectrin during neurulation. J. Embryol. Exp. Morphol. *94:* 73–82.

Salaün, J. (1982). Differentiation of the optic cups from an anophthalmic murine strain, in culture and in intrafoetal grafts. J. Embryol. Exp. Morphol. *67:*71–80.

Salthe, S. N. and J. S. Mecham (1974). Reproductive and courtship patterns. In *Physiology of the Amphibia,* vol. 2 (B. Lofts, ed.). Academic Press, New York, pp. 309–521.

Sanders, W. T., J. R. Parsons, and R. S. Santley (1979). *The Basin of Mexico.* Academic Press, New York.

Sanes, J. R., J. L. R. Rubenstein, and J.-F. Nicolas (1986). Use of a recombinant retrovirus to study post-implantation cell lineage in mouse embryos. EMBO J. *5:*3133–3142.

Sawin, C. T., J. L. Bolaffi, I. P. Callard, P. Bacharach, and I. M. D. Jackson (1978). Induced metamorphosis in *Ambystoma mexicanum:* lack of effect of triiodothyronine on tissue or blood levels of thyrotropin-releasing hormone (TRH). Gen. Comp. Endocrinol. *36:*427–432.

Saxén, L. and S. Toivonen (1962). *Primary Embryonic Induction.* Prentice-Hall, Englewood Cliffs, NJ.

Schapitz, R. (1912). Urgeschlechtszellen von *Amblystoma.* Arch. Mikrosk. Anat. *79:*41–78.

Schlenoff, D. and P. G. Model (1988). The amphibian Mauthner cell is determined during very early neurulation. Dev. Brain Res., *submitted.*

Schmalhausen, I. I. (1968). *The Origin of Terrestrial Vertebrates.* Academic Press, New York.

Schoenwolf, G. C. (1986). On the morphogenesis of the early rudiments of the developing central nervous systems. In *Scanning Electron Microscope Studies of Embryogenesis* G. C. Schoenwolf, ed.). Scanning Electron Microscopy, Inc., Chicago, pp. 97–116.

Schreckenberg, G. M. and A. G. Jacobson (1975). Normal stages of development of the axolotl, *Ambystoma mexicanum.* Dev. Biol. *42:*391–400.

Schultheiss, H. (1980). T_3 and T_4 concentrations during metamorphosis of *Xenopus laevis* and *Rana esculanta* and in the neotenic Mexican axolotl. Gen. Comp. Endocrinol. *40:*372.

Schultze, O. (1887). Untersuchungen über die Reifung und Befruchtung des Amphibieneies. Z. Wiss. Zool. *45:*177–226.

Schwenk, G. C. and E. Hibbard (1977). An autoradiographic study of optic fiber projections from eye grafts in eyeless mutant axolotls. Exp. Neurol. *55:*498–503.

Scott, P. W. (1981). *Axolotls.* T. F. H. Publications, Neptune, NJ.

Seki, T. and S. Kikuyama (1986). Effect of thyrotropin-releasing hormone and dopamine on the *in vitro* secretion of prolactin by the bullfrog pituitary gland. Gen. Comp. Endocrinol. *61:*197–202.

Seldenrijk, R., D. R. W. Hup, P. N. E. de Graan, and F. C. G. van de Veerdonk (1979). Morphological and physiological aspects of melanophores in primary culture from tadpoles of *Xenopus laevis.* Cell Tissue Res. *198:*397–409.

Shaffer, H. B. (1983). Biosystematics of *Ambystoma rosaceum* and *A. tigrinum* in northwestern Mexico. Copeia *1983:*67–78.

Shaffer, H. B. (1984). Evolution in a paedomorphic lineage. I. An electrophoretic analysis of the Mexican ambystomatid salamanders. Evolution *38:*1194–1206.

Shaw, G. (1798). *Gyrinus mexicanus* Shaw. In *Shaw, Naturalist's Miscellany 9:*343–344.

Shaw, G. (1802). *General Zoology or Systematics Natural History,* vol 3. Davidson, London.

Shelton, P. M. J. (1970). The lateral line system at metamorphosis in *Xenopus laevis* (Daudin). J. Embryol. Exp. Morphol. *24:*511–524.

Shelton, P. M. J. (1971). The structure and function of the lateral line system in larval *Xenopus laevis.* J. Exp. Zool. *178:*211–231.

Sicard, R. E. (1985). Leukocytic and immunological influence on regeneration of amphibian forelimbs. In *Regulation of Vertebrate Limb Regeneration* (R. E. Sicard, ed.). Oxford University Press, New York, pp. 128–145.

Signoret, J. (1977). Données récentes sur la cinétique cellulaire au cours de la segmentation du germe d'axolotl. Mém. Soc. Zool. Fr. *41:*123–132.

Signoret, J., R. Briggs, and R. R. Humphrey (1962). Nuclear transplantation in the axolotl. Dev. Biol. *4:*134–164.

Signoret, J. and J. C. David (1986a). Control of the expression of genes for DNA ligase in eukaryotes. Int. Rev. Cytol. *103:*249–279.

Signoret, J. and J. C. David (1986b). DNA ligase activity in axolotl early development: evidence for a multilevel regulation of gene expression. J. Embryol. Exp. Morphol. *97*(Suppl.):85–95.

Signoret, J., J. C. David, J. Lefresne, and C. Houillon (1983). Control of DNA ligase molecular forms in nucleocytoplasmic combinations of axolotl and *Pleurodeles.* Proc. Natl. Acad. Sci. USA, *80:*3368–3371.

Signoret, J. and J. Lefresne (1971). Contribution a l'étude de la segmentation de l'oeuf d'axolotl: I—Définition de la transition blastuléenne. Ann. Embryol. Morphogen. *4:*113–123.

Signoret, J. and J. Lefresne (1975). Mise en évidence d'une RNA synthèse en cours de segmentation chez le germe d'äxolotl. C. R. Acad. Sci. (D), Paris, *281:*301–303.

Signoret, J., J. Lefresne, D. Vinson, and J. C. David (1981). Enzymes involved in DNA replication in axolotl. II—Control of DNA ligase activity during very early development. Dev. Biol. *87:*126–132.

Sims, T. J. (1977). The development of monoamine-containing neurons in the brain and spinal cord of the salamander, *Ambystoma mexicanum.* J. Comp. Neurol. *173:*319–335.

Singer, M. (1952). The influence of the nerve in regeneration of the amphibian extremity. Q. Rev. Biol. *27:*169–200.

Singer, M. (1978). On the nature of the neurotrophic phenomenon in urodele limb regeneration. Am. Zool. *18:*829–841.

Skoblina, M. N. (1963). Characteristics of duration of main stages of embryogenesis in *Ambystoma mexicanum.* In *IV Embryological Conference.* Theses Leningrad State University Press, pp. 172–173 (in Russian).

Skoblina, M. N. (1965). Dimensionless characterization of the duration of the mitotic phases of first cleavage divisions in axolotl. Dokl. Akad. Nauk SSSR *160:*700–703.

Slack, J. M. W. (1976). Determination of polarity in the amphibian limb. Nature *261:*44–46.

Slack, J. M. W. (1977a). Determination of anteroposterior polarity in the axolotl forelimb by an interaction between limb and flank rudiments. J. Embryol. Exp. Morphol. *39:*151–168.

Slack, J. M. W. (1977b). Control of anteroposterior pattern in the axolotl forelimb by a smoothly graded signal. J. Embryol. Exp. Morphol. *39:*169–182.

Slack, J. M. W. (1980). Regulation and potency in the forelimb rudiment of the axolotl embryo. J. Embryol. Exp. Morphol. *57:*203–217.

Slack, J. M. W. (1982). Protein synthesis during limb regeneration in the axolotl. J. Embryol. Exp. Morphol. *70:*241–260.

Slack, J. M. W. (1983). A fail-safe aquarium for aquatic laboratory animals. Lab. Anim. *17:*28–31.

Slack, J. M. W. (1984a). Regional biosynthetic markers in the early amphibian embryo. J. Embryol. Exp. Morphol. *80:*289–319.

Slack, J. M. W. (1984b). *In vitro* development of isolated ectoderm from axolotl gastrulae. J. Embryol. Exp. Morphol. *80:*321–330.

Slack, J. M. W. and D. Forman (1980). An interaction between dorsal and ventral regions of the marginal zone in early amphibian embryos. J. Embryol. Exp. Morphol. *56:*283–299.

Sladecek, F. (1955). Regulative tendencies of the central nervous system during embryogenesis of the axolotl (*Ambystoma mexicanum* Cope). II. Regulation after simultaneous inversion of anteroposterior and mediolateral axes of medullary plate. Acta Soc. Zool. Bohemoslov. *19:*138–149.

Sladecek, F. and J. Lanzová (1959). Cytological mechanism of the production of heteroploids by cold or heat shock in axolotl. Folia Biol. *5:*379–393.

Smith, A. R. and A. M. Crawley (1977). The pattern of cell division during growth of the blastema of regenerating newt forelimbs. J. Embryol. Exp. Morphol. *37:*33–48.

Smith, H. M. (1969a). The Mexican axolotl: some misconceptions and problems. Bioscience *19:*593–597; 615.

Smith, H. M. (1969b). Axolotl, *Ambystoma mexicanum*. Bull. Phila. Herpetol. Soc. *17:*15 (and cover).

Smith, H. M. and R. B. Smith (1971). *Synopsis of the Herpetofauna of Mexico,* vol. 1, *Analysis of the Literature on the Mexican Axolotl.* Eric Lundberg, Augusta, WV.

Smith, H. M. and R. B. Smith (1976). *Synopsis of the Herpetofauna of Mexico,* vol. 4, *Source Analysis and Index for Mexican Amphibians.* John Johnson, North Bennington, VT.

Smith, H. M. and E. H. Taylor (1948). An annotated checklist and key to the amphibia of Mexico. Bull. US Natl. Mus. i–iv:1–118.

Smith, H. M. and E. H. Taylor (1950). Type localities of Mexican reptiles and amphibians. Univ. Kans. Sci. Bull. *33:*313–380.

Smith, J. C., L. Dale, and J. M. W. Slack (1985). Cell lineage labels and region-specific markers in the analysis of inductive interactions. J. Embryol. Exp. Morphol. *89*(Suppl.):317–331.

Smith, J. C. and G. M. Malacinski (1983). The origin of the mesoderm in an anuran, *Xenopus laevis,* and a urodele, *Ambystoma mexicanum.* Dev. Biol. *98:*250–254.

Smith, J. C. and F. M. Watt (1985). Biochemical specificity of *Xenopus* notochord. Differentiation *29:*109–115.

Smith, L. D. and R. E. Ecker (1971). The interaction of steroids with *Rana pipiens* oocytes in the induction of maturation. Dev. Biol. *25:*232–247.

Smith, R. B., D. O. Norris, and H. M. Smith (1971). An analytical survey of the literature on the endocrinology of the axolotl (Amphibia; Caudata). Trans. Kans. Acad. Sci. *73:*319–417.

Smith, R. B. and H. M. Smith (1970a). A review of the literature on osteology of the axolotl. Bull. Phila. Herpetol. Soc. *17:*17–28.

Smith, R. B. and H. M. Smith (1970b). Sources for axolotl myology. Bull. Phila. Herpetol. Soc. *17:*29–35.

Smith, R. B. and H. M. Smith (1970c). An analysis of the literature of the viscera and lymphatic system of the Mexican axolotl. Bull. Phila. Herpetol. Soc. *17:*36–46.

Smith, R. B. and H. M. Smith (1970d). Literature on the blood vascular system of the axolotl. Bull. Phila. Herpetol. Soc. *17:*47–52.

Smith-Gill, S. J. and V. Carver (1981). Biochemical characterization of organ differentiation and maturation. In *Metamorphosis: A Problem in Developmental Biology,* 2nd ed. (L. I. Gilbert and E. Frieden, eds.). Plenum Press, New York, pp. 491–544.

Snieszko, S. F. (1972). Nutritional fish diseases. In *Fish Nutrition* (J. E. Halver, ed.). Academic Press, New York, pp. 403–437.

Solursh, M. and R. S. Reiter (1972). Long-term cell culture of two differentiated cell types from the liver of larval and adult *Xenopus laevis.* Cell Tissue Res. *128:*457–469.

Spehl, G. and J. Polus (1912). Des premiers stades du développement des glandes génitales chez l'Axolotl. Arch. Biol., Paris, *27:*63–90.

Spemann, B. (1931). Über den Anteil von Implantat und Wirtskeim an der Orientierung und Beschaffenheit der induzierten Embryonalanlage. Wilhelm Roux Arch. EntwMech. Org. *123:*389–517.

Spemann, H. (1938). *Embryonic Development and Induction* (reprinted 1962). Hafner, New York.

Spemann, H. and H. Mangold (1924). Über Induktion von Embryonalanlagen durch Implantation artfremder Organisatoren. Arch. Mikrosk. Anat. EntwMech. *100:*599–638.

Sperry, R. W. (1963). Chemoaffinity in the orderly growth of nerve fiber patterns and connections. Proc. Natl. Acad. Sci. USA, *50:*703–710.

Spieth, J. and R. E. Keller (1984). Neural crest cell behavior in white and dark larvae of *Ambystoma mexicanum:* differences in cell morphology, arrangement, and extracellular matrix as related to migration. J. Exp. Zool. *229:*91–107.

Spotte, S. H. (1979). *Fish and Invertebrate Culture: Water Management in Closed Systems,* 2nd ed. Wiley, New York.

Sprules, W. G. (1974). The adaptive significance of paedogenesis in North American species of *Ambystoma* (Amphibia; Caudata): an hypothesis. Can. J. Zool. *52:*393–400.

Stebbins, R. C. (1985). *A Field Guide to Western Reptiles and Amphibians.* Houghton Mifflin, Boston.

Steinberg, M. (1957). A nonnutrient culture medium for amphibian embryonic tissues. In *Carnegie Institution of Washington Year Book 56:*347–348.

Steinberg, M. S. (1970). Does differential adhesion govern self-assembly process in histogenesis? Equilibrium configurations and the emergence of a hierarchy among populations of embryonic cells. J. Exp. Zool. *173:*395–434.

Stephens, D. L., T. J. Miller, L. Silver, D. Zipser, and J. E. Mertz (1981). Easy-to-use equipment for the accurate microinjection of nanoliter volumes into the nuclei of amphibian oocytes. Anal. Biochem. *114:*299–309.

Stephens, N. and N. Holder (1987). The pattern of innervation in serially duplicated axolotl limbs: further evidence for the existence of local pathway cues? Development *100:*479–487.

Stern, C. D. and R. Bellairs (1984). Mitotic activity during somite segmentation in the early chick embryo. Anat. Embryol. *169:*97–102.

Stirling, R. V. and K. Brändle (1982). Expansion of the visual projection to the tectum of axolotls during metamorphosis. Dev. Brain Res. *5:*343–345.

Stocum, D. L. (1985). The role of the skin in urodele limb regeneration. In *Regulation of Vertebrate Limb Regeneration* (R. E. Sicard, ed.). Oxford University Press, New York, pp. 32–53.

Stocum, D. L. and G. E. Dearlove (1972). Epidermal-mesodermal interaction during morphogenesis of the limb regeneration blastema in larval salamanders. J. Exp. Zool. *181:*49–62.

Stocum, D. L. and J. F. Fallon (1982). Control of pattern formation in urodele limb ontogeny: a review and a hypothesis. J. Embryol. Exp. Morphol. *69:*7–36.

Stocum, D. L. and S. D. Thoms (1984). Retinoic acid-induced pattern completion in regenerating double anterior limbs of urodeles. J. Exp. Zool. *232:*207–215.

Stone, L. S. (1922). Experiments on the development of the cranial ganglia and the lateral line sense organs in *Amblystoma punctatum.* J. Exp. Zool. *35:*421–496.

Stone, L. S. (1924). Experiments on the transplantation of placodes of the cranial ganglia in the amphibian embryo. I. Heterotopic transplantations of the ophthalmic placode upon the head of *Amblystoma punctatum.* J. Comp. Neurol. *38:*73–105.

Stone, L. S. (1928a). Primitive lines in *Amblystoma* and their relation to the migratory lateral-line primordia. J. Comp. Neurol. *45:*169–190.

Stone, L. S. (1928b). Experiments on the transplantation of placodes of the cranial ganglia in the amphibian embryo. II. Heterotopic transplantation of the ophthalmic placode upon the head and body of *Amblystoma punctatum.* J. Comp. Neurol. *47:*91–116.

Stone, L. S. (1928c). Experiments on the transplantation of placodes of the cranial ganglia in the amphibian embryo. III. Preauditory and postauditory placodal materials interchanged. J. Comp. Neurol. *47:*117–154.

Stone, L. S. (1929). Experiments on the transplantation of placodes of the cranial ganglia in the amphibian embryo. IV. Heterotopic transplantations of the postauditory placodal material upon the head and body of *Amblystoma punctatum.* J. Comp. Neurol. *48:*311–330.

Stone, L. S. (1931). Induction of the ear by the medulla and its relation to experiments on the lateralis system in Amphibia. Science *74:*577.

Stone, L. S. (1933). The development of lateral-line sense organs in amphibians observed in living and vital-stained preparations. J. Comp. Neurol. *57:*507–540.

Stone, L. S. (1938). Further experimental studies of the development of lateral-line sense organs in amphibians observed in living preparations. J. Comp. Neurol. *68:*83–115.

Stone, L. S. (1964). Return of vision in larval eyes exchanged between *Amblystoma punctatum* and the cave salamander, *Typhlotriton spelaeus.* Invest. Ophthalmol. Vis. Sci. *3:*555–565.

Stoskopf, M. K., J. Arnold, and M. Mason (1987). Aminoglycoside antibiotic levels in the aquatic salamander *Necturus maculosus.* J. Zoo Anim. Med. *18:*81–85.

Straus, N. A. (1971). Comparative DNA renaturation kinetics in amphibians. Proc. Natl. Acad. Sci. USA, *68:*799–802.

Summerbell, D. and J. H. Lewis (1975). Time, place and positional value in the chick limb-bud. J. Embryol. Exp. Morphol. *33:*621–643.

Sunkara, P. S., D. A. Wright, and P. N. Rao (1979). Mitotic factors from mammalian cells induce germinal vesicle breakdown and chromosome condensation in amphibian oocytes. Proc. Natl. Acad. Sci. USA, *76:*2799–2802.

Suzuki, A. S., T. Ueno, and T. Matsusaka (1986). Alteration of cell adhesion system in amphibian ectoderm cells during primary embryonic induction: changes in reaggregation pattern of induced neurectoderm cells and ultrastructural features of the reaggregate. Wilhelm Roux Arch. Dev. Biol. *195:*85–91.

Svistunov, S. A. and V. I. Mitashov (1983). The proliferative activity of the cells of pigment epithelium and regenerating retina in *Ambystoma mexicanum.* Ontogenez *14:*597–606.

Svistunov, S. A. and V. I. Mitashov (1984). An autoradiographic assay of proliferation of the pigment epithelium cells in the axolotl retina. Ontogenez *15:*599–607.

Swainson, William (1839). On the natural history and classification of fishes, amphibians and reptiles. The Cabinet Encyclopaedia *2:*1–452.

Swanberg, L. and D. O. Norris (1972). Peripheral anti-metamorphic action of prolactin in *Bufo cognatus* and *Abystoma tigrinum.* J. Colo. Wyo. Acad. Sci. *7:*66.

Swett, F. H. (1937). Determination of limb-axes. Q. Rev. Biol. *12:*322–339.

Szarski, H., H. M. Smith, and R. B. Smith (1970). Polish contributions to the study of the axolotl, *Abystoma mexicanum.* Bull. Phila. Herpetol. Soc. *17:*4–16.

Taghy-Sadak, Z. and J. P. Vilain (1985). Exocytosis of glycogen during maturation of amphibian oocytes. Gamete Res. *11:*223–236.

Takatsui, A., K. Kohno, and G. Tamura (1975). Inhibition of biosynthesis of polyisoprenol sugars in chick embryo microsomes by tunicamycin. Agric. Biol. Chem. *39:*2089–2091.

Tam, P. P. L. (1981). The control of somitogenesis in mouse embryos. J. Embryol. Exp. Morphol. *65*(Suppl.):103–128.

Tam, P. P. L., S. Meier, and A. G. Jacobson (1982). Differentiation of the metameric pattern in the embryonic axis of the mouse. II. Somitomeric organization of the presomitic mesoderm. Differentiation *29:*109–122.

Tank, P. W. (1981). The ability of localized implants of whole or minced dermis to disrupt pattern formation in the regenerating forelimb of the axolotl. Am. J. Anat. *162:*315–326.

Tank, P. W., B. M. Carlson, and T. G. Connelly (1976). A staging system for forelimb regeneration in the axolotl, *Ambystoma mexicanum*. J. Morphol. *150:*117–128.

Tank, P. W., T. G. Connelly, and F. L. Bookstein (1985). Cellular behavior in the anteroposterior axis of the regenerating forelimb of the axolotl, *Ambystoma mexicanum*. Dev. Biol. *109:*215–223.

Tank, P. W. and N. Holder (1979). The distribution of cells in the upper forelimb of the axolotl, *Ambystoma mexicanum*. J. Exp. Zool. *209:*435–442.

Tank, P. W. and N. Holder (1981). Pattern regulation in the regenerating limbs of urodele amphibians. Q. Rev. Biol. *56:*113–142.

Tassava, R. A. (1983). Limb regeneration to digit stages occurs in well-fed adult newts after hypophysectomy. J. Exp. Zool. *225:*433–441.

Tassava, R. A., L. L. Bennett, and G. D. Zitnik (1974). DNA synthesis without mitosis in amputated denervated forelimbs of larval axolotls. J. Exp. Zool. *190:*111–116.

Tassava, R. A., D. L. Laux, and D. P. Treece (1985). The effects of partial and complete denervation on adult newt forelimb blastema cell-cycle parameters. Differentiation *29:*121–126.

Tassava, R. A. and C. L. Olsen (1985). Neurotrophic influence on cellular proliferation in urodele limb regeneration: in vivo experiments. In *Regulation of Vertebrate Limb Regeneration* (R. E. Sicard, ed.). Oxford University Press, New York, pp. 81–92.

Taurog, A. (1974). Effects of TSH and long-acting thyroid stimulator on thyroid^{131}I-metabolism and metamorphosis of the Mexican axolotl *(Ambystoma mexicanum)*. Gen. Comp. Endocrinol. *24:*257–266.

Taurog, A., C. Olivier, R. L. Eskay, J. C. Porter, and J. M. McKenzie (1974). The role of TRH in the neoteny of the Mexican axolotl *(Ambystoma mexicanum)*. Gen. Comp. Endocrinol. *24:*267–279.

Taylor, E. H. (1938). Concerning Mexican salamanders. Univ. Kans. Sci. Bull. *25:*259–313.

Taylor, E. H. (1940a). New salamanders from Mexico with a discussion of certain known forms. Univ. Kans. Sci. Bull. *26:*407–439.

Taylor, E. H. (1940b). A new *Rhyacosiredon* (Caudata) from western Mexico. Herpetologica *1:*171–176.

Taylor, E. H. (1941). Two new ambystomid salamanders from Chihuahua. Copeia *1941:*143–146.

Taylor, E. H. (1943a). A new ambystomid salamander adapted to brackish water. Copeia *1943:*151–156.

Taylor, E. H. (1943b). Herpetological novelties from Mexico. Univ. Kans. Sci. Bull. *29:*343–361.

Taylor, E. H. (1944). A new ambystomid salamander from the plateau region of Mexico. Univ. Kans. Sci. Bull. *30:*57–61.

Taylor, E. H. and H. M. Smith (1945). Summary of the collections of amphibians made in

Mexico under the Walter Rathbone Bacon Traveling Scholarship. Proc. US Natl. Mus. *95:*521–613.

Thiebaud, P., J. Lefresne, J. Signoret, and J. C. David (1983). Isolation of the messenger RNA for 8S DNA ligase in early developing axolotl egg and its cell free translation. Nucleic Acids Res. *11:*2563–2573.

Thiebaud, P., J. Signoret, J. Lefresne, C. Rimbaut, D. Buffe, and J. C. David (1985). Molecular duality of DNA ligase in axolotl corresponds to distinctive transcriptional information. Exp. Cell Res. *161:*209–218.

Thompson, D. and A. P. Scott (1984). An analysis of recombination data in gynogenetic diploid rainbow trout. Heredity *53:*441–452.

Thorgaard, G. H., F. W. Allendorf, and K. L. Knudsen (1983). Gene–centromere mapping in rainbow trout: high interference over long map distances. Genetics *103:*771–783.

Thornton, C. S. and M. T. Thornton (1965). The regeneration of accessory limb parts following epidermal cap transplantation in urodeles. Experientia *21:*146–148.

Thorsteinsdottir, S. and S. K. Frost (1986). Pigment cell differentiation: the relationship between pterin content, allopurinol treatment, and the melanoid gene in axolotls. Cell Differ. *19:*161–172.

Tickle, C., D. Summerbell, and L. Wolpert (1975). Positional signalling and specification of digits in chick limb morphogenesis. Nature *254:*199–202.

Tiedemann, H. and J. Born (1978). Biological activity of vegetalizing and neuralizing inducing factors after binding to BAC-cellulose and CNBr-Sepharose. Wilhelm Roux Arch. Dev. Biol. *184:*285–299.

Tihen, J. A. (1958). Comments on the osteology and phylogeny of ambystomatid salamanders. Bull. Fla. St. Mus. Biol. Sci. *3:*1–50.

Tihen, J. A. (1969). *Ambystoma.* Cat. Am. Amph. Rept. *75:*1–75.

Tompkins, R. (1970). Biochemical effects of the gene *g* on the development of the axolotl *Ambystoma mexicanum.* Dev. Biol. *22:*59–83.

Toole, B. P. (1981). Glycosaminoglycans in morphogenesis. In *Cell Biology of the Extracellular Matrix* (E. D. Hay, ed.). Plenum Press, New York, pp. 259–294.

Trinkaus, G. P. (1984). *Cells into Organs. The Forces That Shape the Embryo.* Prentice-Hall, Englewood Cliffs, NJ.

Trotier, D. (1986). A patch-clamp analysis of membrane currents in salamander olfactory receptor cells. Pflügers Arch. *407:*589–595.

Trottier, T. M. and J. B. Armstrong (1975). Hormonal stimulation as an aid to artificial insemination in *Ambystoma mexicanum.* Can. J. Zool. *53:*171–173.

Trottier, T. M. and J. B Armstrong (1976). Diploid gynogenesis in the Mexican axolotl. Genetics *83:*783–792.

Trottier, T. M. and J. B. Armstrong (1977). Experimental studies on a mutant gene (*p*) causing premature death of *Ambystoma mexicanum* embryos. J. Embryol. Exp. Morphol. *39:*139–149.

Turley, E. A., C. A. Erickson, and R. P. Tucker (1985). The retention and ultrastructural appearances of various extracellular matrix molecules incorporated into three-dimensional hydrated collagen lattices. Dev. Biol. *109:*347–369.

Turner, D. L. and C. L. Cepko (1987). A common progenitor for neurons and glia persists in rat retina late in development. Nature *328:*131–136.

Twitty, V. C. (1932). Influence of the eye on the growth of its associated structures, studied by means of heteroplastic transplantation. J. Exp. Zool. *61:*333–374.

Twitty, V. C. (1937). Experiments on the phenomenon of paralysis produced by a toxin occuring in *Triturus* embryos. J. Exp. Zool. *76:*67–104.

Twitty, V. C. (1966). *Of Scientists and Salamanders.* Freeman, San Francisco.

Ubbels, G. A., K. Hara, C. H. Koster, and M. W. Kirschner (1983). Evidence for a functional role of the cytoskeleton in determination of dorsoventral axis in *Xenopus laevis* eggs. J. Embrol. Exp. Morphol. *77:*15–37.

Ulshafer, R. J. (1975). *Morphology of the Optic Rudiment in Eyed and Eyeless Axolotls. Master's thesis, Pennsylvania State University.*

Ulshafer, R. J. and E. Hibbard (1976). Morphology of the optic rudiment in eyed and eyeless axolotls. Anat. Rec. *184:*552 (Abstract).

Ulshafer, R. J. and E. Hibbard (1979). An SEM and TEM study of suppression of eye development in eyeless mutant axolotls. Anat. Embryol. *156:*29–35.

Valouch, P., J. Melichna, and F. Sladecek (1971). The number of cells at the beginning of gastrulation depending on the temperature in different species of amphibians. Acta Univ. Carol. Biol. *April*(2):195–205.

Vanable, J. W. (1985). Benzocaine: an excellent amphibian anesthetic. Axolotl Newsletter *14:*19–21.

Van Bambeke, C. (1870). Sur les trous vitellins que présentent les oeufs fécondes des Amphibiens. Bull. Acad. R. Belg. Cl. Sci. *30:*58–71.

Van Duesen, E. B. (1971). Studies on embryonic determination: effects of gene "*e*" on forebrain ectoderm in the axolotl. Am. Zool. *11:*678 (Abstract 312).

Van Duesen, E. B. (1972). *Experimental Studies on a Mutant Gene* (*e*) *Preventing the Differentiation of Eye and Normal Hypothalamus Primordia in the Axolotl.* Doctoral thesis, Indiana University, Bloomington, IN, 71 pp.

Van Deusen, E. (1973). Experimental studies on a mutant gene (*e*) preventing the differentiation of eye and normal hypothalamus primordia in the axolotl. Dev. Biol. *34:*135–158.

Vandorpe, G., G. F. M. Jacobs, and E. R. Kühn (1987). Seasonal changes of the 5′-monodeiodination activity in kidney and skin homogenates of male *Rana ridibunda:* relation to plasma thyroxine (T_4) and testosterone. Gen. Comp. Endocrinol. *68:*163–169.

Van Duijn, C., Jr. (1973). *Diseases of Fishes.* Iliffe Books, London.

Velasco, J. M. (1879). Descripcion, metamorfosis y costumbres de una especie nueva del género *Siredon.* Naturaleza *4:*209–233.

Verhoeff-de Fremery, R. (1980). The maintenance of *Ambystoma mexicanum* in the Hubrecht Laboratory, Utrecht, The Netherlands. Axolotl Newsletter *9:*26–27.

Verhoeff-de Fremery, R. (1984). Survey of the reproductive capacities of three separate age groups of wild-type *Ambystoma mexicanum* (Dutch strain). Axolotl Newsletter *13:*17–20.

Vilain, J. P. (1978). Maturation *in vitro* des ovocytes de *Pleurodeles waltlii* (Amphibien Urodèle). Mem. Soc. Zool. Fr. *41:*93–102.

Vilain, J. P., M. Moreau, and P. Guerrier (1980a). Propionate and cycloheximide reversibly block progesterone-induced calcium surge in *Ambystoma mexicanum* oocytes. Dev. Growth Differ. *22:*25–31.

Vilain, J. P., M. Moreau, and P. Guerrier (1980b). Uncoupling of oocyte–follicle cells triggers reinitiation of meiosis in amphibian oocytes. Dev. Growth Differ. *22:*687–691.

Vogt, W. (1929). Gestaltungsanalyse am Amphibienkeim mit örtlicher Vitalfärbung. II. Gastrulation und Mesodermbildung bei Urodelen und Anuren. Wilhelm Roux Arch. EntwMech. Org. *120:*384–706.

Von Tschudi, J. J. (1839). *Classification der Batrachier* . . . , Neuchât. Mém. Soc. Sci. Nat. *2:*1–100.

Von Woellwarth, C. (1952). Die Induktionsstufen des Gehirns. Wilhelm Roux Arch. EntwMech. Org. *145:*582–668.

Wagler, J. G. (1830). *Naturliches System der Amphibien, mit Vorangehender Classification der Säugethiere und Vögel.* Cotts, Munich.

Wahnschaffe, U., B. Fritzsch, and W. Himstedt (1985). The fine structure of the lateral-line organs of larval *Ichthyophis* (Amphibia; Gymnophiona). J. Morphol. *186:*369–377.

Wakimoto, B. T. (1979). DNA synthesis after polyspermic fertilization in the axolotl. J. Embrol. Exp. Morphol. *52:*39–48.

Wallace, R. A., D. W. Jared, J. N. Dumont, and M. W. Sega (1973). Protein incorporation by isolated amphibian oocytes. III. Optimum incubation conditions. J. Exp. Zool. *184:*321–334.

Waloga, G. and W. L. Pak (1978). Ionic mechanism for the generation of horizontal cell potentials in isolated axolotl retina. J. Gen. Physiol. *71:*69–92.

Warner, A. E. (1973). The electrical properties of the ectoderm in the amphibian embryo during induction and early development of the nervous system. J. Physiol., Lond. *235:*267–286.

Warner, A. E. (1981). The early development of the nervous system. In *Development in the Nervous System* (D. R. Garrod and J. D. Feldman, eds.). Cambridge University Press, Cambridge, pp. 109–127.

Warner, A. E., S. C. Guthrie, and N. B. Gilula (1984). Antibodies to gap-junctional protein selectively disrupt junctional communication in the early amphibian embryo. Nature *311:*127–131.

Wartenberg, H. (1962). Elektronenmikroskopische und histochemische Studien über die Oogenese der Amphibieneizelle. Z. Zellforsch. Mikrosk. Anat. *58:*427–486.

Wasserman, W. J., J. D. Richter, and L. D. Smith (1982). Protein synthesis during maturation-promoting factor and progesterone-induced maturation in *Xenopus* oocytes. Dev. Biol. *89:*152–158.

Webb, J. F. (1985). Evolution and morphological diversity of the lateral line system in fishes. Am. Zool. *25:*19A.

Weinstein, A. (1958). The geometry and mechanics of crossing over. Cold Spring Harbor Symp. Quant. Biol. *23:*177–196.

Weintraub, H., M. Buscaglia, M. Ferrez, S. Weiller, A. Boulet, F. Fabre, and E. E. Baulieu (1982). Mise en évidence d'une activité "MPF" chez *Saccharomyces cerevisiae.* C. R. Acad. Sci. (III), Paris, *295:*787–790.

Weismann, A. (1882). The transformation of the Mexican axolotl into *Amblystoma.* In *Studies in the Theory of Descent* (A. Weismann, ed.). Sampson Low, London, pp. 555–633.

West, J. D. (1984). Cell markers. In *Chimeras in Developmental Biology* (N. Le Douarin and A. McLaren, eds.). Academic Press, London, pp. 39–67.

Weston, J. A. (1970). The migration and differentiation of neural crest cells. Adv. Morphogen. *8:*41–114.

Weston, J. A. (1982). Motile and social behavior of neural crest cells. In *Cell Behaviour* (R. Bellairs, A. Curtis, and G. Dunn, eds.). Cambridge University Press, Cambridge, pp. 429–470.

Weston, J. A. (1983). Regulation of neural crest cell migration and differentiation. In *Cell Interactions and Development* (K. M. Yamada, ed.). Wiley, New York, pp. 153–184.

Weston, J. A., G. Ciment, and J. Girdlestone (1984a). The role of extracellular matrix in neural crest development: a reevaluation. In *The Role of Extracellular Matrix in Development* (R. L. Trelstad, ed.). Alan R. Liss, New York, pp. 433–469.

Weston, J. A., J. Girdlestone, and G. Ciment (1984b). Heterogeneity in neural crest cell populations. In *Cellular and Molecular Biology of Neuronal Development* (I. B. Black, ed.). Plenum Press, New York, pp. 51–62.

Wharton, K. A., K. M. Johansen, T. Xu, and S. Artavanis-Tsakonas (1985). Nucleotide sequence from the neurogenic locus *notch* implies a gene product that shares homology with proteins containing EGF-like repeats. Cell *43:*567–581.

Wiedersheim, R. (1879). Zur Anatomie des *Amblystoma weismanni*. Z. Wiss. Zool. *32*:216–236.

Wiegmann, A.F.A. (1832). *Sirenodon* and *Stegoporus*. In *Handbuch der Zoologie* (A.F.A. Wiegmann and J. F. Ruthe, eds.). Luderitz, Berlin, p. 204.

Wigston, D. J. (1980). Suppression of sprouted synapses in axolotl muscle by transplanted foreign nerves. J. Physiol. *307*:355–366.

Wigston, D. J. (1986). Selective innervation of transplanted limb muscles by regenerating motor axons in the axolotl. J. Neurosci. *6*:2757–2763.

Wigston, D. J. and P. R. Kennedy (1987). Selective reinnervation of transplanted muscles by their original motoneurons in the axolotl. J. Neurosci. *7*:1857–1865.

Wilbur, H. M. (1971). The ecological relationship of the salamander *Ambystoma laterale* to its all-female, gynogenetic associate. Evolution *25*:168–179.

Wilson, C., G. S. Cross, and H. R. Woodland (1986). Tissue-specific expression of actin genes injected into *Xenopus* embryos. Cell *47*:589–599.

Winkelman, E. and A. Winkleman (1970). Experimentelle Unter suchungen zur Regeneration des Telencephalon von *Ambystoma mexicanum* nach resektion beider Hemisphären, Z. Mikrosk.-anat. Forsch. *82*:149–171.

Winklbauer, R. and P. Hausen (1983a). Development of the lateral line system in *Xenopus laevis*. I. Normal development and cell movement in the supraorbital system. J. Embryol. Exp. Morphol. *76*:265–281.

Winklbauer, R. and P. Hausen (1983b). Development of the lateral line system in *Xenopus laevis*. II. Cell multiplication and organ formation in the supraorbital system. J. Embryol. Exp. Morphol. *76*:283–296.

Winklbauer, R. and P. Hausen (1985a). Development of the lateral line system in *Xenopus laevis*. III. Development of the supraorbital system in triploid embryos and larvae. J. Embryol. Exp. Morphol. *88*:183–192.

Winklbauer, R. and P. Hausen (1985b). Development of the lateral line system in *Xenopus laevis*. IV. Pattern formation in the supraorbital system. J. Embryol. Exp. Morphol. *88*:193–207.

Woerdeman, M. W. (1929). Experimentelle Untersuchungen über Lage und Bau der Augenbildenden Bezirke in der Medullarplatte beim Axolotl. Wilhelm Roux Arch. Entw-Mech. Org. *116*:220–241.

Wolpert, L. (1969). Positional information and the spatial pattern of cellular differentiation. J. Theor. Biol. *25*:1–47.

Wolpert, L. (1971). Positional information and pattern formation. Curr. Top. Dev. Biol. *6*:183–224.

Wolpert, L. and W. D. Stein (1984). Positional information and pattern formation. In *Pattern Formation* (G. M. Malacinski and S. V. Bryant, eds.). Macmillan, New York, pp. 3–21.

Wood, J. G. (1863). *The Illustrated Natural History*. Routledge, Warne and Routledge, London.

Wright, D. A. and C. M. Richards (1982). Peptidase isozymes of the leopard frog *Rana pipiens:* properties and genetics. J. Exp. Zool. *221*:283–293.

Wright, M. R. (1947). Regeneration and degeneration experiments on lateral line nerves and sense organs in anurans. J. Exp. Zool. *105*:221–257.

Wright, M. R. (1951). The lateral line system of sense organs. Q. Rev. Biol. *26*:264–280.

Wright, T. R. F. (1970). The genetics of embryogenesis in *Drosophila*. Adv. Genet. *15*:261–395.

Ximénez, F. (1615). Quatro Libros de la Naturaleza, y virtudes de las plantas, y Animales que están recevidos en el uso de Medecina en la Nueva España, y la método, y corrección, y preparación, que para adminstrallas se requiere con lo que el Doctor Francisco Hernández escrivio en lengua latina. México.

Yamada, T. (1937). Der Determinationszustand des Rumpfmesoderms im Molchkeim nach der Gastrulation. Wilhelm Roux Arch. EntwMech. Org. *137*:151–270.

Yamada, T. (1940). Beeinflussung der Differenzierungsleistung des isolierte Mesoderms von Molchkeimen durch zugefügtes Chords-und Neuralmaterial. Okajimas Folia Anat. Jpn. *19*:131–197.

Yamamoto, K. and Kikuyama, S. (1982). Radioimmunoassay of prolactin in plasma of bullfrog tadpoles. Endocrinol. Jpn. *29*:159–167.

Yew, D. T. (1985). Eye enucleation and regeneration of neural retina in axolotl larvae *(Ambystoma mexicanum)*. Anat. Anz. *158*:217–229.

Youn, B. W. and G. M. Malacinski (1981a). Comparative analysis of amphibian somite morphogenesis: cell rearrangement patterns during rosette formation and myoblast fusion. J. Embryol. Exp. Morphol. *66*:1–26.

Youn, B. W. and G. M. Malacinski (1981b). Axial structure development in ultraviolet-irradiated (notochord-defective) amphibian embryos. Dev. Biol. *83*:339–352.

Younglai, E. V., F. Godeau, B. Mulvihill, and E. E. Baulieu (1982). Effects of cholera toxin and actinomycin on synthese of [^{35}S]methionine-labeled proteins during progesterone-induced maturation of *Xenopus laevis* oocytes. Dev. Biol. *91*:36–42.

Zackson, S. L. and M. S. Steinberg (1986). Cranial neural crest cells exhibit directed migration on the pronephric duct pathway: further evidence for an *in vivo* adhesion gradient. Dev. Biol. *117*:342–353.

Zubova, S. E. (1962). Type of variation of development rate in Acipenseridae embryos. Dokl. Akad. Nauk SSSR *145*(3):694–697.

Index